Carbon-Based Materials Applied in Water and Wastewater Treatment

Carbon-Based Materials Applied in Water and Wastewater Treatment

Guest Editor

Athanasia Tolkou

MDPI

Basel • Beijing • Wuhan • Barcelona • Belgrade • Novi Sad • Cluj • Manchester

Guest Editor
Athanasia Tolkou
Department of Chemistry
Democritus University of
Thrace
Kavala
Greece

Editorial Office
MDPI AG
Grosspeteranlage 5
4052 Basel, Switzerland

This is a reprint of the Special Issue, published open access by the journal *C* (ISSN 2311-5629), freely accessible at: https://www.mdpi.com/journal/carbon/special_issues/VDS786TDGA.

For citation purposes, cite each article independently as indicated on the article page online and as indicated below:

Lastname, A.A.; Lastname, B.B. Article Title. *Journal Name* **Year**, *Volume Number*, Page Range.

ISBN 978-3-7258-5267-3 (Hbk)
ISBN 978-3-7258-5268-0 (PDF)
https://doi.org/10.3390/books978-3-7258-5268-0

Contents

About the Editor

Athanasia Tolkou

Athanasia K. Tolkou is an Assistant Professor of Wastewater Treatment, in the Department of Chemistry, Democritus University of Thrace, GR-65404 Kavala, Greece. Dr. Athanasia Tolkou is a Chemist, having graduated from the Aristotle University of Thessaloniki (AUTH), and holds a PhD in Chemistry, in the field of wastewater treatment. She is now working at the Department of Chemistry, School of Science, Democritus University of Thrace (DUTH), Kavala, Greece, as an Assistant Professor. Her research interests include wastewater treatment, municipal, industrial and hazardous wastewater treatment technologies (coagulation/flocculation, adsorption, ozonation, membrane bioreactors), the recovery of nutrients (phosphates, nitrates) from wastewater, and the removal of ions, heavy metals, and dyes from water and wastewater. She has gained research experience at the Technical University of Hamburg (Germany), Laboratory of Water Resources and Water Supply and at the University of Insubria, Varese (Italy) Department of Theoretical and Applied Sciences. She has also gained teaching experience at the Department of Chemistry of the University Maria Curie-Skłodowska, Lublin, Poland. Her scientific work has been published in more than 72 Papers in international journals, and she has published four chapters in scientific Books. Currently (2025), her work is widely recognized, with more than 1500 citations (h-index 25) in Scopus and 2000 citations (h-index of 27) in Google Scholar. She has served as the Guest Editor of 6 Special Issues and she is a reviewer at over 40 international scientific journals (published by Elsevier, Springer, Wiley, Taylor & Francis, MDPI, etc.). She has also participated in about 20 research projects related to her field.

C Journal of Carbon Research

MDPI

Editorial

Editorial for Special Issue "Carbon-Based Materials Applied in Water and Wastewater Treatment"

Athanasia K. Tolkou

Hephaestus Laboratory, School of Chemistry, Faculty of Sciences, Democritus University of Thrace, GR-65404 Kavala, Greece; atolkou@chem.duth.gr; Tel.: +30-2510-462-351

In the past decade, carbon nanostructures have emerged as one of the most rapidly advancing areas of research [1–3]. Novel carbon-based materials—such as activated carbon, carbon nanotubes, fullerenes, and graphene—exhibit superior mechanical, electrical, chemical, and optical properties, making them highly promising candidates for water and wastewater treatment applications [4–7]. Given that over two billion people worldwide face water scarcity, access to clean water remains one of the most critical global challenges [8–10]. At the same time, wastewater, often regarded as spent water, can be viewed as a valuable resource when properly treated and reused [11–13]. In addition, carbon-based nanomaterials—including activated carbon, single- and multi-walled carbon nanotubes [14], carbon quantum dots, and graphene or reduced graphene—have recently emerged as promising photocatalysts for removing organic and inorganic pollutants from water, owing to their abundance, high efficiency and selectivity, cost-effectiveness, and recyclability [15–18]. This Special Issue of *C—Journal of Carbon Research* compiles and highlights cutting-edge research on the design and application of carbon materials in water and wastewater treatment, emphasizing their remarkable adsorption capacity, high pollutant removal efficiency, multifunctional properties, structural stability, cost-effectiveness, and environmental sustainability. Collectively, these articles present a comprehensive overview of recent progress in developing innovative treatment strategies for aquatic environments through the use of carbon-based materials. The contributions included in this Special Issue are summarized as follows:

- Tasić et al. [19] present a study that highlights the potential of porous cellulose-based carbon fibers as effective adsorbents for chlorpyrifos removal form water.
- Ouyang et al. [20] developed "carbon-magnetic" Ni@SiC/CNF (C/Ni, C/SiC) composites, synthesized by electrostatic spinning, that efficiently degraded methylene blue under microwave radiation.
- An et al. [21] fabricated wood-derived carbon aerogel/$Mg(OH)_2$ bio-composites that were applied in relation to the adsorption and separation of cadmium ions.
- Guagliano et al. [22] produced chestnut waste-derived Fe-based photocatalysts for diclofenac degradation through a hybrid adsorption/photocatalytic process.
- Vamvuka et al. [23] explored magnesium-, zinc-, or calcium-modified nut residue biochar activated by nitrogen/steam for the adsorption of arsenate from wastewaters.
- Sengupta and Hussain [24] reported a review regarding advancements in graphene-based technologies for removing antibiotics from wastewater between 2016 and 2024.
- Li et al. [25] presented the research progress in relation to graphene-based adsorbents for wastewater treatment. Focus was given to the preparation, adsorption properties, and mechanisms of inorganic and organic pollutants.

- Guo et al. [26] synthesized TiO_2-modified g-C_3N_4 composites using one-pot calcination for the photocatalytic purification of orange-II-dye-polluted wastewater.
- Rahimi et al. [27] developed and fully characterized biomass-derived carbons for the removal of Cu^{2+} and Pb^{2+} from aqueous solutions. The biomass-based carbon adsorbents were produced from olive stones and almond shells using various activating agents including KOH, H_3PO_4, and $ZnCl_2$.
- Tavukcuoglu et al. [28] demonstrated a monohydrogen phosphate-selective carbon composite membrane electrode for soil water samples. This electrode is developed for on-site and real-time applications using a silver polyglutaraldehyde phosphate and carbon nanotube (CNT) matrix.
- Ledesma et al. [29] investigated the possible reuse of activated carbons used as filters in steam cycle treatment in a nuclear power plant.
- Puente Torres et al. [30] applied an innovative X-Ray absorption technology for the improved monitoring of the degradation and oxidation of granular activated carbon filters used in hospital water treatment systems.

As shown by the published papers contained in this Special Issue, recent advances in graphene synthesis have enabled the development of efficient graphene-based adsorbents for the rapid removal of heavy metals, dyes, pharmaceuticals, and other pollutants from wastewater. Their performance can be further enhanced through functionalization with specific groups or the incorporation of nanomaterials. For instance, functionalized graphene materials, as well as graphene combined with metal oxides or polymers, show improved efficiency in antibiotic removal via adsorption and photocatalytic mechanisms. Moreover, biomass-based carbon adsorbents produced from agricultural waste, i.e., olive stones or chestnut waste and cellulose-based porous carbon fibers, are promising materials for the removal of heavy metals and pharmaceutical substances from waste. Finally, carbon-based materials were applied for the photocatalytic purification of water and wastewater.

We sincerely thank all contributing authors and the reviewers for their excellent efforts and insights. We also extend our appreciation to the editorial team of *C* for their continuous support. We hope that this Special Issue will serve as a valuable reference for researchers in the fields of carbon-based materials applied in water and wastewater treatment.

Funding: This research received no external funding.

Conflicts of Interest: The author declares no conflicts of interest.

References

1. Saleh, M.; Gul, A.; Nasir, A.; Moses, T.O.; Nural, Y.; Yabalak, E. Comprehensive Review of Carbon-Based Nanostructures: Properties, Synthesis, Characterization, and Cross-Disciplinary Applications. *J. Ind. Eng. Chem.* **2025**, *146*, 176–212. [CrossRef]
2. Methods, F.S.; Barhoum, A.; Deshmukh, K. *Handbook of Functionalized Carbon Nanostructures*; Springer: Berlin/Heidelberg, Germany, 2024; ISBN 9783031321498.
3. Slepičková Kasálková, N.; Slepička, P.; Švorčík, V. Carbon Nanostructures, Nanolayers, and Their Composites. *Nanomaterials* **2021**, *11*, 2368. [CrossRef] [PubMed]
4. Tolkou, A.K.; Trikkaliotis, D.G.; Kyzas, G.Z.; Katsoyiannis, I.A.; Deliyanni, E.A. Simultaneous Removal of As(III) and Fluoride Ions from Water Using Manganese Oxide Supported on Graphene Nanostructures (GO-MnO_2). *Sustainability* **2023**, *15*, 1179. [CrossRef]
5. Baby, R.; Saifullah, B.; Hussein, M.Z. Carbon Nanomaterials for the Treatment of Heavy Metal-Contaminated Water and Environmental Remediation. *Nanoscale Res. Lett.* **2019**, *14*, 583. [CrossRef]
6. Rao, N.; Singh, R.; Bashambu, L. Carbon-Based Nanomaterials: Synthesis and Prospective Applications. *Mater. Today Proc.* **2021**, *44*, 608–614. [CrossRef]
7. Siraj, A.; Adugna, H.; Lemessa, R.; Deso, L. Carbon-Based Nanomaterials for Water Treatment: A Comprehensive Review of Recent Advances and Mechanisms. *Chem. Eng. J. Adv.* **2025**, *23*, 100834. [CrossRef]

8. Yang, Y.C.E.; Son, K.; Hung, F.; Tidwell, V. Impact of Climate Change on Adaptive Management Decisions in the Face of Water Scarcity. *J. Hydrol.* **2020**, *588*, 125015. [CrossRef]
9. Tzanakakis, V.A.; Paranychianakis, N.V.; Angelakis, A.N. Water Supply and Water Scarcity. *Water* **2020**, *12*, 2324. [CrossRef]
10. Shemer, H.; Wald, S.; Semiat, R. Challenges and Solutions for Global Water Scarcity. *Membranes* **2023**, *13*, 612. [CrossRef]
11. Al-Hazmi, H.E.; Mohammadi, A.; Hejna, A.; Majtacz, J.; Esmaeili, A.; Habibzadeh, S.; Saeb, M.R.; Badawi, M.; Lima, E.C.; Mąkinia, J. Wastewater Reuse in Agriculture: Prospects and Challenges. *Environ. Res.* **2023**, *236*, 116711. [CrossRef]
12. Christou, A.; Beretsou, V.G.; Iakovides, I.C.; Karaolia, P.; Michael, C.; Benmarhnia, T.; Chefetz, B.; Donner, E.; Gawlik, B.M.; Lee, Y.; et al. Sustainable Wastewater Reuse for Agriculture. *Nat. Rev. Earth Environ.* **2024**, *5*, 504–521. [CrossRef]
13. Salgot, M.; Folch, M. Wastewater Treatment and Water Reuse. *Curr. Opin. Environ. Sci. Health* **2018**, *2*, 64–74. [CrossRef]
14. Asgari, M.; Lohrasbi, E. Comparison of Single-Walled and Multiwalled Carbon Nanotubes Durability as Pt Support in Gas Diffusion Electrodes. *ISRN Electrochem.* **2013**, *2013*, 1–7. [CrossRef]
15. Soni, S. Carbon-Based Nanoparticles as Visible-Light Photocatalysts. In *Carbon-Based Nanomaterials for Green Applications*; Wiley IEE Press: Hoboken, NJ, USA, 2025; pp. 533–551.
16. Mohapatra, L.; Cheon, D.; Yoo, S.H. Carbon-Based Nanomaterials for Catalytic Wastewater Treatment: A Review. *Molecules* **2023**, *28*, 1805. [CrossRef] [PubMed]
17. Devi, M.; Praharaj, S.; Rout, D. 6-Carbon-Based Nanomaterials for Photocatalytic Application. In *Water Treatment Using Engineered Carbon Nanotubes*; Hadi Dehghani, M., Karri, R.R., Mubarak, N.M., Eds.; Elsevier: Amsterdam, The Netherlands, 2024; pp. 153–178, ISBN 978-0-443-18524-3.
18. Xu, Y.; Li, S.; Chen, M.; Zhang, J.; Rosei, F. Carbon-Based Nanostructures for Emerging Photocatalysis: CO_2 Reduction, N_2 Fixation, and Organic Conversion. *Trends Chem.* **2022**, *4*, 984–1004. [CrossRef]
19. Tasić, T.; Milanković, V.; Unterweger, C.; Fürst, C.; Breitenbach, S.; Pašti, I.A.; Lazarević-Pašti, T. Highly Porous Cellulose-Based Carbon Fibers as Effective Adsorbents for Chlorpyrifos Removal: Insights and Applications. *C* **2024**, *10*, 58. [CrossRef]
20. Ouyang, H.; Liu, J.; Li, C.; Bao, L.; Shen, T.; Li, Y. Synthesis of Ni@SiC/CNFs Composite and Its Microwave-Induced Catalytic Activity. *C* **2024**, *10*, 72. [CrossRef]
21. An, R.; Liu, J.; Ma, H.; Yan, Y.; Guo, Y.; Pan, Q.; Li, S. Fabrication of Wood-Derived Carbon Aerogel/$Mg(OH)_2$ Bio-Composite and Its High Performance for Adsorption and Separation of Cadmium Ions. *C* **2025**, *11*, 32. [CrossRef]
22. Guagliano, M.; Bahamonde, A.; Bellotto, M.; Cristiani, C.; Finocchio, E.; Gasco, A.; Muelas-Ramos, V.; Jiménez-Bautista, K.; de los Ríos, C.; Hermosilla, D. Chestnut Waste-Derived Fe-Based Photocatalyst for Diclofenac Degradation. *C* **2025**, *11*, 38. [CrossRef]
23. Vamvuka, D.; Sdoukou, E.; Stratakis, A.; Pentari, D. Enhanced Adsorption of Arsenate from Contaminated Waters by Magnesium-, Zinc- or Calcium-Modified Biochar—Modeling and Mechanisms. *C* **2024**, *10*, 61. [CrossRef]
24. Sengupta, J.; Hussain, C.M. Advanced Graphene-Based Technologies for Antibiotic Removal from Wastewater: A Review (2016–2024). *C* **2024**, *10*, 92. [CrossRef]
25. Li, G.; Du, R.; Cao, Z.; Li, C.; Xue, J.; Ma, X.; Wang, S. Research Progress in Graphene-Based Adsorbents for Wastewater Treatment: Preparation, Adsorption Properties and Mechanisms for Inorganic and Organic Pollutants. *C* **2024**, *10*, 78. [CrossRef]
26. Guo, B.; Uzzaman, M.; Furukawa, M.; Tateishi, I.; Katsumata, H.; Kaneco, S. Photocatalytic Purification of Orange-II-Dye-Polluted Wastewater Using TiO_2-Modified g-C_3N_4 Composite. *C* **2024**, *10*, 103. [CrossRef]
27. Rahimi, V.; Pimentel, C.H.; Gómez-Díaz, D.; Freire, M.S.; Lazzari, M.; González-Álvarez, J. Development and Characterization of Biomass-Derived Carbons for the Removal of Cu^{2+} and Pb^{2+} from Aqueous Solutions. *C* **2025**, *11*, 2. [CrossRef]
28. Tavukcuoglu, O.; Erci, V.; Ciftci, F.; Isildak, I.; Kasapoglu, M.Z. A New Monohydrogen Phosphate-Selective Carbon Composite Membrane Electrode for Soil Water Samples. *C* **2025**, *11*, 18. [CrossRef]
29. Ledesma Cano, B.; Rodríguez, E.M.; González González, J.F.; Nogales-Delgado, S. Reuse of Activated Carbons from Filters for Water Treatment Derived from the Steam Cycle of a Nuclear Power Plant. *C* **2025**, *11*, 19. [CrossRef]
30. Puente Torres, J.; Crespo Sariol, H.; Mariño Peacok, T.; Haeldermans, T.; Reggers, G.; Yperman, J.; Adriaensens, P.; Carleer, R.; Vandamme, D. Innovative X-Ray Absorption Technology for Improved Monitoring of the Degradation and Oxidation of Granular Activated Carbon Filters Used in Hospital Water Treatment Systems. *C* **2025**, *11*, 30. [CrossRef]

C Journal of Carbon Research

MDPI

Article

Chestnut Waste-Derived Fe-Based Photocatalyst for Diclofenac Degradation

Marianna Guagliano [1,*], Ana Bahamonde [2], Maurizio Bellotto [3], Cinzia Cristiani [1], Elisabetta Finocchio [4], Antonio Gasco [5], Virginia Muelas-Ramos [5], Karla Jiménez-Bautista [5], Christian de los Ríos [5] and Daphne Hermosilla [5]

1 Dipartimento di Chimica, Materiali e Ingegneria Chimica "G. Natta" (CMIC), Politecnico di Milano, Piazza Leonardo da Vinci 32, 20133 Milano, Italy; cinzia.cristiani@polimi.it
2 Instituto de Catálisis y Petroleoquímica, Consejo Superior de Investigaciones Científicas, Calle de Marie Curie 2, 28049 Madrid, Spain; abahamonde@icp.csic.es
3 Opigeo Srl, Via dell'Industria, 13, 36040 Grisignano di Zocco VI, Italy; maurizio.bellotto@opigeo.eu
4 Dipartimento di Ingegneria Civile, Chimica e Ambientale (DICCA), Università di Genova, via Opera Pia 15, 16145 Genova, Italy; elisabetta.finocchio@unige.it
5 Escuela Técnica Superior de Ingeniería de Montes, Forestal y del Medio Natural, Universidad Politécnica de Madrid, Calle José Antonio Novais 10, 28040 Madrid, Spain; antonio.gasco@upm.es (A.G.); virginia.muelas@upm.es (V.M.-R.); karla.jimenez@upm.es (K.J.-B.); christian.delosrios@upm.es (C.d.l.R.); daphne.hermosilla@upm.es (D.H.)
* Correspondence: marianna.guagliano@polimi.it

Abstract: This study aims to demonstrate the feasibility of the use of chestnut waste as a green and circular material for developing iron-based photocatalysts for non-steroidal anti-inflammatory drug (NSAID) photodegradation. Four Fe-based catalysts and two pristine biochars were obtained upon a pyrolysis process at 500 and 700 °C and fully characterised. Due to the applied synthesis, iron is present in the form of isotropic grains of magnetite (Fe_3O_4), quite homogeneously dispersed onto the biochar. The textural properties of all the materials are mainly determined by the pyrolytic temperature, which results in macroporous materials at 500 °C and microporous ones at 700 °C. Fe-based catalysts were tested in Diclofenac (DFC) photodegradation. DFC removal was the result of both adsorption and photocatalytic reactions. Despite the good yield in DFC removal (80–100%), the formation of degradation by-products can partially invalidate the good effectiveness of this approach. However, the encouraging results of this study represent a step forward for the possible development of waste-derived biochar-based catalysts for in-field application.

Keywords: chestnuts shells; pyrolysis; Diclofenac degradation; adsorption; photocatalysis

1. Introduction

Non-steroidal anti-inflammatory drugs (NSAIDs) are extensively used due to their antipain effect in various diseases and their availability without a medical prescription. However, NSAID consumption is followed by excretion through faecal matter and urine, causing adverse effects on the environment and the ecosystem, if not properly removed from the water before its reintroduction in the aquatic environment [1]. As they are so widespread, the efficient removal of NSAIDs from contaminated waters is a global problem [2], as demonstrated by the numerous and diverse approaches proposed by researchers from several countries [3–6]. Different techniques have been proposed for the efficient removal of NSAIDs, for instance, electrochemical methods, sonochemical processes, and photocatalytic degradation; however, their total elimination is still an open point [7]. Among others, adsorption by different sorbents has been reported to be able to remove large amounts of NSAIDs from contaminated water, and different sorbents have been studied, such as activated carbon, ligninolytic enzymes, graphene-based adsorbents, and molecularly imprinted polymers [8,9].

Diclofenac (DFC), one of the NSAIDs, is often detected in wastewater and surface water, due to its high consumption by both humans and animals, its improper use, and its disposal. European Commission regulations stated 0.1–0.01 $\mu g\ L^{-1}$ and 75–7.5 $\mu g\ L^{-1}$ as the maximum DFC concentrations in order to avoid chronic and acute toxicity for inland and coastal surface waters, respectively [10]; therefore, the degradation of DFC is one of the most extensively debated and assessed problems within the scientific community [10].

Different strategies are available for DFC removal, such as adsorption, membrane separation, and advanced oxidation processes (AOPs) [10]. Among these, AOPs represent an emerging strategy to remove organic contaminants from wastewater [10]. They consist of non-selective processes that exploit the action of radical species, such as hydroxyl radicals ($OH^{\bullet}$), superoxide radical anions ($O_2^{-\bullet}$), a hydroperoxyl radical ($HO^{2\bullet}$), or a sulphate radical ($SO_4^{-\bullet}$), to degrade and mineralize the organic molecules into CO_2, H_2O, and inorganic elements [11]. Typically, the formation of radicals is facilitated by the activation of an oxidant agent, such as H_2O_2 or persulphate, in the presence of a metal oxide that acts as a catalyst (e.g., iron, copper, zinc, cobalt, manganese) [12,13].

The photo-Fenton reaction, which occurs in the presence of Fe^{2+}, H_2O_2, and UV light in an acidic condition (pH = 3), has been reported as a good strategy [14]. Despite the high level of effectiveness of this methodology, it presents some drawbacks, as it is performed in a homogeneous phase, harsh conditions, such as strong acid pH, with the formation of Fe-containing sludges [15].

To overcome these limitations, the use of heterogeneous catalysts, such as metal-supported ones, has been proposed [15,16]. For this purpose, several materials have been reported as supports, for instance, clays, Layered Double Hydroxides (LDHs), zeolites, graphene, Metal–Organic Frameworks (MOFs), and activated carbon [15]. Although these supports are valid and efficient, new, low-cost, and "greener" materials, often generated by residual waste, are highly requested to achieve the sustainability goals and the transition to circularity. Accordingly, biochar, being the by-product of bio-energy production [17,18], is gaining a lot of attention.

Biochar's characteristics are diverse, including a high surface area, micro-, macro-, and meso-porosity, as well as the presence of oxygenated functional groups at its surface, which make biochar an exceptionally versatile material. This versatility includes the possibility of catalytic applications [11,19,20]; indeed, the morphological and surface properties of biochar are proper for its efficient functioning as a catalytic support [20]. In this context, efforts are reported in the literature for the development of biochar-based catalysts for different applications [11,21–23], including biodiesel production, removal of tar, electrochemical reactions, microbial fuel cell electrodes, photocatalytic degradation, and advanced oxidation processes. In this regard, in the literature, there is still a debate about the catalytic activity that pristine biochar may present due to its diverse functional structural characteristics, such as its relatively high surface area, its pore structure, its inorganic components, and the presence of Persistent Free Radicals (PFRs) [24–28].

The characteristics of biochar strongly depend on the original biomass and the applied thermochemical process used to create it. Therefore, different biomass sources have been experimented with, such as sugarcane, maize straw, rice straw, peanut shells, herb residues, rice hulls, and Taihu Blue Algae, mainly by adopting pyrolysis conditions [12,29–33]. Similarly, several preparation routes have been used to synthetize biochar-based catalysts, such as co-precipitation, hydrothermal and sol/gel reactions, and self-assembly methods [28]. Furthermore, to improve the catalyst performances, the combination of different metals has also been proposed, to obtain, for instance, bi-metallic-supported catalysts, such as $ZnFe_2O_4$/biochar [34], $CuFeO_2$/biochar [35], $MnFe_2O_4$/chitosan modified biochar [36], biochar decorated with a core–shell of nano-Zero Valent Iron (nZVI)/Fe_3O_4 [37], and $LaFeO_3$/lignin biochar [38].

These catalysts were proposed for the degradation of organic molecules such as antibiotics (tetracycline [35], ofloxacin [38]), and dyes (rhodamine B, methyl orange, methylene blue [34,36,37]). The main parameters that can affect the degradation of the organic molecules were ascribed to the dose of catalyst and H_2O_2, the pH of the medium, and the light source [35–37]. For example, a higher dosage of catalyst and H_2O_2 could result in a decreased degradation of the pollutant probably due to the high resistance for light to penetrate the solution, and to the generation of undesirable decomposition of active radicals [35,37].

Castanea sativa is considered a valuable tree in the agricultural and forestry economy. Indeed, in 2023, about 382,916 tons of chestnuts were produced and processed in South Europe (France, Greece, Italy, Spain, Portugal, and Turkey), with Italy and Spain providing almost 64% of chestnut production [39]. The agro-residue resulting from chestnut processing (removal of outer shells and peeling of inner shells) generates large quantities of wastes, which are usually burned [40]. According to EUROSTAT data, in 2021, primary production and manufacturing of food products accounted for 9% and 21% of food wastes, respectively [41], and the disposal of these wastes represents a global issue, since the majority is currently incinerated or landfilled in soil, causing the generation of greenhouse gases (GHGs) like carbon dioxide (CO_2), nitrous oxide (N_2O), and methane (CH_4), which poses a threat to the human and the natural environment [42]. An efficient valorisation of these copious wastes, and, more generally, of lignocellulosic agricultural residues, is represented by their thermochemical conversion, namely pyrolysis and/or gasification [43]. However, to the best of the authors' knowledge, there is a limited number of studies regarding the utilisation of chestnut shells to produce biochar for catalytic application, and it is a relatively unexplored field within the scientific community [44–46].

The aim and the novelty of this work rely on the possibility of reusing and re-evaluating a waste material, produced in large amounts, i.e., chestnut waste, in the perspective of circular economy and sustainability, rather than in the need of replacing biochars derived from other different biomasses.

In this context, the main goal of this work was to investigate the behaviour of iron-based catalysts, prepared by the simply mixing of the chestnut shells (outer pericarp) with an aqueous solution of the iron precursor ($FeSO_4{\cdot}7H_2O$), subject them directly to a pyrolysis process, and apply them in the degradation process.

Both the adsorption and the photocatalytic degradation of Diclofenac was evaluated for a DFC initial concentration of 10 mg L^{-1}; this initial concentration was selected to exploit both the adsorption and the photodegradation processes.

The physico-chemical properties, the adsorption capacity, the photodegradation activity, as well as the stability of the Fe-based catalysts were analysed by considering the iron content and the temperatures of heat treatment (i.e., 500 °C and 700 °C).

2. Experimental

2.1. Materials

Diclofenac (DFC), $C_{14}H_{11}C_{12}NO_2$, (CAS 15307-86-5), a pure commercial powder from Sigma-Aldrich (St. Louis, MO, USA), was chosen as representative of Non-steroidal Anti-inflammatory Drug (NSAID) molecules [47]. DFC is a weak acid (pka = 4.5), characterised by two aromatic rings, two chloride (Cl) atoms, and a negative charge for pH > pKa [48].

$FeSO_4$ 7 H_2O was supplied by Sigma-Aldrich (St. Louis, MO, USA).

Ultrapure water (Type I) was used to prepare all solutions.

2.2. Preparation of the Biochar-Supported Iron Catalyst

Fe-chestnut-biochar-based catalysts were synthesised according to Muelas-Ramos et al. [49]. Briefly, washed chestnut shells were dried at 60 °C in an oven (Digitheat—TFT, J.P. Selecta, Abrera (Barcelona), Spain) for 12 h. To separate the outer and the inner pericarp, the chestnut shells were treated in a blender (MC300, Moulinex, Bagnolet, France), and the pericarps were grounded (<1 mm) before the activation with Fe [50]. Then, 24.89 g of $FeSO_4$ 7 H_2O were dissolved in 80 mL of ultrapure water, and 10 g of outer pericarps were added to the solution to obtain the desired 1/2 Fe/pericarps ratio. The suspension was kept at 80 °C for 2 h, continuously mixing, and, finally, dried at 105 °C.

The dried solids were pyrolysed at two different temperatures, 500 and 700 °C, for a dwelling time of 2 h, and a heating rate of 10 $°C \cdot min^{-1}$ (AAF 11/3 Muffle, Carbolite, Hope, Hope Valley, Derbyshire, UK), according to a modified Burbano et al. [51] protocol, previously developed for lab-scale application. The pyrolysed powders were divided into two batches and washed with ultrapure water five or twelve times by resuspending and centrifuging (Didacen II, Orto Alresa, Spain) the suspensions for any washing step. The washing steps are required to avoid iron leaching during the photocatalytic reactions [49]; five and twelve washing steps were applied, considering them representative of a smaller and larger number of washings, respectively.

At the end of the washing process, the materials were dried at 105 °C before testing. Iron leaching upon washing was determined by applying the 1,10-phenantroline colorimetric method [52].

For the sake of comparison, pristine biochar, i.e., without Fe addition and washing, was also prepared following the same pyrolytic procedure reported above.

Four catalysts and two pristine biochars were prepared in six different batches. The acronyms of the samples, as well as the parameters of the synthesis, are summarised in Table 1.

Table 1. Acronyms and synthesis operating paraments of pristine biochar and Fe-biochar catalysts.

Sample	Material	Temperature (°C)	Washing Number
B5 B7	Pristine biochar	500 700	none
Fe-B5W5 Fe-B5W12	Fe-biochar catalyst	500	5 12
Fe-B7W5 Fe-B7W12	Fe-biochar catalyst	700	5 12

2.3. Materials Characterisation

The textures of the biochar and of the catalysts were analysed using N_2 adsorption–desorption isotherms at −196 °C (Micromeritics ASAP 2420 apparatus (Norcross, GA, USA)); the samples were previously outgassed overnight at 120 °C to a residual pressure lower than 10^{-4} Pa. The specific surface area (S_{BET}) was calculated according to the Brunauer–Emmett–Teller (BET) equation [53], the micropore volume (V_{micro}) and the external surface (S_{ext}) were calculated according to the t-plot method.

Meso- and macropores volume was determined by mercury intrusion porosimetry (Micromeritics Autopore IV apparatus, Norcross, GA, USA), which also allows for the materials' skeletal density determination [41].

Elemental Analysis (EA) was performed by Costech ECS mod.4010 instrument (Avenue Stanford, Santa Clarita, CA, USA), which allows for the detection of carbon (C), nitrogen (N), hydrogen (H), and sulphur (S) simultaneously.

Inductively Coupled Plasma–Optical Emission Spectroscopy (ICP-OES) was performed with a PerkinElmer OPTIMA 7000 DV spectrometer (PerkinElmer, Waltham, MA, USA). The average of three measurements is reported, and the estimated error is within 3%.

Total Organic Carbon (TOC) was determined using the Shimadzu TOC-V CSH model (Kyoto, Japan) with a Non-purgeable Organic Carbon (NPOC) method.

X-Ray Powder Diffraction (XRPD) patterns were collected in the 4–90 2θ° range, $0.02° \cdot s^{-1}$ scan rate, and 50 s of counting time, by a PANalytical X'Pert PRO X-ray polycrystal diffractometer (Almelo, The Netherlands), using a Ni-filtered Cu Kα radiation and an ultrafast X'Celerator detector. The patterns were analysed via the Rietveld method to derive the microstructural parameters of the magnetite phase from the measured line broadening. The instrumental line broadening has been determined via X-ray tracing of the instrumental optical path, considering the characteristics of the source and all the present optical elements. The Profex software 5.4.1 [54] has been used for the analysis.

Thermogravimetric (TG-DTG) analyses were performed in nitrogen in the temperature range of 30–800 °C at a heating rate of 5 °C/min (Perkin Elmer Q600 DTA-TG apparatus, Waltham, MA, USA).

FTIR-ATR analysis of the sample powders has been performed (100 scans, resolution 4 cm^{-1}, background air) using a Thermo Nicolet 380 Instrument (Waltam, MA, USA) equipped with a diamond window for Attenuated Total Reflectance (ATR) analysis.

Scanning Electron Microscopy and Energy Dispersive X-ray (SEM-EDX) analyses were performed with a Zeiss EVO 50 EP (Zeiss, Jena, Germany) combined with an Oxford INCA energy 2000 spectrometer (Oxford Instruments, Abingdon, UK). The SEM-EDX equipment was operated at a high electron tension (EHT) voltage of 15 and 20 kV, a probe current of 120 and 300 pA, and at high vacuum (about 10^{-4} Pa) conditions.

2.4. Adsorption and Photocatalytic Experiments

The photocatalytic activity tests of Fe-B5W5, Fe-B5W12, Fe-B7W5, and Fe-B7W12 were performed in batch, using 250 mL glass bottles, shaking in a Digital Orbital Shaker at 150 rpm (Edmund Buhler KL 2, Bodelshausen, Tübingen, Germany) at room temperature and atmospheric pressure.

In a typical experiment, 0.1 g of catalyst was suspended in 100 mL of deionised water at constant pH 4–4.5. A 10 $mg \cdot L^{-1}$ DFC solution was added to the suspension, together with the stoichiometric amount of H_2O_2 (35.2 mg L^{-1}) for the complete oxidation of DFC to CO_2 and H_2O.

Subsequently, the photoreactor was illuminated with UV light at 365 nm (I = 250 mA; total photon flux of about 1.0×10^{21} photon$\cdot m^{-2} \cdot s^{-1}$)a custom assembled lamp (10 LEDs CUN6GB1A supplied by Seoul Vyosis, Republic of Korea) was used to perform the photocatalytic process. The advancement of the reaction was followed by collecting samples of the suspension at different reaction times, namely 10, 20, 30, 45, 60, 90, and 120 min by a syringe equipped with a 0.22 μm PTFE filter (Branchia, Labbox, Spain).

DFC concentration was analysed by High-performance Liquid Chromatography (HPLC) (12600II QuatAgilent 1260 Infinity II, Agilent Technologies (Santa Clara, CA, USA) equipped with a C18 column (120 EC-C18, Poroshell, Santa Clara, CA, USA) with an estimated error within 4.5–5.5%. A mixture of 80% HPLC-grade methanol (Sigma-Aldrich, St. Louis, MO, USA) and 20% Ultrapure Type I H_2O (Wasserlab, Spain) was used as the mobile phase; a constant flow of 0.8 mL·min^{-1} was applied.

The DFC detection wavelength was fixed at 222 nm, i.e., the maximum absorption value for DFC.

The evolution of Diclofenac degradation was expressed as molar conversion (X), calculated according to Equation (1):

$$\text{X DFC}(\%) = \frac{(C_0 - C_t)}{C_0} * 100 \quad (1)$$

where C_0 (mg L^{-1}) is the DFC initial concentration and C_t (mg L^{-1}) is DFC concentration measured at time t during the photocatalytic experiment.

The consumption of H_2O_2 was analysed according to the titanium sulphate spectrophotometric Pobiner's method [55].

The consumption of hydrogen peroxide is calculated according to Equation (2):

$$100 - XH_2O_2(\%) = 100 - \frac{(C_{0,H2O2} - C_{t,H2O2})}{C_{0,H2O2}} * 100 \quad (2)$$

where $C_{0,\,H2O2}$ (mg L^{-1}) is the H_2O_2 initial concentration and $C_{t,\,H2O2}$ (mg L^{-1}) is the concentration of H_2O_2 at time t during the photocatalytic experiment.

Lixiviated iron, at 60 and 120 min of irradiation times, was determined according to Tamura et al. [52].

Adsorption tests were performed applying the same protocol as the photocatalytic tests, without the addition of H_2O_2 and without illumination; the amount of adsorbed DFC was expressed as molar conversion (X), calculated according to Equation (1).

3. Results and Discussion

3.1. Materials Characterisation

3.1.1. Pristine Biochar

Nitrogen adsorption–desorption isotherms of the pristine biochar pyrolysed at 500 °C (B5) and 700 °C (B7), respectively, are shown in Figure S1.

Both materials present a type IV isotherm, and an H3 hysteresis loop, characteristics of adsorbents with a narrow distribution of uniform pores [56]. The textural parameters of the samples are summarised in Table 2 (recalling that S_{BET}: BET surface area; S_{ext} and S_{micro}: external and micropores' surface; V_{Tot}: total pores' volume; V_{micro}, V_{meso}, and V_{macro}: pores' volume; d_{meso} and d_{macro}: meso- and macropores diameters).

Table 2. Textural Properties of B5 and B7 biochar.

Sample	S_{BET} (m^2 g^{-1})	S_{ext} (m^2 g^{-1})	S_{micro} (m^2 g^{-1})	V_{micro} (cm^3 g^{-1})	V_{meso} (cm^3 g^{-1})	V_{macro} (cm^3 g^{-1})	V_{Tot} (cm^3 g^{-1})	d_{meso} (nm)	d_{macro} (nm)
B5	3.5	0.3	3.2	0.000	0.011	0.179	0.190	-	1030
B7	343.2	4.3	338.9	0.134	0.009	0.213	0.355	3.8	894

Noteworthy differences can be observed between B5 and B7 textures (Table 2). B5 shows a very low S_{BET} (3.5 m^2 g^{-1}), which is the result of the presence of large macropores (1030 nm) accounting for 94.1% of the total volume. Moreover, macropores are accompanied by a very limited number of mesopores, whose dimensions could not be measured, accounting for 5.8% of the total volume.

On the contrary, B7 is characterised by a very high S_{BET} value (343.2 $m^2\ g^{-1}$), where the contribution of the micropores (37.6%) determines both the high surface area and the high total pore volume (0.355 ($cm^3\ g^{-1}$). Furthermore, a large number (59.9%) of macropores (894 nm) and a very limited number (2.5%) of mesopores (3.8 nm) contribute to the complex texture of this sample. According to the literature [57,58], the high surface area of B7 is the result of the pyrolysis temperature applied to this sample (700 °C), enough to decompose the organic matter, thus leading to the formation of micropores [59]. Moreover, in the B7 sample, the removal of the pore-blocking substances also promotes the accessibility to the external surface area (4.3 $m^2\ g^{-1}$ for B7 vs. 0.3 $m^2\ g^{-1}$ for B5) [60].

The skeletal density of the materials was also determined: 1.263 g mL^{-1} was found for B5, while 1.244 g mL^{-1} was found for B7, and despite the different textures, B5 and B7 evidenced very close skeletal density; therefore, this finding suggests that only a modest amount of residual undecomposed biomass should be remained in B5 following decomposition.

Pristine biochars were also characterised by X-Ray Powder Diffraction (XRPD) (Figure 1).

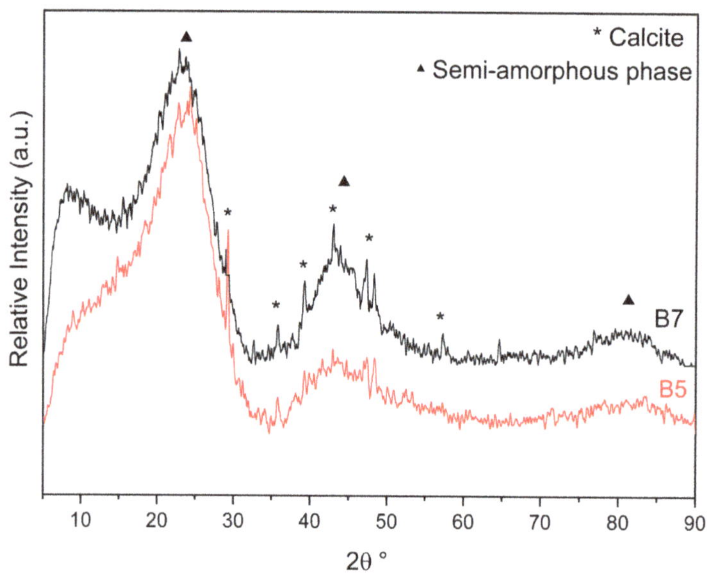

Figure 1. XRPD patterns of B5 (red) and B7 (black) (stars: $CaCO_3$ calcite [JCPDS 05-0586], triangles: semi-amorphous phase).

B5 and B7 materials showed similar patterns (Figure 1), characterised by broad reflections at about 25 (main one), 45, and 85 2θ° of a semi-amorphous phase (Figure 1, triangles), accompanied by the reflections (Figure 1, stars) of a crystalline phase. The semi-amorphous features are consistent with a carbon structure containing randomly oriented aromatic carbon layers [61,62], which, in the case of B7, apparently presents a more ordered organisation with a high spacing distance. The crystalline phase is associated with a polycrystalline $CaCO_3$ calcite phase [JCPDS 05-0586] formed during the pyrolysis process from the natural inorganic components of the biomass cells [63]. In view of the higher pyrolytic temperature, the calcite phase is more evident in the B7 sample, as it is better crystallised.

Infra-red analysis in Attenuated Total Reflectance mode (FT-IR-ATR) of the pristine biochar is compared in Figure 2 in the region of 3500–500 cm^{-1}. Indeed, the two materials show completely different features, which can be related to the different thermal treatments.

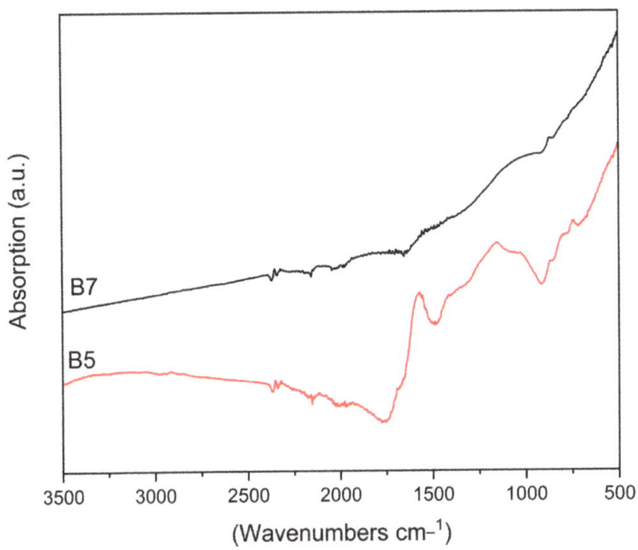

Figure 2. FT-IR ATR spectra of pristine biochar B5 (red), B7 (black).

Bands at 751, 874, and 1571 cm^{-1} in the spectrum of sample B5 are assigned to carbonate species [64], as evidenced by XRPD analysis. The broad envelope of bands in the spectral region of 1200–1000 cm^{-1} can be attributed to the overlap of CC and CO stretching modes, in addition to ring vibration and COH deformation modes, which are characteristics of the residual cellulosic material [65], and lignin matrix. Clearly, this material is still characterised by a high amount of oxygen-containing functional groups arising from biomass degradation. In the spectrum of sample B7, all these features disappear, and, only very weak bands, if any, due to residual carbonate species, are detected. The overall spectrum is flattened and consistent with the spectrum of carbonaceous materials reported in the literature, showing that the carbon matrix and the surface structure have been settled following pyrolysis at 700 °C [66].

As expected, a higher pyrolysis temperature leads to the removal of most of the oxygen-containing functional groups.

Thermogravimetric (TG) analysis, performed in N_2, on B5 and B7 are reported in Figure 3.

Once more, a markedly different behaviour has been found for B5 and B7 (Figure 3); in fact, in the B5 sample, a large weight loss, accounting for 100% of the mass, is observed between 400 and 550 °C. According to the literature, such a weight loss could be ascribed to a further pyrolysis process of the residual lignocellulosic material, not fully decomposed during the pyrolytic treatment at 500 °C of this sample [67,68]. The absence of calcite decomposition, expected between 700 and 800 °C (790 °C for highly pure and crystalline material [69]), can be due either to the limited amount of this phase, below the detection limit of the technique, or to an anticipation of the decomposition, due to the presence of polycrystals and impurities. Calcite decomposition, thus, could be overlapped by the strong thermal phenomenon between 400 and 550 °C.

On the contrary, no weight losses associated with biomass decomposition are manifest in B7, confirming that the pyrolysis temperature of this sample was high enough to fully decompose the biomass [57]. Moreover, in B7, no calcite decomposition was detected, thus suggesting that the low content of this phase, in agreement with IR data, is the most probable explanation for the findings.

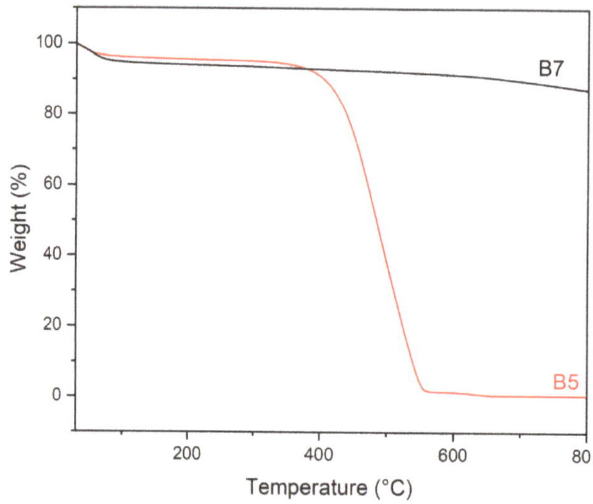

Figure 3. TG curves in N_2 of B5 (red) and B7 (black).

The morphology of the samples was assessed by Scanning Electron Microscopy (SEM) (Figure 4).

Figure 4. SEM images, at different magnifications (250 and 5000×), of (**a**) B5 and (**b**) B7. (Yellow arrows: fibres).

B5 and B7 exhibit quite similar morphology, characterised by an irregular and coarse surface, irrespective of the pyrolysis temperature (Figure 4a,b). The typical biochar porous structure, characterised by quite regular channels [70], is not clearly discernible in this case, even at high magnification (Figure 4a,b, left); moreover, the presence of fibres is clearly detectable. These fibres may correspond to lignocellulosic components of the original biomass only partially decomposed [71], suggesting that, even if quite effective, the pyrolysis at 500 and 700° is not enough to totally decompose all the biomass components, in particular, lignin [72]. These results are in line with TG analysis of pristine biochar, where a further biomass decomposition was observed during the analysis; however, these residues are present in a very limited amount and do not change the bulk morphology, texture, or composition.

The elemental composition of B5 and B7 was evaluated by both Elemental Analysis (EA) and Energy Dispersive X-ray spectroscopy (EDX) (Table 3).

Table 3. Elemental and EDX analysis of B5 and B7 (* by EA, ** by EDX).

Sample	H/C *	C/N *	O/C **
	%		
B5	0.04	92.3	0.3
B7	0.01	109.3	0.1

The higher carbon content of B7 (82%) compared to B5 (72%) is due to the higher degree of carbonisation of the first sample, in view of the higher pyrolysis temperature. This picture is also confirmed by the lower H/C and O/C ratio of B7 (Table 3).

The higher C/N ratio of CB7 (109.3) as compared to CB5 (92.3) still confirms the effect of the different employed temperatures; indeed, it is reported that N diminishes at a faster rate than C with the increase in pyrolysis temperature [73].

Other elements, mainly Mg, Al, K, Ca, and P, natural components of the pyrolysed biomass cells, were also detected by EDX analysis (not reported), but in a very limited amount (0.1–0.8%); moreover, no sulphur or iron were detected in these samples.

3.1.2. Fe-Based Catalysts

A similar characterisation has been performed on the Fe-supported catalysts, Fe-B5W5, Fe-B5W12, Fe-B7W5, and Fe-B7W12 (recalling that W5 and W12 correspond to samples washed 5 and 12 times).

After the introduction of iron, the texture of all the samples became more complex than the original biochar, suggesting that either the synthesis process or the presence of iron may exert some effects on the texture of the pristine systems.

In the case of B5-based catalysts, a considerable increase in the surface area and in the total pore volume occurred. Such an increase is mainly related to both micro- and mesopores formation upon the catalysts' synthesis and the washing process (Table 4), and it can be explained once more with B5 pyrolysis temperature (500 °C).

Table 4. Textural properties of the Fe-based catalysts, pristine B5 and B7 reported for comparison. (Abbreviations as in Table 2).

Sample	S_{BET} ($m^2\ g^{-1}$)	S_{ext} ($m^2\ g^{-1}$)	S_{micro} ($m^2\ g^{-1}$)	V_{micro} ($cm^3\ g^{-1}$)	$V_{meso.}$ ($cm^3\ g^{-1}$)	V_{macro} ($cm^3\ g^{-1}$)	V_{Tot} ($cm^3\ g^{-1}$)	d_{meso} (nm)	d_{macro} (nm)
B5	3.5	0.3	3.2	0.000	0.011	0.179	0.190	-	1030
Fe-B5W5	55.0	29.0	26.1	0.005	0.077	0.274	0.356	3.8	58, 3099
Fe-B5W12	135.8	31.8	104.0	0.041	0.064	0.328	0.433	3.8	84, 2792
B7	343.2	4.3	338.9	0.134	0.009	0.213	0.355	3.8	894
Fe-B7W5	134.9	25.5	109.4	0.045	0.054	0.358	0.457	3.8	106, 2840
Fe-B7W12	162.2	29.2	133.0	0.057	0.062	0.385	0.504	3.8	202, 2840

As already discussed, this temperature is not enough to totally decompose the biomass, whose residues remain in the pores. When the iron solution is present in the mixture, an acid environment is generated, that, together with the pyrolytic process, could favour biomass decomposition. This could result in the emptying of the pores, thereby generating a new porosity in Fe-B5W5. The further pores' volume and surface area increase, observed in Fe-B5W12, could be the result of the larger number of washing steps, which, additionally, favour the pores' emptying (Table 4). Accordingly, Fe-B5W12 texture appears to be more the result of emptying the already existing pores' than the formation of new ones.

On the contrary, in B7-based catalysts, the textural properties can be explained by considering two concurrent opposite effects. The decreased surface area in Fe-B7W5, with respect to pristine B7, can be the result of the occupation of the pores, mainly the micropores, by the iron phase (Table 4). This hypothesis is supported by the halving of the micropores' volume in Fe-B7W5 and Fe-B7W12 (Table 4). On the contrary, the increase in surface area after 12 washes could be the consequence of the partial emptying of the pores due to the removal of the iron during washing.

Iron release upon washing has been qualitatively checked by colorimetric determination [52], and the presence of iron was found in the washing water both after 5 and 12 washing steps. However, considering the textural evolution of B5-based catalysts, it can be assumed that the effect of the pores emptying, due to biomass removal, may be predominant over the effect of iron removal during washing.

Irrespective of the pyrolysis temperature and the number of washing cycles, all catalysts are characterised by very close skeletal densities, in the range 1.703–1.837 g mL^{-1}, and, in any case, higher than those of pristine B5 and B7 due to the presence of iron. Considering that the pristine biochars and the Fe-based catalysts have been prepared in four different batches, the constancy of skeletal densities can be taken as an indication of the reproducibility of the preparation procedure.

XRPD patterns of the catalysts in the 2 θ° range of 10–90 are plotted in Figure 5.

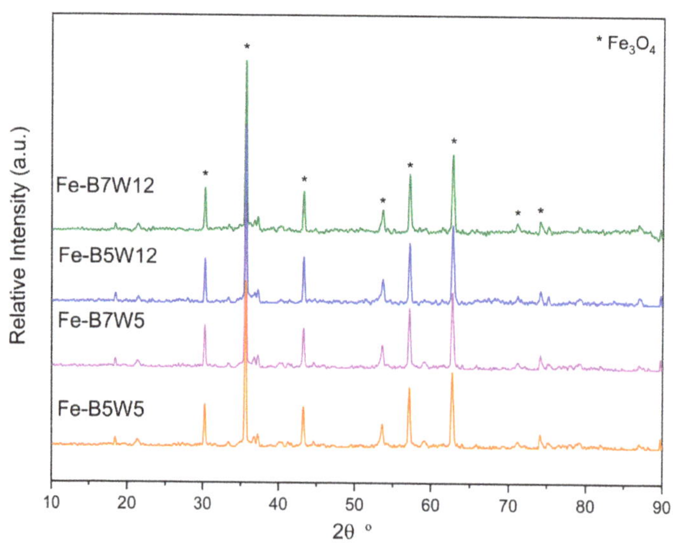

Figure 5. XRPD patterns of Fe-B5W5 (orange), Fe-B5W12 (blue), Fe-B7W5 (violet), and Fe-B7W12 (green). (Stars: Fe_3O_4, magnetite, [JCPDS 19–629]).

The Fe-supported catalysts, irrespective of the number of washing cycles, presented mainly the reflections corresponding to Fe_3O_4, magnetite [JCPDS 19–629], as confirmed by the magnetic properties of the samples (Figure S2); moreover, the partial reduction of Fe(III) to Fe(II) is a confirmation of the effectiveness of the applied pyrolytic process.

The reflections attributed to the calcite phase, present in the pristine biochar, are no longer detected in the catalysts' samples. The disappearance of calcite is possibly due to the formation of $CaSO_4$ during the impregnation, via a double exchange reaction with $FeSO_4$, followed by the $CaSO_4$ dissolution when the washing process is applied. This behaviour is justified by the higher strength of the sulphuric acid, compared with the carbonic one, and by the higher $CaSO_4$ solubility (2.4 gL^{-1} at 25 °C), compared with that of $CaCO_3$ (0.015 g L^{-1} at 25 °C) [74]. Furthermore, the acid environment could also have decomposed the calcite phase.

XRPD patterns have been analysed by Rietveld analysis (plots not reported), and results are summarised in Table 5.

Table 5. Rietveld analysis of the catalysts (* calculated on (1, 1, 1) and (1, 0, 0) directions).

Sample	a_0 Fe_3O_4 (Å)	Microstrain	Grain Size (nm) *	Rwp (%)
Fe-B5W5	8.3905 (8)	0.0024 (2)	64 (5)	2.82
Fe-B7W5	8.3966 (4)	0.00124 (8)	151 (13)	2.78
Fe-B5W12	8.3964 (4)	0.00133 (8)	170 (16)	2.80
Fe-B7W12	8.3963 (4)	0.00138 (9)	173 (17)	2.76

Except for the Fe-B5W5 sample, all the catalysts present very close features, with cell parameters in line with reference Fe_3O_4 [JCPDS 19–629] (Table 5). The difference observed in both lattice parameters and grain size for sample Fe-B5W5 can be related to the presence of Fe vacancies in the structure [75], associated with incomplete pyrolysis and different redox conditions during Fe-B5W5 preparation. Both iron vacancies and preparation conditions can account for the low grain size of this sample (Table 5); on the contrary, Fe-B5W12, Fe-B7W5, and Fe-B7W12 showed isotropic grains of similar dimensions, in the range of 150–170 nm. For all the samples, in view of the very limited microstrains, one can suppose that the Fe_3O_4 phase is not perturbated when interacting with the biochar structure.

Moreover, despite the fact that no direct comparison can be performed among the samples, as they have been prepared in different batches, the slight grain size increase, observed in the samples upon 12 washing, irrespective of the pyrolytic temperature (Table 5), can be related to a sort of ripening effect, due to the dissolution and recrystallisation of the active phase during the washing process.

Iron content in the samples was determined by Inductively Coupled Plasma–Optical Emission Spectroscopy (ICP-OES). Iron contents of 24.6 and 22.5% (w/w) were found for Fe-B7W5 and Fe-B7W12, respectively, and of 14.6 and 13.3% (w/w) were detected for Fe-B5W5 and Fe-B5W12, respectively.

Since the samples were prepared from the same initial iron content, the differences in iron content of the catalysts after washing, can be explained by referring to the morphology of the systems. In fact, the higher iron content of the sample pyrolysed at 700° seems to be linked to the presence of Fe_3O_4 possibly allocated in the smaller pores, such as the micropores, which can trap the active phase, thus limiting the leaching during washing. Instead, in the case of the B5-based systems, Fe_3O_4 is accommodated in the macropores, therefore, more accessible, thus its removal is favoured during washing. The slightly lower percentage of iron in the samples washed 12 times, compared to those washed only 5 times, highlights how the number of washes, in addition to the texture, becomes a determining factor in establishing the content of the active phase in the catalyst.

In Figure 6, the FT-IR spectra in the ATR mode of the Fe-supported catalysts are reported.

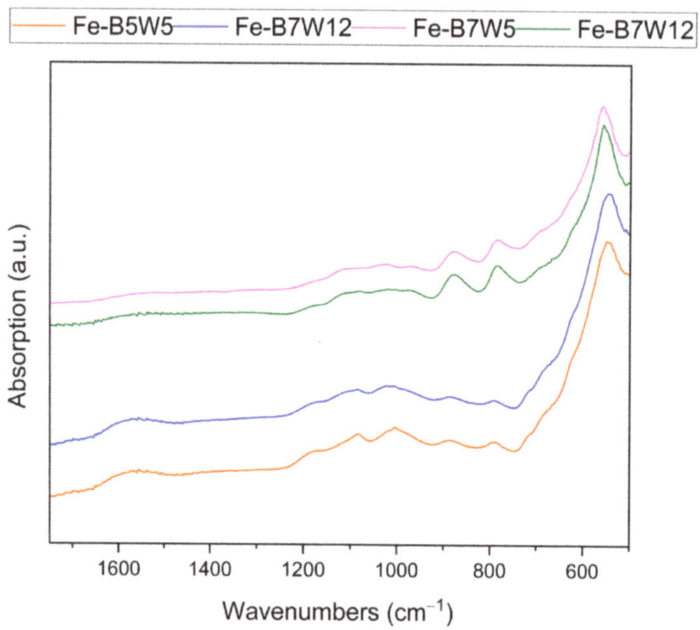

Figure 6. FT-IR ATR spectra of Fe-B5W5 (orange), Fe-B5W12 (blue), Fe-B7W5 (violet), and Fe-B7W12 (green).

All four samples exhibit a band at approximately 574 cm^{-1}, indicative of Fe-O bonding in Fe_3O_4 [76]. As reported by Gadgeel et al. [77], the other characteristic bands of Fe_3O_4 at 694, 627, and 896 cm^{-1} are no longer observed in our spectra; otherwise, two bands at 781 and 878 cm^{-1}, more intense in B7 sample, are evident. These bands, close to those of magnetite but not exactly corresponding, suggest a possible interaction of Fe_3O_4 with the substrate, hence, the consequent bands' perturbation.

Several weak bands in the region of 1200–900 cm^{-1} can be assigned to residual oxygen-containing groups and, mainly, to SO vibrational modes of sulphur-containing species, namely ionic bulklike sulphates, with a possible contribution of sulphites formed by the decomposition of sulphates [78].

TG analysis (Figure S3) showed, for all the samples, a limited weight loss of about 20–25% w/w, where specific decomposition phenomena are hardly distinguishable; for this reason, DTG curves have also been analysed (Figure 7).

The broad peaks of the DTG curves can be associated with dehydration and thermal phenomena occurring in both Fe_3O_4 and residual $FeSO_4$ $7H_2O$.

The features observed below 150 °C may account for water loss in Fe_3O_4 [79], while those in the range of 200–400 °C could be related to water release from residual $FeSO_4$, better evident in B5-based samples [80]. This hypothesis is also confirmed by the presence of a further thermal phenomenon at higher temperatures in B5-based samples. Indeed, $FeSO_4$ decomposition to give Fe_2O_3 is reported as a two-step process in the temperature range of 525–710 °C [79]; thus, DTG findings are in line with the incomplete pyrolysis observed in B5-based catalysts. Finally, the phenomenon observed starting from 650 °C could be associated with the oxidation of Fe(II) to Fe(III) in Fe_3O_4 to give Fe_2O_3 [81].

The morphology and the elemental composition of the Fe-supported catalysts were assessed by SEM-EDX (Figure 8a,b).

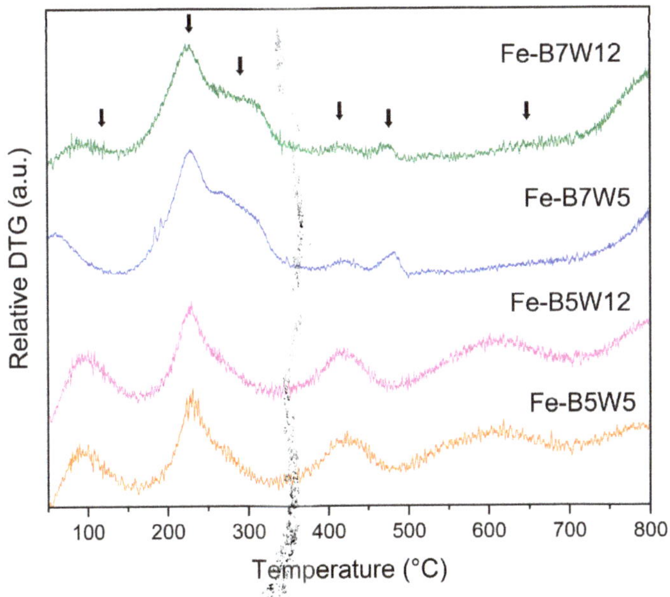

Figure 7. DTG analysis of Fe-B5W5 (orange), Fe-B5W12 (blue), Fe-B7W5 (violet), and Fe-B7W12 (green). (Arrows: thermal phenomena).

Despite changes observed in textural properties, all the materials show a morphology close to that of the pristine biochar, suggesting that neither the catalyst preparation nor the washing process apparently modified it, and the main difference with respect to the pristine B5 and B7 is the presence of iron, accompanied by sulphur.

Iron is more homogenously distributed in the Fe-B7-based samples, irrespective of the number of washing cycles (Figure 8b); on the contrary, in Fe-B5W5 and Fe-B5W12, Fe-containing clusters, clearly evident in the figure, are allocated in specific areas of the sample (Figure 8a). However, at the eye's inspection, these clusters appear to be a mere aggregation of smaller particles than large crystals formation.

The different dispersion of iron in the catalysts may be related to both the catalysts' synthetic procedure and the different textural properties obtained during the thermal treatment. As a matter of fact, the pyrolysis of the biomass and the catalyst production occurs one pot; thus, to understand the reason for the different Fe dispersion in B5- and B7-based samples, the textural evolution of pristine B5 and B7 must be recalled.

As already discussed, very different surface areas (3.5 and 343.2 m^2 g^{-1} for B5 and B7, respectively) and porosity (such as, 94.1% of macropores and 5.8 % of mesopores in B5, and 59.9% of macropores and 37.6% of micropore in B7) were found when the biomass was pyrolysed without the presence of Fe (Table 4).

When the biomass and the iron sulphate mixture are pyrolysed together, the texture of the biochar and the Fe-containing phase are contemporary formed; therefore, Fe_3O_4, while forming, could be allocated in the pores.

In the case of B7-based catalysts, where pores of smaller dimensions are present, the iron-containing phase, "trapped" inside the pores, is not free to move, and aggregate; thus, it gives rise to a more homogeneous iron distribution (Figure 8b). Instead, in the Fe-B5-based catalysts, in view of the macroporosity of the system, the iron phase, being allocated in larger pores, is free to move and aggregate (Figure 8a), so that a lower iron dispersion is found.

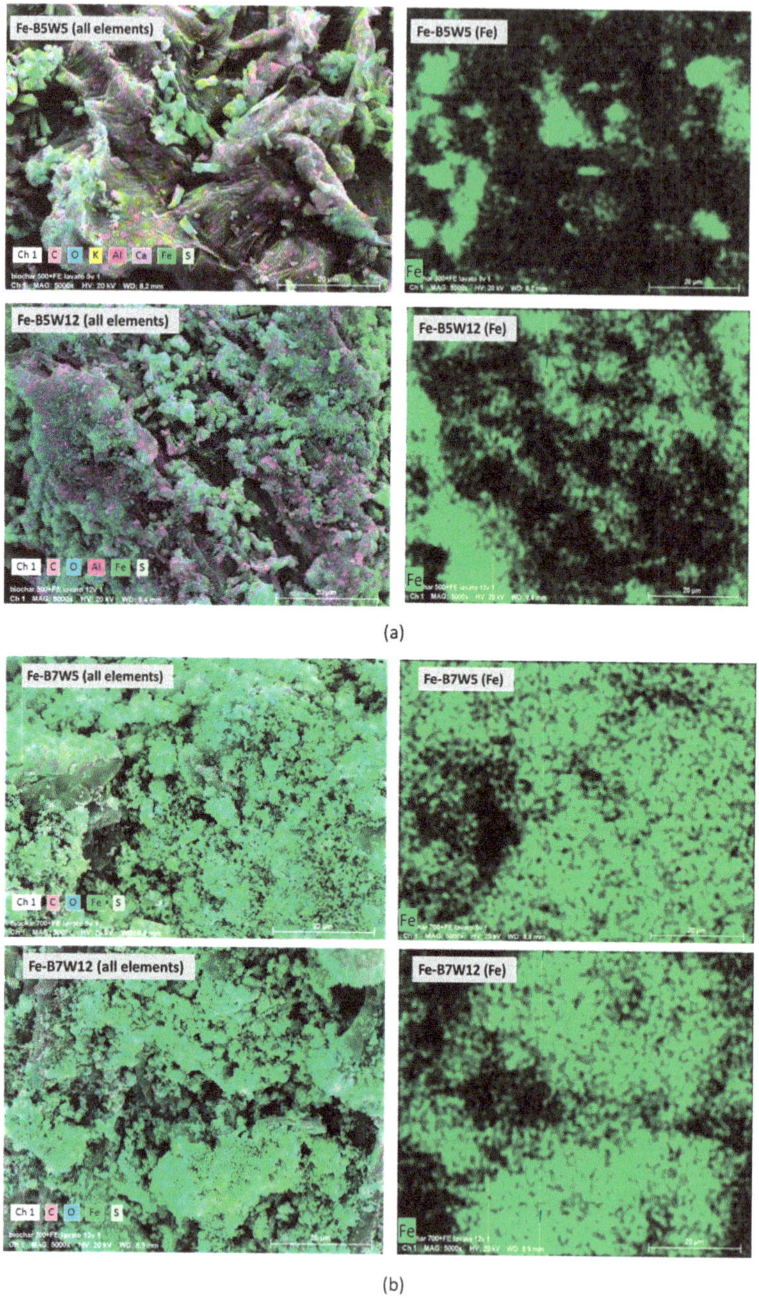

Figure 8. SEM-EDX analysis of (**a**) Fe-B5W5 and Fe-B5W12 and (**b**) Fe-B7W5 and Fe-B7W12.

Moreover, in the case of the samples pyrolysed at 500 °C, the residual functional surface groups, observed by FTIR analysis, are available to preferentially interact with the iron phase, thus explaining iron clustering in preferential areas of Fe-B5W5 and Fe-B5W12 (Figure 8a). In B7-based systems, such a preferential interaction could hardly occur in view of the limited number of functional groups retained at 700 °C.

Finally, considering the very low microstrain of Fe_3O_4, calculated by Rietveld analysis for all the catalysts (Table 5), it can be speculated that the iron-containing clusters are only weakly interacting with the biochar structure.

3.2. Photocatalytic and Adsorption Tests

The photocatalytic activities of Fe-B5W5, Fe-B5W12, Fe-B7W5, and Fe-B7W12 were first investigated as a strategy for the removal of Diclofenac (DFC) from a solution with an initial concentration of 10 mg L^{-1}. Experiments were performed according to the methodological approach described in Section 2.4. The outcomes of the photocatalytic activity of the Fe-supported catalysts are reported in Figure 9a, while the corresponding H_2O_2 decomposition reaction is plotted in Figure 9b.

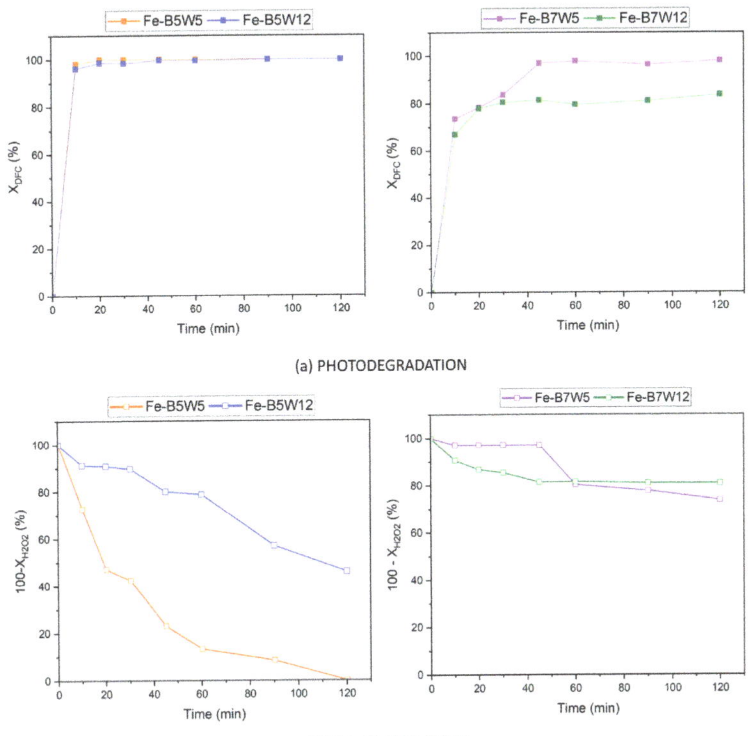

Figure 9. DFC removal by Fe-B5 and Fe-B7-based catalysts: (**a**) photocatalytic activity and (**b**) H_2O_2 decomposition. (Operating conditions: 10 mg L^{-1} DFC, 10–120 min, 0.1 g catalyst, UV light at 365 nm).

Total DFC removal was achieved by all the catalysts, except for Fe-B7W12, whose yield was not higher than 80% (Figure 9a). However, the total conversion of hydrogen peroxide was observed only for the Fe-B5W5 sample (Figure 9b), suggesting that the other Fe-based catalysts are less efficient in hydrogen peroxide decomposition to hydroxyl radicals; as a matter of fact, hydroxyls are directly involved in the photocatalytic process; thus the H_2O_2 conversion is an indication of the photocatalytic reaction efficiency [82].

The photodegradation reaction was irrespective of both iron content and surface areas, since the better performing catalysts (Fe-B5W5 and Fe-B5W12) retained the lower Fe content and were not characterized by the higher surface areas.

Instead, the photocatalytic activity seems related to the temperature of pyrolysis; indeed, the higher efficiency was obtained for the samples pyrolysed at 500 °C, and it could be related to the presence of the surface functional groups, as suggested by the FTIR analysis of these samples.

Notably, in the case of Fe-B7W12, the iron loss upon washing, could be partially compensated by the high dispersion of the iron phase, thereby preventing a potential marked decrease in photoactivity.

Moreover, iron leaching was experimentally verified at the end of the photocatalytic experiments. A very limited, practically negligible, iron leaching of 0–0.005 mg L^{-1} was measured; therefore, if can be concluded that all the iron loss has occurred during the washing process.

Furthermore, in the samples obtained at 500 °C, the photodegradation is faster (100% in 10 min) than in the samples obtained at 700° (80–100% in 40 min), maybe due to the different catalysts' texture.

TOC analysis was also performed to evaluate whether DFC was completely mineralised to CO_2 and H_2O, or if any by-products were generated during the photocatalytic process. A residual TOC of about 5–5.7 mg L^{-1} was observed at the end of the process, indicating that part of the DFC was not fully degraded, and some organic by-products remained in the medium, contributing to the detected TOC.

The degradation mixtures can be highly complex and dependent on the catalysts' nature. The degradation products have been fully studied in the literature, both in homogeneous [83] and heterogeneous [84] processes. The by-products can be, for instance, the result of hydroxylation, decarboxylation, and oxidation processes, as well as those resulting from the degradation of recalcitrant organic acids, acetic, maleic, oxalic, and formic [84].

The potential application of biochar as a catalyst in advanced oxidation processes (AOPs) is closely linked to its physico-chemical characteristics. The oxygen-containing functional groups on the surface of biochar—primarily carboxyl, carbonyl, hydroxyl, and ester groups—play a key role in determining the catalytic, adsorptive, and redox properties. Indeed, these functional groups can activate the oxidising agents, the hydrogen peroxide in this work, by donating electrons. This electron transfer leads to the formation of reactive species, including hydroxyl radicals [85].

In addition to oxidant activation, the oxygen-containing groups at the biochar surface can adsorb pharmaceutical contaminants like Diclofenac; thus increasing the probability of interactions between this pollutant and the reactive radicals, thereby enhancing the degradation efficiency of the process [86].

Furthermore, as a carbon-based material, biochar can contain structural defects such as curvatures, vacancies, and edge defects. These defects also facilitate the activation of the oxidant and the subsequent formation of the radical species. Indeed, the adsorption of the oxidant onto the biochar surface, via sigma (σ) bonds and interactions with delocalised π-electrons, enables electron transfer from the biochar to the oxidant, resulting in the generation of reactive radicals, such as hydroxyl ($^{\bullet}HO$).

Finally, the iron active species present at the catalyst surface also can generate different reactive species, including, $^{\bullet}HO$ and $O_2^{\bullet-}$, which contribute to the Diclofenac degradation in the presence of light and hydrogen peroxide.

Figure 10 presents a simplified mechanism for Diclofenac degradation by the use of Fe-biochar-based catalysts proposed on the basis of the literature. When photons interact with the active phase at the catalyst surface, they photoexcite electrons, promoting them from the valence band to the conduction band, thus generating positively charged holes (h^+). Both the excited electrons (e^-) and the holes (h^+) can activate hydrogen peroxide, dissolved oxygen, hydroxyl ions, and/or the iron active phase present at the catalyst surface, leading to the formation of reactive species. Such species, for instance hydroxyl radicals (•OH) or superoxide radicals ($O_2^{\bullet-}$), are able to degrade Diclofenac, possibly until its complete mineralisation.

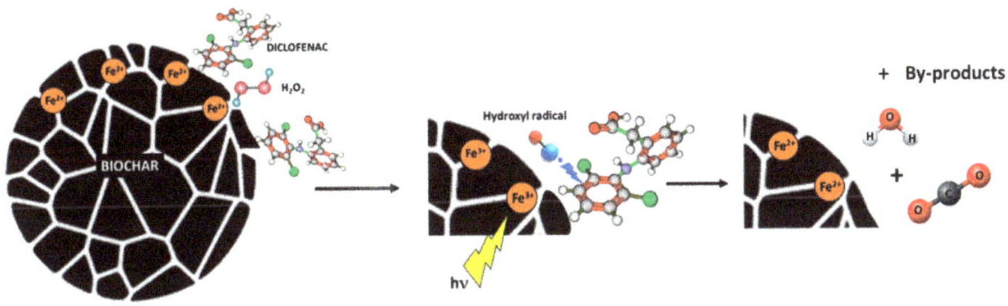

Figure 10. Simplified reaction pathway of the Diclofenac degradation by the Fe-biochar catalysts.

On the samples with the best photocatalytic performances, namely, Fe-B5W5 and Fe-B7W5, the adsorption capacity towards a DFC solution at an initial concentration of 10 mg L^{-1} was also investigated, as adsorption is the most probable reaction occurring in the system, other than photocatalysis.

As a matter of fact, it is difficult to distinguish between these two processes, and certain considerations should be carried out regarding the H_2O_2 decomposition reaction, given that the adsorption process occurs in the absence of H_2O_2.

For sake of comparison, the adsorption capability of pristine B5 and B7 was also assessed, and the results are reported in Figure 11.

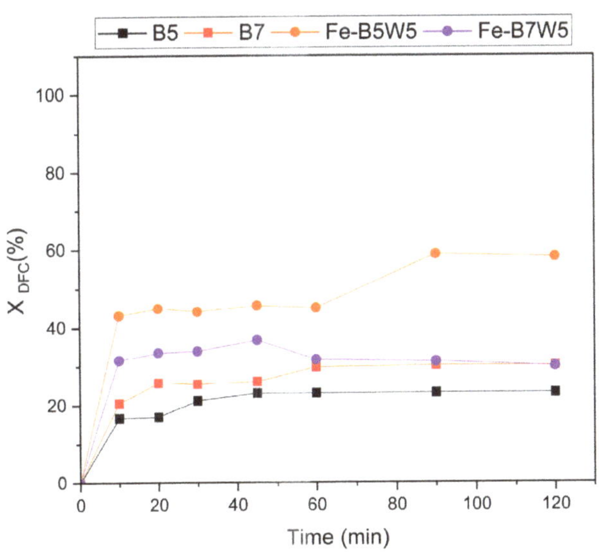

Figure 11. Absorption of DFC by B5, B7, Fe-B5W5, and Fe-B7W5. (Operating conditions: 10 mg L^{-1} DFC, 10–120 min, 0.1 g catalyst).

DFC removal by adsorption in the range of 20–60% was achieved by all the samples. However, Fe-B5W5 (Figure 11) showed the highest removal efficiency (60%), while B5 the lowest one (20%) at the end of the experiment (120 min). On the contrary, B7 and Fe-B7W12 achieved comparable DFC adsorptions of about 30% (Figure 11).

These results suggest that the adsorption capacity of B5 and B7, and of the corresponding catalysts, is not solely related to their textural properties, as B7, the biochar with the highest S_{BET} (343 m^2 g^{-1}), did not give the highest DFC adsorption. Instead, the high DFC adsorption observed in Fe-B5W5 could be explained with the presence of a larger number of surface groups, for instance, hydroxyls or aromatic rings (detected by FTIR) able to coordinate DFC better [87]. Similarly, Bano et al. [88] found a maximum DFC adsorption of 40.2% when the initial DFC concentration was set at 20 mg L^{-1} and the adsorbent dose at 0.1 g L^{-1}, and, a maximum DFC removal efficiency equal to 97.8% was obtained by increasing the adsorbent dose.

The improvement in pharmaceutical removal by biochar-based photocatalysts was attributed to the synergistic effect between adsorption and photodegradation, as the photodegradation degrade the pollutants more effectively in comparison to adsorbent and photocatalyst used separately. Biochar, thanks to its porous structure and adsorption capacity, facilitates the distribution of the active phase, iron in this case, thus improving the contact between reagents and contaminants. Moreover, biochar acting as a support for iron improves the stability and the efficiency of the catalyst [89].

A possible mechanism of interaction of DFC, while adsorbed, can be hypothesised only on the bases of the literature report, as data here reported were not performed targeted to mechanistic studies, and devoted experiments would be required on purpose. However, on the basis of literature indications, it can be assumed that the formation of π-π interactions or hydrogen bonds, proposed in the literature, are also active in our samples [87,90]. As a matter of fact, π-π interactions should be favoured by the presence of the observed aromatic rings in the layers of both B5 and B7 biochars, being them able to interact with the DFC molecules. Moreover, at the catalyst surfaces Fe ions could also behave as Lewis acid sites, coordinating the electron-rich regions of the DFC molecule, and finally H-bonds between the sorbent and DFC could be responsible for the above interaction. Finally, pH value during adsorption may also exert an influence on the adsorption process [91,92], by favouring interaction among species with opposite charges [91,93].

It can be concluded that, in the case of both Fe-B5W5 and Fe-B7W12, the observed total DFC removal could be mainly attributed to the photodegradation process, while in the case of Fe-B5W12 and Fe-B7W12, the observed DFC removal could be the combined effect of both adsorption and photocatalysis. In any case, the occurrence of the photodegradation process is fundamental for DFC total removal, because no removal higher than 60% can be reached by adsorption.

4. Conclusions

The feasibility of using chestnut waste as a green and circular material for developing iron-based photocatalysts has been demonstrated in this study. Specifically, chestnut waste-derived materials were used to produce iron-based biochar catalysts capable of removing a pharmaceutical pollutant, such as Diclofenac, through a hybrid adsorption/photocatalytic process.

Four Fe-based catalysts, namely Fe-B5W5, Fe-B5W12, Fe-B7W5, and Fe-B7W12 along with two pristine biochars, B5 and B7, were prepared via pyrolysis at 500 °C and 700 °C, respectively, and fully characterised.

The physico-chemical properties of the materials were primarily determined by the thermal treatment, leading to very different textural properties, such as, in the case of pristine biochar (B5 and B7) for which very different surface areas of 3.5 m^2 g^{-1} (B5) and 343.2 m^2 g^{-1} (B7) were measured. Differences in textural properties were still preserved by the catalysts, for which a broad surface area range, 55–162 m^2 g^{-1}, was found depending on both pyrolysis and washing conditions.

After the synthesis and the repeated washing cycles, the Fe-based catalysts retained an iron content in the range of 15–20%, determined by EDX analysis. Iron was incorporated as isotropic magnetite (Fe_3O_4) grains, characterised by the presence of Fe vacancies in the structure. Iron is more homogenously distributed in the Fe-B7-based samples, irrespective of the number of washing cycles. This different dispersion of iron was associated with the synthesis route of the materials, where the Fe_3O_4 formation is prone to the microporosity of the biochar. Finally, considering the very low microstrain of Fe_3O_4, calculated by Rietveld analysis, it can be speculated that the iron-containing clusters are only weakly interacting with the biochar structure.

The pyrolysis temperature also influences the number of the surface sites, thus resulting in a larger number of them when the samples were obtained at the lower temperature (i.e., 500 °C); as a result, carbonates, CO, CC, and COH, are the groups available for DFC adsorption.

All the samples exhibited good adsorption and photocatalytic efficiencies, in the range of 84–100%, and the synergistic effect between these two mechanisms enabled a high—often complete—removal of Diclofenac (DFC).

Finally, the adsorption and photocatalytic properties appeared to be irrespective of the surface area of the materials, since the more active catalyst, Fe-B5W5, was characterised by the lower surface area (55 m^2 g^{-1}); therefore, the nature and the number of the residual functional groups play a pivotal role in the DFC removal efficiency.

Despite the high DFC removal efficiency, the formation of degradation by-products was observed, which may partially compromise the overall effectiveness of the process. Nevertheless, the promising results of this study highlight the potential for further development and possible in-field application of these systems.

Supplementary Materials: The following supporting information can be downloaded at: https://www.mdpi.com/article/10.3390/c11020038/s1, Figure S1: N_2 adsorption–desorption isotherms of biochar materials (B5, B7) and their corresponding Fe-based catalysts, Figure S2: Fe-B5W5 catalyst with magnetic properties, Figure S3: TG analysis of Fe-B5W5 (orange line), Fe-B5W12 (blue line) Fe-B7W5 (violet line), and F-B7W12 (green line) catalysts.

Author Contributions: Conceptualisation, A.G., D.H. and A.B.; methodology, M.G., C.d.l.R. and K.J.-B.; validation, M.G., A.G. and D.H.; formal analysis, M.G., C.d.l.R. and K.J.-B.; investigation, M.G., C.C., M.B., E.F. and A.B.; resources, A.B., E.F., M.B. and C.C.; data curation, M.G.; writing—original draft preparation, M.G., C.C, A.B., M.B. and E.F.; writing—review and editing, M.G., A.B., E.F., M.B., A.G., D.H., K.J.-B., V.M.-R., C.d.l.R. and C.C.; visualisation, M.G. and C.C.; supervision, A.G. and D.H. All authors have read and agreed to the published version of the manuscript.

Funding: The research activities were possible thanks to the financial support received under Project PID2021-124021OB-I00 (URBRAINTREAT), funded by MCIN/AEI/10.13039/501100011033 and "ERDF: A way of making Europe".

Data Availability Statement: Data are available upon request from the authors.

Acknowledgments: The research activities were possible thanks to the financial support received under the European project PON FSE REACT-EU, and the resources allocated by Ministerial Decree No. 1061 of 10 August 2021.

Conflicts of Interest: Author Maurizio Bellotto was employed by the company Opigeo s.r.l. The remaining authors declare that the research was conducted in the absence of any commercial or financial relationships that could be construed as potential conflicts of interest.

Abbreviations

The following abbreviations are used in this manuscript:

B5	Pristine biochar pyrolysed at 500 °C
B7	Pristine biochar pyrolysed at 700 °C
Fe-B5W5	Fe-biochar catalyst pyrolysed at 500 °C and washed 5 times
Fe-B5W12	Fe-biochar catalyst pyrolysed at 500 °C and washed 12 times
Fe-B7W5	Fe-biochar catalyst pyrolysed at 700 °C and washed 5 times
Fe-B7W12	Fe-biochar catalyst pyrolysed at 700 °C and washed 12 times
EA	Elemental Analysis
ICP-OES	Inductively Coupled Plasma–Optical Emission Spectroscopy
TOC	Total Organic Carbon
XRPD	X-Ray Powder Diffraction
TG-DTG	Thermogravimetric analyses
SEM-EDX	Scanning Electron Microscopy and Energy Dispersive X-Ray
AOP	Advanced Oxidation process
NSAID	Non-steroidal Anti-inflammatory Drug
DFC	Diclofenac

References

1. Lakshmi, S.D.; Geetha, B.V.; Vibha, M. From Prescription to Pollution: The Ecological Consequences of NSAIDs in Aquatic Ecosystems. *Toxicol. Rep.* **2024**, *13*, 101775. [CrossRef]
2. Tran, N.H.; Reinhard, M.; Gin, K.Y.-H. Occurrence and Fate of Emerging Contaminants in Municipal Wastewater Treatment Plants from Different Geographical Regions—A Review. *Water Res.* **2018**, *133*, 182–207. [CrossRef]
3. Zorita, S.; Mårtensson, L.; Mathiasson, L. Occurrence and Removal of Pharmaceuticals in a Municipal Sewage Treatment System in the South of Sweden. *Sci. Total Environ.* **2009**, *407*, 2760–2770. [CrossRef] [PubMed]
4. Kermia, A.E.B.; Fouial-Djebbar, D.; Trari, M. Occurrence, Fate and Removal Efficiencies of Pharmaceuticals in Wastewater Treatment Plants (WWTPs) Discharging in the Coastal Environment of Algiers. *Comptes Rendus Chim.* **2016**, *19*, 963–970. [CrossRef]
5. Sun, Q.; Li, M.; Ma, C.; Chen, X.; Xie, X.; Yu, C.-P. Seasonal and Spatial Variations of PPCP Occurrence, Removal and Mass Loading in Three Wastewater Treatment Plants Located in Different Urbanization Areas in Xiamen, China. *Environ. Pollut.* **2016**, *208*, 371–381. [CrossRef]
6. Kosma, C.I.; Lambropoulou, D.A.; Albanis, T.A. Investigation of PPCPs in Wastewater Treatment Plants in Greece: Occurrence, Removal and Environmental Risk Assessment. *Sci. Total Environ.* **2014**, *466–467*, 421–438. [CrossRef]
7. Mlunguza, N.Y.; Ncube, S.; Nokwethemba Mahlambi, P.; Chimuka, L.; Madikizela, L.M. Adsorbents and Removal Strategies of Non-Steroidal Anti-Inflammatory Drugs from Contaminated Water Bodies. *J. Environ. Chem. Eng.* **2019**, *7*, 103142. [CrossRef]
8. De Andrade, J.R.; Oliveira, M.F.; Da Silva, M.G.C.; Vieira, M.G.A. Adsorption of Pharmaceuticals from Water and Wastewater Using Nonconventional Low-Cost Materials: A Review. *Ind. Eng. Chem. Res.* **2018**, *57*, 3103–3127. [CrossRef]
9. Gautam, M.K.; Mondal, T.; Nath, R.; Mahajon, B.; Chincholikar, M.; Bose, A.; Das, D.; Das, R.; Mondal, S. Harnessing Activated Hydrochars: A Novel Approach for Pharmaceutical Contaminant Removal. *C* **2024**, *10*, 8. [CrossRef]
10. Alessandretti, I.; Rigueto, C.V.T.; Nazari, M.T.; Rosseto, M.; Dettmer, A. Removal of Diclofenac from Wastewater: A Comprehensive Review of Detection, Characteristics and Tertiary Treatment Techniques. *J. Environ. Chem. Eng.* **2021**, *9*, 106743. [CrossRef]
11. Shan, R.; Han, J.; Gu, J.; Yuan, H.; Luo, B.; Chen, Y. A Review of Recent Developments in Catalytic Applications of Biochar-Based Materials. *Resour. Conserv. Recycl.* **2020**, *162*, 105036. [CrossRef]
12. Zhou, X.; Zhu, Y.; Niu, Q.; Zeng, G.; Lai, C.; Liu, S.; Huang, D.; Qin, L.; Liu, X.; Li, B.; et al. New Notion of Biochar: A Review on the Mechanism of Biochar Applications in Advannced Oxidation Processes. *Chem. Eng. J.* **2021**, *416*, 129027. [CrossRef]
13. Lai, C.; Huang, F.; Zeng, G.; Huang, D.; Qin, L.; Cheng, M.; Zhang, C.; Li, B.; Yi, H.; Liu, S.; et al. Fabrication of Novel Magnetic $MnFe_2O_4$/Bio-Char Composite and Heterogeneous Photo-Fenton Degradation of Tetracycline in near Neutral pH. *Chemosphere* **2019**, *224*, 910–921. [CrossRef] [PubMed]
14. Hermosilla, D.; Cortijo, M.; Huang, C.P. The Role of Iron on the Degradation and Mineralization of Organic Compounds Using Conventional Fenton and Photo-Fenton Processes. *Chem. Eng. J.* **2009**, *155*, 637–646. [CrossRef]
15. Thomas, N.; Dionysiou, D.D.; Pillai, S.C. Heterogeneous Fenton Catalysts: A Review of Recent Advances. *J. Hazard. Mater.* **2021**, *404*, 124082. [CrossRef]

16. Gou, Y.; Chen, P.; Yang, L.; Li, S.; Peng, L.; Song, S.; Xu, Y. Degradation of Fluoroquinolones in Homogeneous and Heterogeneous Photo-Fenton Processes: A Review. *Chemosphere* **2021**, *270*, 129481. [CrossRef] [PubMed]
17. Ramos, R.; Abdelkader-Fernández, V.K.; Matos, R.; Peixoto, A.F.; Fernandes, D.M. Metal-Supported Biochar Catalysts for Sustainable Biorefinery, Electrocatalysis, and Energy Storage Applications: A Review. *Catalysts* **2022**, *12*, 207. [CrossRef]
18. Amalina, F.; Razak, A.S.A.; Krishnan, S.; Zularisam, A.W.; Nasrullah, M. A Comprehensive Assessment of the Method for Producing Biochar, Its Characterization, Stability, and Potential Applications in Regenerative Economic Sustainability—A Review. *Clean. Mater.* **2022**, *3*, 100045. [CrossRef]
19. Wang, J.; Wang, S. Preparation, Modification and Environmental Application of Biochar: A Review. *J. Clean. Prod.* **2019**, *227*, 1002–1022. [CrossRef]
20. Okiemute Akpasi, S.; Michael Smarte Anekwe, I.; Adedeji, J.; Lewis Kiambi, S. Biochar Development as a Catalyst and Its Application. In *Biochar—Productive Technologies, Properties and Applications*; Bartoli, M., Giorcelli, M., Tagliaferro, A., Eds.; IntechOpen: London, UK, 2023; ISBN 978-1-80356-251-3.
21. Lee, J.; Kim, K.-H.; Kwon, E.E. Biochar as a Catalyst. *Renew. Sustain. Energy Rev.* **2017**, *77*, 70–79. [CrossRef]
22. Lyu, H.; Zhang, Q.; Shen, B. Application of Biochar and Its Composites in Catalysis. *Chemosphere* **2020**, *240*, 124842. [CrossRef]
23. Yuan, X.; Cao, Y.; Li, J.; Patel, A.K.; Dong, C.-D.; Jin, X.; Gu, C.; Yip, A.C.K.; Tsang, D.C.W.; Ok, Y.S. Recent Advancements and Challenges in Emerging Applications of Biochar-Based Catalysts. *Biotechnol. Adv.* **2023**, *67*, 108181. [CrossRef] [PubMed]
24. Devi, P.; Dalai, A.K.; Chaurasia, S.P. Activity and Stability of Biochar in Hydrogen Peroxide Based Oxidation System for Degradation of Naphthenic Acid. *Chemosphere* **2020**, *241*, 125007. [CrossRef]
25. Fang, G.; Liu, C.; Wang, Y.; Dionysiou, D.D.; Zhou, D. Photogeneration of Reactive Oxygen Species from Biochar Suspension for Diethyl Phthalate Degradation. *Appl. Catal. B Environ.* **2017**, *214*, 34–45. [CrossRef]
26. Zhou, X.; Zeng, Z.; Zeng, G.; Lai, C.; Xiao, R.; Liu, S.; Huang, D.; Qin, L.; Liu, X.; Li, B.; et al. Insight into the Mechanism of Persulfate Activated by Bone Char: Unraveling the Role of Functional Structure of Biochar. *Chem. Eng. J.* **2020**, *401*, 126127. [CrossRef]
27. Zhu, N.; Li, C.; Bu, L.; Tang, C.; Wang, S.; Duan, P.; Yao, L.; Tang, J.; Dionysiou, D.D.; Wu, Y. Bismuth Impregnated Biochar for Efficient Estrone Degradation: The Synergistic Effect between Biochar and Bi/Bi_2O_3 for a High Photocatalytic Performance. *J. Hazard. Mater.* **2020**, *384*, 121258. [CrossRef] [PubMed]
28. Subramaniam, M.N.; Wu, Z.; Goh, P.S.; Zhou, S. The State-of-the-Art Development of Biochar Based Photocatalyst for Removal of Various Organic Pollutants in Wastewater. *J. Clean. Prod.* **2023**, *429*, 139487. [CrossRef]
29. Park, J.-H.; Wang, J.J.; Xiao, R.; Tafti, N.; DeLaune, R.D.; Seo, D.-C. Degradation of Orange G by Fenton-like Reaction with Fe-Impregnated Biochar Catalyst. *Bioresour. Technol.* **2018**, *249*, 368–376. [CrossRef]
30. Huang, Y.; Gao, M.; Deng, Y.; Khan, Z.H.; Liu, X.; Song, Z.; Qiu, W. Efficient Oxidation and Adsorption of As(III) and As(V) in Water Using a Fenton-like Reagent, (Ferrihydrite)-Loaded Biochar. *Sci. Total Environ.* **2020**, *715*, 136957. [CrossRef]
31. Yi, Y.; Tu, G.; Zhao, D.; Tsang, P.E.; Fang, Z. Pyrolysis of Different Biomass Pre-Impregnated with Steel Pickling Waste Liquor to Prepare Magnetic Biochars and Their Use for the Degradation of Metronidazole. *Bioresour. Technol.* **2019**, *289*, 121613. [CrossRef]
32. Wang, H.; Wang, H.; Zhao, H.; Yan, Q. Adsorption and Fenton-like Removal of Chelated Nickel from Zn-Ni Alloy Electroplating Wastewater Using Activated Biochar Composite Derived from Taihu Blue Algae. *Chem. Eng. J.* **2020**, *379*, 122372. [CrossRef]
33. Yan, J.; Qian, L.; Gao, W.; Chen, Y.; Ouyang, D.; Chen, M. Enhanced Fenton-like Degradation of Trichloroethylene by Hydrogen Peroxide Activated with Nanoscale Zero Valent Iron Loaded on Biochar. *Sci. Rep.* **2017**, *7*, 43051. [CrossRef] [PubMed]
34. Welter, N.; Leichtweis, J.; Silvestri, S.; Sánchez, P.I.Z.; Mejía, A.C.C.; Carissimi, E. Preparation of a New Green Composite Based on Chitin Biochar and $ZnFe_2O_4$ for Photo-Fenton Degradation of Rhodamine B. *J. Alloys Compd.* **2022**, *901*, 163758. [CrossRef]
35. Xin, S.; Ma, B.; Liu, G.; Ma, X.; Zhang, C.; Ma, X.; Gao, M.; Xin, Y. Enhanced Heterogeneous Photo-Fenton-like Degradation of Tetracycline over $CuFeO_2$/Biochar Catalyst through Accelerating Electron Transfer under Visible Light. *J. Environ. Manag.* **2021**, *285*, 112093. [CrossRef]
36. Wang, Z.; Li, Y.; Xie, X.; Wang, Z. Bifunctional $MnFe_2O_4$/Chitosan Modified Biochar Composite for Enhanced Methyl Orange Removal Based on Adsorption and Photo-Fenton Process. *Colloids Surf. A Physicochem. Eng. Asp.* **2021**, *613*, 126104. [CrossRef]
37. Bashir, A.; Pandith, A.H.; Qureashi, A.; Malik, L.A.; Gani, M.; Perez, J.M. Catalytic Propensity of Biochar Decorated with Core-Shell nZVI@Fe_3O_4: A Sustainable Photo-Fenton Catalysis of Methylene Blue Dye and Reduction of 4-Nitrophenol. *J. Environ. Chem. Eng.* **2022**, *10*, 107401. [CrossRef]
38. Chen, X.; Zhang, M.; Qin, H.; Zhou, J.; Shen, Q.; Wang, K.; Chen, W.; Liu, M.; Li, N. Synergy Effect between Adsorption and Heterogeneous Photo-Fenton-like Catalysis on $LaFeO_3$/Lignin-Biochar Composites for High Efficiency Degradation of Ofloxacin under Visible Light. *Sep. Purif. Technol.* **2022**, *280*, 119751. [CrossRef]
39. FAOSTAT. Available online: https://www.fao.org/faostat/en/#data/QCL (accessed on 15 November 2024).
40. De Vasconcelos, M.C.; Bennett, R.N.; Rosa, E.A.; Ferreira-Cardoso, J.V. Composition of European Chestnut (*Castanea sativa* Mill.) and Association with Health Effects: Fresh and Processed Products. *J. Sci. Food Agric.* **2010**, *90*, 1578–1589. [CrossRef]

41. Eurostat. Food Waste and Food Waste Prevention-Estimates. Available online: https://ec.europa.eu/eurostat/statistics-explained/index.php?title=Food_waste_and_food_waste_prevention_-_estimates (accessed on 15 January 2024).
42. Foong, S.Y.; Liew, R.K.; Yang, Y.; Cheng, Y.W.; Yek, P.N.Y.; Wan Mahari, W.A.; Lee, X.Y.; Han, C.S.; Vo, D.-V.N.; Van Le, Q.; et al. Valorization of Biomass Waste to Engineered Activated Biochar by Microwave Pyrolysis: Progress, Challenges, and Future Directions. *Chem. Eng. J.* **2020**, *389*, 124401. [CrossRef]
43. Blasi, A.; Verardi, A.; Lopresto, C.G.; Siciliano, S.; Sangiorgio, P. Lignocellulosic Agricultural Waste Valorization to Obtain Valuable Products: An Overview. *Recycling* **2023**, *8*, 61. [CrossRef]
44. Singh, A.; Himanshu, M.; Verma, B.; Singh, R.; Lal, B.; Syed, A.; Elgorban, A.M.; Wong, L.S.; Srivastava, N. Evaluation of Sustainability of Fabrication Process and Characterization Studies of Activated Carbon Nanocatalyst from Waste Chestnut Peels. *J. Mol. Struct.* **2025**, *1321*, 139810. [CrossRef]
45. Nguyen, T.-K.-T.; Nguyen, T.-B.; Chen, C.-W.; Chen, W.-H.; Bui, X.-T.; Lam, S.S.; Dong, C.-D. Boosting Acetaminophen Degradation in Water by Peracetic Acid Activation: A Novel Approach Using Chestnut Shell-Derived Biochar at Varied Pyrolysis Temperatures. *Environ. Res.* **2024**, *252*, 119143. [CrossRef] [PubMed]
46. Zhao, L.; Zhang, H.; Dai, Z.; Zhang, A.-Y.; Yin, J.; Peng, S.; Liang, H. Recycling Chestnut Shell for Superior Peroxymonosulfate Activation in Contaminants Degradation via the Synergistic Radical/Non-Radical Mechanisms. *J. Hazard. Mater.* **2022**, *430*, 128471. [CrossRef] [PubMed]
47. Altman, R.; Bosch, B.; Brune, K.; Patrignani, P.; Young, C. Advances in NSAID Development: Evolution of Diclofenac Products Using Pharmaceutical Technology. *Drugs* **2015**, *75*, 859–877. [CrossRef] [PubMed]
48. De Oliveira, T.; Guégan, R.; Thiebault, T.; Milbeau, C.L.; Muller, F.; Teixeira, V.; Giovanela, M.; Boussafir, M. Adsorption of Diclofenac onto Organoclays: Effects of Surfactant and Environmental (pH and Temperature) Conditions. *J. Hazard. Mater.* **2017**, *323*, 558–566. [CrossRef]
49. Muelas-Ramos, V.; Gascó, A.; Salvatierra, M.; De Los Ríos, C.; Jiménez-Bautista, K.; Merayo, N.; Bahamonde, A.; Hermosilla, D. Assessing the Application of a Biochar-Supported Iron Oxide Catalyst to the Treatment of Imidacloprid by Photo-Fenton Technologies. *Catal. Today* **2024**, *438*, 114782. [CrossRef]
50. Shen, T.; Zhang, F.; Yang, S.; Wang, Y.; Liu, H.; Wang, H.; Hu, J. Comprehensive Study on the Pyrolysis Process of Chestnut Processing Waste (Chestnut Shells): Kinetic Triplet, Thermodynamic, in-Situ Monitoring of Evolved Gasses and Analysis Biochar. *Fuel* **2023**, *331*, 125944. [CrossRef]
51. Burbano, A.A.; Medina, G.A.M.; Sánchez, F.H.; Lassalle, V.L.; Horst, M.F.; Gascó, G.; Méndez, A. Influence of Post-Pyrolysis Treatment on Physicochemical Properties and Acid Medium Stability of Magnetic Carbon Nanocomposites. *Biomass Conv. Bioref.* **2024**, *14*, 27871–27884. [CrossRef]
52. Tamura, H.; Goto, K.; Yotsuyanagi, T.; Nagayama, M. Spectrophotometric determination of iron(II) with 1,10-phenanthroline in the presence of large amounts of iron(III). *Talanta* **1974**, *21*, 314–318. [CrossRef]
53. Brunauer, S.; Emmett, P.H.; Teller, E. Adsorption of Gases in Multimolecular Layers. *J. Am. Chem. Soc.* **1938**, *60*, 309–319. [CrossRef]
54. Doebelin, N.; Kleeberg, R. *Profex*: A Graphical User Interface for the Rietveld Refinement Program *BGMN*. *J. Appl. Crystallogr.* **2015**, *48*, 1573–1580. [CrossRef]
55. Pobiner, H. Determination of Hydroperoxides in Hydrocarbon by Conversion to Hydrogen Peroxide and Measurement by Titanium Complexing. *Anal. Chem.* **1961**, *33*, 1423–1426. [CrossRef]
56. Yang, Q.; Sun, Y.; Sun, W.; Qin, Z.; Liu, H.; Ma, Y.; Wang, X. Cellulose Derived Biochar: Preparation, Characterization and Benzo[a]Pyrene Adsorption Capacity. *Grain Oil Sci. Technol.* **2021**, *4*, 182–190. [CrossRef]
57. Tomczyk, A.; Sokołowska, Z.; Boguta, P. Biochar Physicochemical Properties: Pyrolysis Temperature and Feedstock Kind Effects. *Rev. Environ. Sci. Biotechnol.* **2020**, *19*, 191–215. [CrossRef]
58. Chen, B.; Chen, Z. Sorption of Naphthalene and 1-Naphthol by Biochars of Orange Peels with Different Pyrolytic Temperatures. *Chemosphere* **2009**, *76*, 127–133. [CrossRef]
59. Katyal, S.; Thambimuthu, K.; Valix, M. Carbonisation of Bagasse in a Fixed Bed Reactor: Influence of Process Variables on Char Yield and Characteristics. *Renew. Energy* **2003**, *28*, 713–725. [CrossRef]
60. Rafiq, M.K.; Bachmann, R.T.; Rafiq, M.T.; Shang, Z.; Joseph, S.; Long, R. Influence of Pyrolysis Temperature on Physico-Chemical Properties of Corn Stover (*Zea mays* L.) Biochar and Feasibility for Carbon Capture and Energy Balance. *PLoS ONE* **2016**, *11*, e0156894. [CrossRef]
61. Da Luz Corrêa, A.P.; Bastos, R.R.C.; Rocha Filho, G.N.D.; Zamian, J.R.; Conceição, L.R.V.D. Preparation of Sulfonated Carbon-Based Catalysts from Murumuru Kernel Shell and Their Performance in the Esterification Reaction. *RSC Adv.* **2020**, *10*, 20245–20256. [CrossRef]
62. Rodríguez-Sánchez, S.; Ruiz, B.; Martínez-Blanco, D.; Sánchez-Arenillas, M.; Diez, M.A.; Suárez-Ruiz, I.; Marco, J.F.; Blanco, J.; Fuente, E. Sustainable Thermochemical Single-Step Process To Obtain Magnetic Activated Carbons from Chestnut Industrial Wastes. *ACS Sustain. Chem. Eng.* **2019**, *7*, 17293–17305. [CrossRef]

63. Rahman, M.A.; Oomori, T. Structure, Crystallization and Mineral Composition of Sclerites in the Alcyonarian Coral. *J. Cryst. Growth* **2008**, *310*, 3528–3534. [CrossRef]
64. Bekiaris, G.; Peltre, C.; Jensen, L.S.; Bruun, S. Using FTIR-Photoacoustic Spectroscopy for Phosphorus Speciation Analysis of Biochars. *Spectrochim. Acta Part A Mol. Biomol. Spectrosc.* **2016**, *168*, 29–36. [CrossRef]
65. Ivanova, N.V.; Korolenko, E.A.; Korolik, E.V.; Zhbankov, R.G. IR Spectrum of Cellulose. *J. Appl. Spectrosc.* **1989**, *51*, 847–851. [CrossRef]
66. Yang, R.; Liu, G.; Li, M.; Zhang, J.; Hao, X. Preparation and N_2, CO_2 and H_2 Adsorption of Super Activated Carbon Derived from Biomass Source Hemp (*Cannabis sativa* L.) Stem. *Microporous Mesoporous Mater.* **2012**, *158*, 108–116. [CrossRef]
67. Jeyasubramanian, K.; Thangagiri, B.; Sakthivel, A.; Dhaveethu Raja, J.; Seenivasan, S.; Vallinayagam, P.; Madhavan, D.; Malathi Devi, S.; Rathika, B. A Complete Review on Biochar: Production, Property, Multifaceted Applications, Interaction Mechanism and Computational Approach. *Fuel* **2021**, *292*, 120243. [CrossRef]
68. De Lima, R.S.; De Paiva E Silva Zanta, C.L.; Meili, L.; Dos Santos Lins, P.V.; De Souza Dos Santos, G.E.; Tonholo, J. Fenton-Based Processes for the Regeneration of Biochar from Syagrus Coronata Biomass Used as Dye Adsorbent. *Desalin. Water Treat.* **2019**, *162*, 391–398. [CrossRef]
69. Karunadasa, K.S.P.; Manoratne, C.H.; Pitawala, H.M.T.G.A.; Rajapakse, R.M.G. Thermal Decomposition of Calcium Carbonate (Calcite Polymorph) as Examined by in-Situ High-Temperature X-Ray Powder Diffraction. *J. Phys. Chem. Solids* **2019**, *134*, 21–28. [CrossRef]
70. Thies, J.E.; Rillig, M.C. Characteristics of Biochar: Biological Properties. In *Biochar for Environmental Management*; Routledge: London, UK, 2012.
71. Hong, X.; Fang, C.; Hui, K.S.; Hui, K.N.; Zhuang, H.; Liu, W.; Shan, S. Influence of Interfering Anions on Cu^{2+} and Zn^{2+} Ions Removal on Chestnut Outer Shell-Derived Hydrochars in Aqueous Solution. *RSC Adv.* **2017**, *7*, 51199–51205. [CrossRef]
72. López-Beceiro, J.; Díaz-Díaz, A.M.; Álvarez-García, A.; Tarrío-Saavedra, J.; Naya, S.; Artiaga, R. The Complexity of Lignin Thermal Degradation in the Isothermal Context. *Processes* **2021**, *9*, 1154. [CrossRef]
73. Nkomo, N.; Odindo, A.O.; Musazura, W.; Missengue, R. Optimising Pyrolysis Conditions for High-Quality Biochar Production Using Black Soldier Fly Larvae Faecal-Derived Residue as Feedstock. *Heliyon* **2021**, *7*, e07025. [CrossRef]
74. Patnaik, P. *Handbook of Inorganic Chemicals*; McGraw-Hill Handbooks; McGraw-Hill: New York, NY, USA, 2003; ISBN 978-0-07-049439-8.
75. Cervellino, A.; Frison, R.; Cernuto, G.; Guagliardi, A.; Masciocchi, N. Lattice Parameters and Site Occupancy Factors of Magnetite–Maghemite Core–Shell Nanoparticles. A Critical Study. *J. Appl. Crystallogr.* **2014**, *47*, 1755–1761. [CrossRef]
76. Safavi, F.S.; Ebrahimipour, S.Y.; Fatemi, S.J.; Mohammadi, P.; Shamspur, T. Green Synthesis of Silver Nanoparticles and Their Immobilization on Magnetic Biochar for the Removal of Tetracycline and Enrofloxacin. *Biomass Conv. Biorefin.* **2025**. [CrossRef]
77. Gadgeel, A.A.; Mhaske, S.T.; Duerr, C.; Liu, K.L. In-Situ Preparation and Characterization of Aconitic Acid Capped Fe_3O_4 Nanoparticle by Using Citric Acid as a Reducing Agent. *J. Inorg. Organomet. Polym. Mater.* **2019**, *29*, 1688–1700. [CrossRef]
78. Zhao, L.; Bi, S.; Pei, J.; Li, X.; Yu, R.; Zhao, J.; Martyniuk, C.J. Adsorption Performance of SO_2 over $ZnAl_2O_4$ Nanospheres. *J. Ind. Eng. Chem.* **2016**, *41*, 151–157. [CrossRef]
79. Joshi, R.; Singh, B.P.; Ningthoujam, R.S. Confirmation of Highly Stable 10 Nm Sized Fe_3O_4 Nanoparticle Formation at Room Temperature and Understanding of Heat-Generation under AC Magnetic Fields for Potential Application in Hyperthermia. *AIP Adv.* **2020**, *10*, 105033. [CrossRef]
80. Masset, P.; Poinso, J.Y.; Poignet, J.C. TG/DTA/MS Study of the Thermal Decomposition of $FeSO_4{\cdot}6H_2O$. *J. Therm. Anal. Calorim.* **2006**, *83*, 457–462. [CrossRef]
81. Chaklader, A.C.D.; Blair, G.R. Differential thermal study of FeO and Fe_3O_4. *J. Therm. Anal.* **1970**, *2*, 165–179. [CrossRef]
82. Gallego-Ramírez, C.; Chica, E.; Rubio-Clemente, A. Combination of Biochar and Advanced Oxidation Processes for the Sustainable Elimination of Pharmaceuticals in Water. *Sustainability* **2024**, *16*, 10761. [CrossRef]
83. Iovino, P.; Chianese, S.; Canzano, S.; Prisciandaro, M.; Musmarra, D. Photodegradation of Diclofenac in Wastewaters. *Desalination Water Treat.* **2017**, *61*, 293–297. [CrossRef]
84. Pérez-Estrada, L.A.; Malato, S.; Gernjak, W.; Agüera, A.; Thurman, E.M.; Ferrer, I.; Fernández-Alba, A.R. Photo-Fenton Degradation of Diclofenac: Identification of Main Intermediates and Degradation Pathway. *Environ. Sci. Technol.* **2005**, *39*, 8300–8306. [CrossRef]
85. Jiang, T.; Wang, B.; Gao, B.; Cheng, N.; Feng, Q.; Chen, M.; Wang, S. Degradation of Organic Pollutants from Water by Biochar-Assisted Advanced Oxidation Processes: Mechanisms and Applications. *J. Hazard. Mater.* **2023**, *442*, 130075. [CrossRef]
86. Inyang, M.; Dickenson, E. The Potential Role of Biochar in the Removal of Organic and Microbial Contaminants from Potable and Reuse Water: A Review. *Chemosphere* **2015**, *134*, 232–240. [CrossRef] [PubMed]
87. Moreno-Castilla, C. Adsorption of Organic Molecules from Aqueous Solutions on Carbon Materials. *Carbon* **2004**, *42*, 83–94. [CrossRef]

88. Bano, A.; Aziz, M.K.; Mishra, R.; Dave, H.; Prasad, B.; Kumari, M.; Dubey, D.; Meili, L.; Shah, M.P.; Prasad, K.S. Response Surface Methodology–Based Optimisation of Adsorption of Diclofenac and Treatment of Pharmaceutical Effluent Using Combined Coagulation-Adsorption onto nFe_2O_3 Decorated Water Chestnut Shells Biochar. *Environ. Sci. Pollut. Res.* **2024**, *31*, 55317–55335. [CrossRef]
89. Zahid, M.; Khan, Z.U.H.; Sun, J.; Muhammad, N.; Sabahat, S.; Shah, N.S.; Iqbal, J. Biochar-Derived Photocatalysts for Pharmaceutical Waste Removal, a Sustainable Approach to Water Purification. *Appl. Surf. Sci. Adv.* **2025**, *26*, 100721. [CrossRef]
90. Chauhan, S.; Shafi, T.; Dubey, B.K.; Chowdhury, S. Biochar-Mediated Removal of Pharmaceutical Compounds from Aqueous Matrices via Adsorption. *Waste Dispos. Sustain. Energy* **2023**, *5*, 37–62. [CrossRef]
91. Lonappan, L.; Rouissi, T.; Kaur Brar, S.; Verma, M.; Surampalli, R.Y. An Insight into the Adsorption of Diclofenac on Different Biochars: Mechanisms, Surface Chemistry, and Thermodynamics. *Bioresour. Technol.* **2018**, *249*, 386–394. [CrossRef] [PubMed]
92. Sun, S.; Hu, Y.; Xu, M.; Cheng, F.; Zhang, H.; Li, Z. Photo-Fenton Degradation of Carbamazepine and Ibuprofen by Iron-Based Metal-Organic Framework under Alkaline Condition. *J. Hazard. Mater.* **2022**, *424*, 127698. [CrossRef]
93. Kaur, K.; Kaur, R.; Kaur, H. A Systematic Review of Lignocellulosic Biomass for Remediation of Environmental Pollutants. *Appl. Surf. Sci. Adv.* **2024**, *19*, 100547. [CrossRef]

C Journal of Carbon Research

MDPI

Article

Fabrication of Wood-Derived Carbon Aerogel/$Mg(OH)_2$ Bio-Composite and Its High Performance for Adsorption and Separation of Cadmium Ions

Ran An [1], Jinyue Liu [1], Haomiao Ma [1], Yuqing Yan [1], Yuanru Guo [1,*], Qingjiang Pan [2] and Shujun Li [1,*]

[1] Engineering Research Center of Advanced Wooden Materials (Ministry of Education), College of Material Science and Engineering, Northeast Forestry University, Harbin 150040, China; anran20000905@163.com (R.A.); ljy121266@163.com (J.L.); nefumhm@163.com (H.M.); yyq20010208@163.com (Y.Y.)

[2] Key Laboratory of Functional Inorganic Material Chemistry (Ministry of Education), School of Chemistry and Materials Science, Heilongjiang University, Harbin 150080, China; panqingjiang@hlju.edu.cn

* Correspondence: guoyr@nefu.edu.cn (Y.G.); lishujun@nefu.edu.cn (S.L.)

Abstract: To address the need for reducing carbon emissions and enhancing the sustainable utilization of non-fossil resources, a one-step calcination strategy has been developed to fabricate hierarchical carbon aerogels from balsa wood. The resulting wood-derived carbon aerogels (WCA) were functionalized with $Mg(OH)_2$ to boost their environmental remediation potential. Comprehensive characterization using XRD, FT-IR, XPS, and SEM confirmed that the optimized WCA/$Mg(OH)_2$ composite (WCAMg) retained a three-dimensional hierarchical porous structure, and $Mg(OH)_2$ nanosheets were attached to it. The adsorption performance of WCAMg composites towards Cd^{2+} was systematically investigated through controlled experiments, which focused on three critical variables ($Mg(OH)_2$ loading content, initial Cd^{2+} concentration and solution ionic strength). The functionalized WCAMg demonstrated a maximum Cd^{2+} adsorption capacity of 351.1 mg g^{-1}—a tenfold improvement over pristine WCA. Combined with exceptional adsorption efficiency, this biomass-derived composite offers an eco-friendly, cost-effective solution for heavy metal ion remediation. Its scalable fabrication from renewable resources aligns with sustainable water treatment objectives, presenting the advantage of pollution mitigation.

Keywords: balsa wood; carbon aerogel; $Mg(OH)_2$; Cd^{2+} contamination; adsorption

1. Introduction

With the rapid advancement of industrialization, the widespread use of heavy metal ions has emerged as a global environmental challenge [1,2]. Among these contaminants, cadmium ions (Cd^{2+}) have gained significant attention due to their extensive utilization in battery manufacturing, electroplating processes, and pigment production [3,4], owing to their favorable electrochemical properties. However, cadmium's high toxicity and bioaccumulative potential in the human body pose severe ecological and public health risks [5]. Chronic exposure to cadmium-contaminated environments has been linked to renal dysfunction, osteoporosis, carcinogenesis, and various systemic disorders. Effective wastewater treatment prior to discharge is critically required to safeguard ecological integrity and public health safety.

To address Cd^{2+} contamination, researchers have devoted themselves to various conventional remediation strategies, including chemical precipitation [6,7], ion exchange [8,9], membrane separation [10,11], and adsorption [12,13]. Adsorption has become a particularly promising solution due to its operational simplicity, cost-effectiveness, and high removal efficiency. Various materials have been applied as adsorbents in treating water, such as activated carbon [14–17], zeolites [18,19], and clay minerals [20,21]. Of them, carbon materials are widely used because of their low preparation cost and good adsorption property. For example, Xu and his co-workers prepared three-dimensional N-doped carbon (NC) polyhedrons through pyrolysis carbonization and KOH activation of zeolitic imidazolate framework-8 (ZIF-8) precursors. The resulting NC material exhibited an ultra-high specific surface area of 3041 $m^2\ g^{-1}$ and achieved a maximum adsorption capacity of 370.2 mg g^{-1} for Cd^{2+} [22]. In a separate study, Wei and his colleagues synthesized magnetic carbon nanopolymers (Fe_3O_4/C@PM) using suspension polymerization; these materials demonstrated a maximum Cd^{2+} adsorption capacity of 250.7 mg g^{-1} [23].

Since most carbon-based materials exist in powder form, their recovery after adsorption is challenging, limiting their practical applications. To solve this problem, researchers have developed magnetic–carbon composites for efficient post-treatment recovery [24]. Carbon aerogels (CAs) with three-dimensional (3D) porous architectures present another remediation strategy. Their inherent structural integrity enables bulk removal from aqueous systems post-adsorption, significantly simplifying separation processes. Many materials, such as MOF, graphene, carbon nanotubes, polyacrylonitrile, polyimide, and chitosan have been applied to engineer CAs. For instance, Zhang's group prepared a modified graphene oxide composite aerogel with a saturated Cd^{2+} adsorption capacity of 174.85 mg g^{-1} [25].

Being a natural material, wood possesses vertical channels and a cellulose skeleton that can form multistage channels after carbonization. These characteristics render itself a highly suitable raw material candidate for the synthesis of CA. Due to its renewable property and 3D–porous structure, wood-derived carbon aerogel (WCA) serves as an excellent substrate for combining with other functional materials. The resulted composite aerogel can be utilized as binder-free supercapacitor electrode materials [26], flexible electrode materials [27], advanced packaging materials [28], electromagnetic interference shielding materials, and flame-retardant materials [29].

In this study, wood-derived carbon aerogel (WCA) was synthesized via balsa wood calcination. With it as a substrate, $Mg(OH)_2$ nanoparticles were loaded. The WCA/$Mg(OH)_2$ composite was fabricated (designated as WCAMg). The Cd^{2+} adsorption performance was thoroughly investigated. The influence of $Mg(OH)_2$ loading amount on adsorption was tested, and the adsorption mechanism was proposed. As the cadmium complexes were adsorbed and embedded into the 3D structure of WCAMg, it enabled facile separation and recovery from wastewater. These WCAMg materials provide a novel water treatment strategy that eliminates secondary pollution risks, and offer potential solutions for heavy metal pollution remediation and recovery.

2. Experimental

2.1. Materials and Chemicals

The balsa wood comes from Papua New Guinea. Magnesium chloride hexahydrate ($MgCl_2{\cdot}6H_2O$, AR, purity $\geq$ 99.0%, and CAS 7791-18-6), nitric acid (HNO_3, AR, 65–68% w/w, and CAS 7697-37-2), and sodium hydroxide (NaOH, AR, purity $\geq$ 99.0%, and CAS 1310-73-2) were purchased from Tianli Chemical Reagent Manufacturing Co., Ltd. (Tianjin, China). Acetic acid (CH_3COOH, AR, 99.5% w/w, and CAS 64-19-7) was purchased from Tianjin Ruijinte Chemical Co., Ltd., Tianjin, China. Calcium chloride ($CaCl_2$, AR, purity $\geq$ 99.0%, and CAS 10043-52-4), sodium chloride (NaCl, AR, purity $\geq$ 99.0%, and CAS 7647-14-5), and potassium chloride (KCl, AR, purity $\geq$ 99.0%, and CAS 7447-40-7) were purchased from Tianjin Comin Chemical Reagent Co., Ltd., Tianjin, China. Cadmium nitrate tetrahydrate ($Cd(NO_3)_2{\cdot}4H_2O$, AR, purity $\geq$ 99.0%, and CAS 10325-94-7) was purchased from Aladdin Biochemical Technology Co., Ltd., Shanghai, China. All the solutions were prepared using deionized water.

2.2. Preparation of WCA

Balsa wood was firstly cut into pieces with the size of 10 mm $\times$ 10 mm $\times$ 20 mm, and dried at 102 $\pm$ 3 °C for 24 h. Then, the wood pieces were put into a beaker with 70 mL of 5% $NaClO_2$ solution for delignification at 80 °C for 10 h. After that, delignified wood pieces were transferred into a beaker with deionized water. After heating at 90 °C for 5 h, the delignified wood pieces were taken out and frozen at −18 °C for 4 h in a polystyrene cup. Then, they were frozen dry in a freeze dryer for 48 h to obtain delignified wood aerogel. The same method was also adopted in references [30–32].

The delignified wood aerogel was calcined at 600 °C in nitrogen atmosphere for 3 h and WCA was prepared.

2.3. Preparation of WCAMg

$MgCl_2{\cdot}6H_2O$ served as the precursor in the synthesis of $Mg(OH)_2$. Two pieces of WCA (≈0.08 g) were immersed into 10 mL 1.5 mol L^{-1} $MgCl_2$ solution for ultrasonic treatment at 80 °C (40 KHz, ultrasonic power of 360 W). After 30 min, the WCAs were taken out and vacuumed in a vacuum dryer at −0.1 MPa for 1 h, and then put into 10 mL 1 M L^{-1} NaOH solution for 30 min at 80 °C. The WCA samples were removed from the solution, and evacuated to −0.1 MPa vacuum pressure.

After repeating the procedure three times, the WCAMg composite was collected, subsequently rinsed with deionized water, and vacuum-dried at 50 °C for 6 h. The loaded content is calculated by the change of mass of WCA before and after $Mg(OH)_2$ loading according to Equation (1), as follows:

$$R = \frac{M_2 - M_1}{M_1} \times 100\% \tag{1}$$

R (%) is the loading rate of the prepared sample; M_2 (g) is the mass of the WCAMg sample after the loading of $Mg(OH)_2$; and M_1 (g) is the mass of the WCA sample before the loading of $Mg(OH)_2$.

The $MgCl_2$ concentration variations (0.75, 1.5, and 3.0 mol L^{-1}) were systematically evaluated as critical experimental parameters for $Mg(OH)_2$ loading. Accordingly, the synthesized WCAMg composites were labeled WCAMg-1, -2, and -3 to reflect their respective $MgCl_2$ precursor concentrations.

2.4. Adsorption Experiments

Adsorption experiments were conducted using 200 mL cadmium solutions (200 mg L^{-1} Cd^{2+}, pH 6) with WCAMg specimens (≈0.09 g) under ambient conditions. At predetermined intervals, the solutions were extracted for a cadmium concentration analysis via atomic absorption spectroscopy.

The Langmuir (Equation (2)) and Freundlich (Equation (3)) models were applied to analyze the adsorption mechanism [33], as follows:

$$\frac{c_e}{q_e} = \frac{1}{K_L q_{max}} + \frac{c_e}{q_{max}} \tag{2}$$

$$lgq_e = lgK_F + \frac{1}{n} lgc_e \tag{3}$$

where q_e (mg g^{-1}) represents the adsorption capacity at equilibrium, and c_e (mg L^{-1}) represents equilibrium concentration. The binding energy of the whole system is generally expressed by Langmuir's constant K_L (L g^{-1}), and the q_{max} is the maximum adsorption capacity. The Freundlich constant is K_F (mg g^{-1} $L^{1/n}$ $mg^{-1/n}$) and n is the adsorption strength.

The adsorption kinetics were modeled using three classical approaches: pseudo-first-order (PFO, Equation (4)), pseudo-second-order (PSO, Equation (5)), and intra-particle diffusion (IPD, Equation (6)), with their respective formulations defined as follows [34]:

PFO model:

$$\ln(q_e - q_t) = lnq_e - k_1 t \tag{4}$$

PSO model:

$$\frac{t}{q_t} = \frac{1}{k_2 q_e^2} + \frac{t}{q_e} \tag{5}$$

IPD model:

$$q_t = K_{id} t^{\frac{1}{2}} + C \tag{6}$$

where k_1 (h^{-1}) and k_2 (g mg^{-1} h^{-1}) are the PFO and PSO adsorption rate constant; K_{id} (mg g^{-1} $h^{-1/2}$) is the IPD rate constant; q_e (mg g^{-1}) and q_t (mg g^{-1}) are the adsorption capacities at equilibrium and time t; and C (mg g^{-1}) is the boundary layer thickness correlation constant for the IPD model.

2.5. Characterizations

A Fourier transform infrared spectrometer (FT-IR, PerkinElmer, Waltham, MA, USA) was used to test the corresponding functional groups of the samples. X-ray diffraction (XRD, Rigaku D/Max-RC, Tokyo, Japan) with Cu $K\alpha$ radiation (wavelength of 1.5418 Å) was used to analyze the structure of the samples. Scanning electron microscopy (SEM, Hitachi S-4800, Tokyo, Japan) was applied to observe the morphology of the composites. The chemical compositions of the samples were measured by energy dispersive X-ray spectroscopy (EDS, Oxford Instruments ULTIM Max 65, Abingdon, UK) and X-ray photoelectron spectroscopy (XPS, K-Alpha, XPS, Thermo Fisher Scientific K-Alpha, Waltham, MA, USA). Cd^{2+} was quantitatively analyzed by an atomic adsorption spectrometer (AAS, PerkinElmer PinAAcle 900H, Waltham, MA, USA).

3. Results and Discussion

3.1. Structure and Morphology of WCAMg

The structure of WCAMg and WCA were characterized by XRD, and the results are shown in Figure 1a. For WCA, there is a broad peak located at 20~30° [35], indicating the amorphous structure of calcined WCA. The XRD plot for WCAMg shows some differences: diffraction peaks were observed at 2θ values of 18.5°, 37.9°, 50.8°, and 58.6°, which correspond to the (001), (101), (102), and (110) lattice planes of hexagonal $Mg(OH)_2$ (JCPDS# 44-1482). The XRD analysis confirms that WCAMg comprises two distinct crystalline phases: WCA and $Mg(OH)_2$. It demonstrates successful $Mg(OH)_2$ attachment into the WCA matrix.

The FT-IR spectra of WCAMg samples are shown in Figure 1b. For $Mg(OH)_2$, a distinct and sharp peak was observed at 3699 cm^{-1}, attributed to the -OH stretching vibration in $Mg(OH)_2$. Meanwhile, the peak at 1433 cm^{-1} was associated with the O-H bending vibration of adsorbed water. Once loaded with $Mg(OH)_2$, the samples of WCAMg also show these two adsorption peaks. However, the intensity of the -OH stretching vibration at 3699 cm^{-1} exhibits a distinct concentration-dependent trend: an increased $MgCl_2$ concentration correlates directly with an enhanced peak intensity. This indicated that elevated $MgCl_2$ concentrations significantly enhanced $Mg(OH)_2$ loading efficiency on WCA. Meanwhile, the peak at 1433 cm^{-1} in WCAMg shows significant enhancement compared to pure $Mg(OH)_2$. Two key effects are indicated: (i) improved water adsorption amount, and (ii) the emergence of additional functional groups (e.g., C-O) within the composite aerogel structure. Moreover, the characteristic peaks observed at 860, 1562, and 1704 cm^{-1} are related to the C-H bending vibration, C=C stretching vibration, and C=O stretching vibration of WCA, respectively [36].

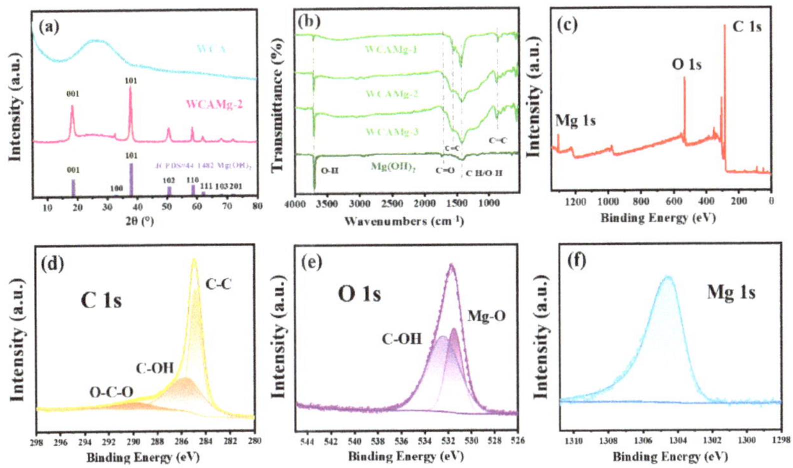

Figure 1. (**a**) XRD patterns of WCA, WCAMg-2, and $Mg(OH)_2$, (**b**) FT-IR spectra of $Mg(OH)_2$ and WCAMg-n, and XPS spectra of the composite WCAMg-2, (**c**) survey, (**d**) C 1s, (**e**) O 1s, and (**f**) Mg 1s.

One can note that WCAMg-3 exhibits the largest loaded amount of $Mg(OH)_2$. However, the loaded amount does not linearly correlate with the initial feed ratio. So, WCAMg-2 was chosen for further experiments due to its optimal balance between performance and cost-effectiveness.

The XPS was applied to analyze the elemental environment of WCAMg. The XPS survey of the composite aerogel (Figure 1c) shows three main peaks, which correspond to C 1s, O 1s, and Mg 1s. Figure 1d is the C 1s spectrum, and three peaks at 284.8, 285.6, and 289.8 eV can be fitted. They are binding energies of C-C/C-H, C-OH, and C-O-C originated from WCA. The spectrum for O 1s in Figure 1e can be fitted into two peaks at 531.3 and 532.4 eV. They are attributed to the binding energies of Mg-O in $Mg(OH)_2$ and C-O/C-OH in WCA, respectively. The Mg 1s spectrum was depicted in Figure 1f, and a distinct peak was observed at a binding energy of 1303.0 eV, which is related to the binding energy of Mg-O in $Mg(OH)_2$ [37]. The XPS results further confirm the successful up-loading of $Mg(OH)_2$ into the WCA.

The morphology of WCA and WCAMg samples were examined using SEM. As illustrated in Figure 2a, WCA exhibits an irregular fibrous architecture containing interconnected cavities and channels. Morphological features are characteristic of biomass-derived carbon aerogels. However, observations of WCAMg samples (Figure 2b–d) find that surfaces of WCAs were covered. After the deposition of $Mg(OH)_2$ nanoparticles, plate-like nanostructures with the size of about 200 nm were determined via SEM (Figure 2e–g). This result not only confirms the successful in situ crystallization of $Mg(OH)_2$, but also demonstrates its effective anchoring on the WCA surface through interfacial bonding. WCAMg-1 displays aggregated $Mg(OH)_2$ nanosheets that distribute across the WCA. While increasing the $MgCl_2$ concentration, the prepared WCAMg-2 and WCAMg-3 show a morphology of progressive surface coverage by vertically aligned $Mg(OH)_2$ nanosheets. These results also give the evidence that a high concentration of $MgCl_2$ would result in a high load capacity of $Mg(OH)_2$, agreeing with the discussion above.

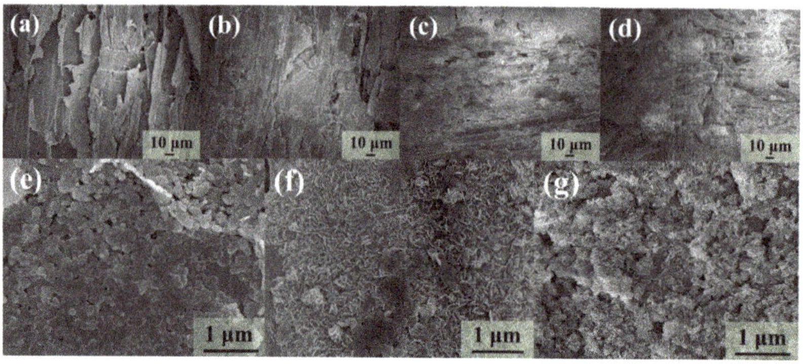

Figure 2. SEM images of (**a**) WCA, (**b**,**e**) WCAMg-1, (**c**,**f**) WCAMg-2, and (**d**,**g**) WCAMg-3.

3.2. Absorption Performance

The adsorption performance of WCAMg for Cd^{2+} was fully studied. The systematic evaluation of Cd^{2+} adsorption efficiency revealed a significant enhancement in WCAMg samples (Figure 3a). It can be observed that pristine WCA alone has an adsorption capacity of 33.30 mg g^{-1}. Interestingly, the three composites WCAMg-n (n = 1, 2, and 3) give sequentially increasing capacities of 225.8, 351.1, and 355.9 mg g^{-1}. As shown in Figure 3b, their $Mg(OH)_2$ loading ratios were calculated to be 42.38% (±1.09 wt%), 56.82% (±1.38 wt%), and 60.27% (±0.87 wt%) according to Equation (1), respectively. It further confirms that the adsorption capacity depends on the loaded amount of $Mg(OH)_2$. Notably, WCAMg-2 has the best cost–benefit ratio. It achieves a 10-times-higher adsorption capacity than WCA. This suggests that 1.5 mol L^{-1} $MgCl_2$ is the optimal concentration for loading $Mg(OH)_2$.

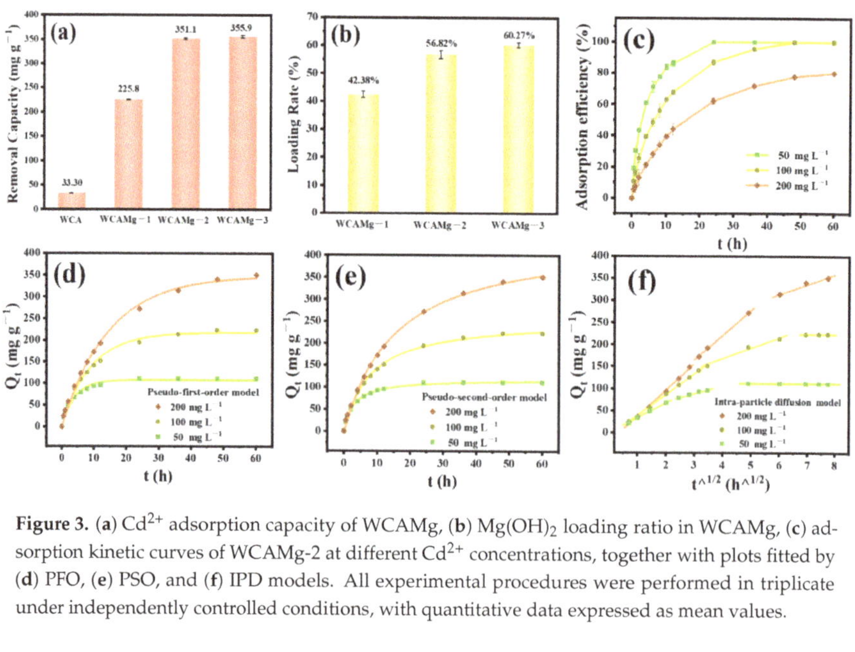

Figure 3. (**a**) Cd^{2+} adsorption capacity of WCAMg, (**b**) $Mg(OH)_2$ loading ratio in WCAMg, (**c**) adsorption kinetic curves of WCAMg-2 at different Cd^{2+} concentrations, together with plots fitted by (**d**) PFO, (**e**) PSO, and (**f**) IPD models. All experimental procedures were performed in triplicate under independently controlled conditions, with quantitative data expressed as mean values.

Table 1 presents adsorption capacities of carbon aerogels reported in previous studies and in this work. Compared to conventional easily recoverable adsorbents, our composite WCAMg not only exhibits a superior Cd^{2+} adsorption capacity, but also demonstrates significant cost-effectiveness.

Table 1. Comparison of Cd^{2+} adsorption capacity of different carbon aerogels.

Carbon Aerogel Materials	Cd^{2+} Adsorption Capacity (mg g^{-1})	References
WCAMg	351.1	This study
MCA (Fe_3O_4)	143.88	[38]
SA/GO-βCD	174.85	[25]
rGO/Fe_3O_4/CNTs	67.96	[24]
MCA ($MnFe_2O_4$)	73.63	[39]
GO/PDA-MMT/CMCS	317.46	[40]

The kinetic curves of WCAMg-2 to various Cd^{2+} concentrations are shown in Figure 3c. To 50 and 100 mg L^{-1} Cd^{2+} solution, adsorption equilibriums were reached within 12 h and 24 h, respectively, and the adsorption efficiencies are almost 99.9%. Whereas, for the 200 mg L^{-1} Cd^{2+} solution, it took 48 h to get to the adsorption equilibrium, and the adsorption efficiency is 80.1%. The q_e were determined as 110.9, 223.6, and 351.1 mg g^{-1} at initial Cd^{2+} concentrations of 50, 100, and 200 mg L^{-1}, respectively (Table 2). For systems with a concentration less than 200 mg L^{-1}, the post-adsorption analysis confirmed a cadmium concentration reduction to <0.1 mg L^{-1} (Cd^{2+} removal efficiency > 99.9%). This meets the Class I discharge standards per China's Integrated Wastewater Discharge Standard (GB 8978-1996) [41]. These findings indicate the exceptional adsorption efficiency of WCAMg for industrial wastewater remediation.

Table 2. The parameters fitted by PF/PSO models for Cd^{2+} adsorption.

	PFO Model				PSO Model		
C_0 (mg L^{-1})	$q_{e,\,exp}$ (mg g^{-1})	$q_{e,\,cal}$ (mg g^{-1})	k_1 (h^{-1})	R^2	$q_{e,\,cal}$ (mg g^{-1})	k_2 (g mg^{-1} h^{-1})	R^2
50	110.9	107.8	0.2419	0.9773	118.5	0.002890	0.9965
100	223.6	218.5	0.1104	0.9882	256.5	0.0004935	0.9968
200	351.1	349.4	0.0691	0.9955	445.5	0.0001457	0.9984

To study the adsorption mechanism, PFO and PSO models were applied to fit adsorption kinetic curves (Figure 3d,e). The relevant fitting parameters are shown in Table 2. Both the PFO and PSO models fit the adsorption process well. The PSO model has higher R^2 values (0.9965~0.9984). This indicates that the process is predominantly governed by the chemical adsorption [42]. The calculated k_2 of Cd^{2+} exhibits a clear concentration-dependent pattern, measuring 0.002890 g mg^{-1} h^{-1} at 50 mg L^{-1} Cd^{2+}, 0.0004935 g mg^{-1} h^{-1} at 100 mg L^{-1} Cd^{2+}, and 0.0001457 g mg^{-1} h^{-1} at 200 mg L^{-1} Cd^{2+}. These values suggest an inverse correlation between initial cadmium concentration and adsorption kinetics, where lower Cd^{2+} concentrations correspond to significantly faster contaminant removal rates.

To elucidate the rate-limiting mechanism that governs cadmium adsorption, we analyzed the kinetic data using the Weber–Morris IPD model. As illustrated in Figure 3f, the Cd^{2+} adsorption profile on WCAMg exhibits triphasic behavior. Three sequential regimes were characterized: (I) instantaneous surface adsorption, (II) gradual pore diffusion, and (III) adsorptive equilibrium. The K_p (Table 3) indicates a consistent concentration-dependent progression: $K_{p1} > K_{p2} > K_{p3}$. It finds that (1) boundary layer diffusion dominates initial rapid capture, (2) gradual penetration through mesopores controls mid-term kinetics, and (3) final saturation occurs through micropore filling. Notably, elevated Cd^{2+} concentrations accelerate diffusion rates across all the phases (K_p increased 65.93% in Phase 1 and 229.3% in Phase 2 at 200 mg L^{-1} vs. 50 mg L^{-1}). This confirms that the concentration gradient-driven mass transfer serves as the principal transport mechanism.

Table 3. The parameters fitted by IPD model for Cd^{2+} adsorption.

	Phase 1			Phase 2			Phase 3		
C_0 (mg L^{-1})	R_1^2	C_1 (mg g^{-1})	K_{p1} (mg g^{-1} $h^{-1/2}$)	R_2^2	C_2 (mg g^{-1})	K_{p2} (mg g^{-1} $h^{-1/2}$)	R_3^2	C_3 (mg g^{-1})	K_{p3} (mg g^{-1} $h^{-1/2}$)
50	0.9973	−3.069	35.57	0.9767	36.99	17.25	1	110.9	0
100	0.9991	−10.67	48.13	0.9778	67.36	24.89	1	223.6	0
200	0.9951	−21.81	59.02	0.9987	−5.745	56.81	0.9580	189.0	21.24

To investigate the effect of initial concentration on adsorption capacity, experiments were carried out. They started with different initial concentrations of Cd^{2+} solutions by WCAMg-2. The results are shown in Figure 4 along with the fitted plots.

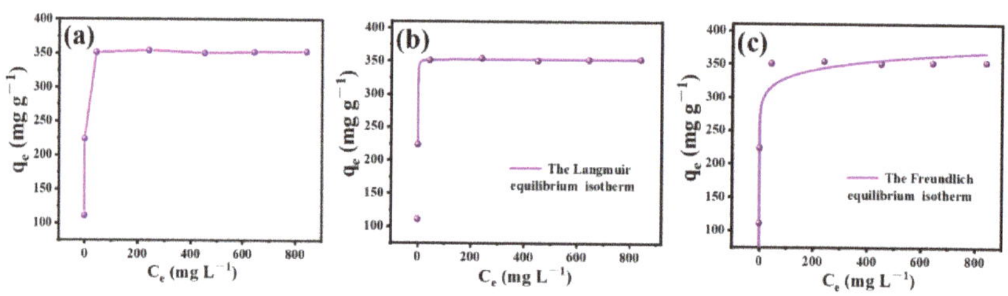

Figure 4. (**a**) Adsorption isotherm of WCAMg-2 for Cd^{2+}, and experimental data with (**b**) Langmuir fitting curve and (**c**) Freundlich fitting curve.

From the adsorption isotherm (Figure 4a), one can see that q_{max} of 351.1 mg g^{-1} was obtained when the concentration of Cd^{2+} was 200 mg L^{-1}. The adsorption isotherms were fitted using Langmuir and Freundlich models (Figure 4b,c). The results are shown in Table 4, and the R^2 values for these two models are 0.9994 and 0.7417, respectively. Clearly, the Langmuir model describes the removal process better, indicating a monolayer adsorption mechanism. The q_{max} calculated by fitting results from the Langmuir model is about 352.5 mg g^{-1}, which is consistent with the experimentally determined result.

Table 4. The parameters fitted by Langmuir and Freundlich models for Cd^{2+} removal isotherm.

	Langmuir Model			Freundlich Model		
	q_{max} (mg g^{-1})	K_L (L g^{-1})	R^2	n	K_F (mg g^{-1} $L^{1/n}$ $mg^{-1/n}$)	R^2
WCAMg-2	352.5	8.675	0.9994	0.05020	261.5	0.7417

3.3. *Adsorption Mechanism*

To comprehensively elucidate the adsorption mechanism, the sample of WCAMg after adsorbing Cd^{2+} was recovered and characterized. The spent sample is marked as WCAMg-Cd. Its XRD patterns in Figure 5a show that recovered Cd^{2+} exists as $Cd(OH)_2$ (JCPDS# 31-0228) and $CdCO_3$ (JCPDS# 42-1342). It agrees with the previous studies [37,43].

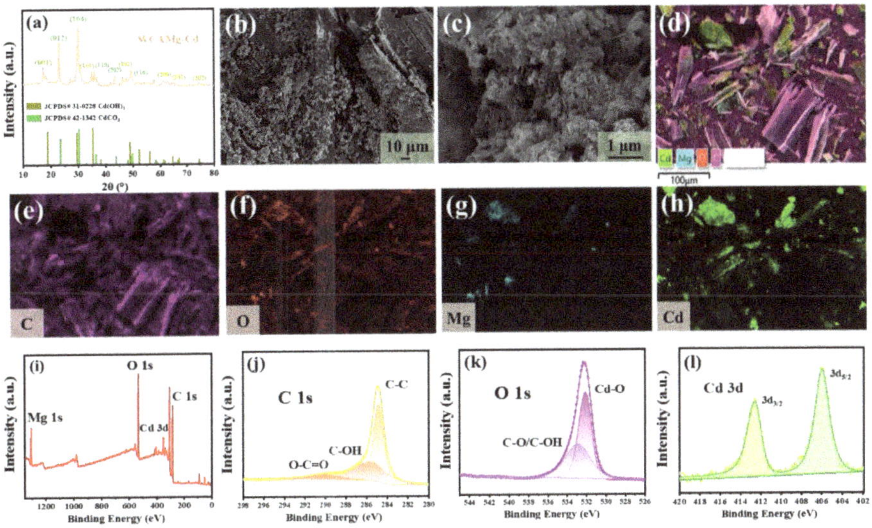

Figure 5. (**a**) XRD patterns, (**b**,**c**) SEM images, (**d**–**h**) EDS mapping images, and (**i**–**l**) XPS spectra of WCAMg-Cd.

From SEM images (Figure 5b,c), one can observe particles instead of nanosheets on the surface of WCAMg-Cd. The EDS mapping (Figure 5d–h) evidences that Cd^{2+} has been adsorbed onto WCAMg. The XPS was applied to analyze the elemental environment of WCAMg-Cd. It is found in Figure 5i that Cd element appears in the survey spectrum other than Mg, C, and O. It further indicates the successful adsorption of Cd^{2+} by WCAMg. For the Cd 3d spectrum (Figure 5l), a doublet peak appears at binding energies of 405.9 eV and 412.6 eV. This corresponds to the $3d_{5/2}$ and $3d_{3/2}$ states of Cd^{2+} in $CdCO_3/Cd(OH)_2$ [44,45]. In the XPS spectrum of C 1s (Figure 5j), a peak at 289.9 eV was fitted, which is the binding energy of $CO_3{}^{2-}$ [46]. This result also indicates the formation of $CdCO_3$ post-adsorption.

According to these results, the adsorption mechanism is summarized. Weak basic $Mg(OH)_2$ is dissolved into Mg^{2+} and OH^- in solution, and Cd^{2+} would precipitate by the released OH^- to give $Cd(OH)_2$ [47]. It is described by the reaction in Equation (7), as follows:

$$Mg(OH)_2 + Cd^{2+} \rightarrow Cd(OH)_2 + Mg^{2+} \quad (7)$$

Meanwhile, CO_2 from air dissolves in water to form H_2CO_3. The released OH^- ions subsequently react with H_2CO_3 to generate $CO_3{}^{2-}$, which would precipitate Cd^{2+} as $CdCO_3$. The involved reactions are shown in Equations (8)–(12). Notably, the pH of solution is kept at 6 during the adsorption.

$$CO_2\ (air) + H_2O \rightarrow H_2CO_3 \quad (8)$$

$$H_2CO_3 \rightarrow CO_3{}^{2-} + 2H^+ \quad (9)$$

$$Mg(OH)_2 \rightarrow Mg^{2+} + 2OH^- \quad (10)$$

$$H^+ + OH^- \rightarrow H_2O \quad (11)$$

$$Cd^{2+} + CO_3{}^{2-} \rightarrow CdCO_3\downarrow \quad (12)$$

3.4. Competitive Ion Interference Effects and Environmental Performance in Natural Water

Since industrial effluents typically contain high salinity that may compromise the adsorption efficiency of WCAMg, we have systematically investigated the interference effects of four common co-existing ions (K^+, Na^+, Ca^{2+}, and Mg^{2+}). The results are presented in Figure 6a–d. Distinct structure–activity relationships are revealed. Monovalent cations (K^+/Na^+) exhibit no significant inhibitory effects on Cd^{2+} adsorption across the tested concentration range (100–800 mg L^{-1}). In contrast, divalent cations (Ca^{2+}/Mg^{2+}) show concentration-dependent inhibitory effects, where the adsorption capacity decreases proportionally as their concentration increases.

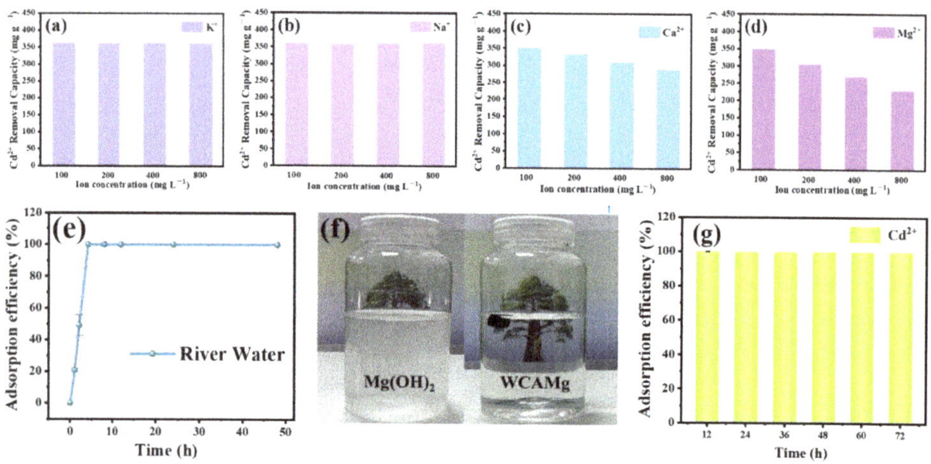

Figure 6. The effects of (**a**) K^+, (**b**) Na^+, (**c**) Ca^{2+}, and (**d**) Mg^{2+} on the adsorption capacity of Cd^{2+} by WCAMg-2, (**e**) the adsorption efficiency of WCAMg for treating river water, (**f**) comparison diagram of pure $Mg(OH)_2$ and WCAMg after adsorption, and (**g**) durability test for WCAMg-2.

It may be caused by the following reason. Mg^{2+} induces competitive precipitation through $Mg(OH)_2$ formation (reducing its solubility equilibrium), thereby suppressing Cd^{2+} up-taken efficiency. Conversely, Ca^{2+} interference primarily originates from OH^- depletion effects, where elevated cation concentrations progressively diminish adsorption capacity via hydroxyl ion competition.

To evaluate the environmental applicability of WCAMg as a water purification material, a field-simulated adsorption experiment was carried out using Songhua River water (Harbin, China). It was amended with 1 mg L^{-1} Cd^{2+} through the controlled addition of cadmium nitrate tetrahydrate ($Cd(NO_3)_2{\cdot}4H_2O$). As demonstrated in Figure 6e, WCAMg achieves exceptional remediation efficiency over 99.99% Cd^{2+} removal within 4 h. Crucially, the treated water maintains visually clear (Figure 6f), whereas parallel tests with conventional $Mg(OH)_2$ produces turbid suspensions due to uncontrolled $Cd(OH)_2/CdCO_3$ precipitation. This marked visual contrast confirms WCAMg's unique microstructural advantage in effectively encapsulating cadmium complexes within its carbon aerogel architecture. Thereby, it enables the prevention of secondary particulate pollution and simplifies the recovery.

To assess the durability of the composite WCAMg, six consecutive adsorption cycles were conducted using 100 mL of 20 mg L^{-1} Cd^{2+} aqueous solution. Upon completing each adsorption cycle, the adsorbent was taken out with tweezers and re-immersed in fresh Cd^{2+} solution for subsequent remediation phases. Remarkably, the composite maintains almost 100% retention efficiency throughout six cycles (Figure 6g). This demonstrates the exceptional structural stability and confirms its potential as a long-lasting reagent for heavy metal removal in water purification applications.

4. Conclusions

A novel balsa wood-derived carbon aerogel (WCA) has been prepared in this study. Further functionalization with $Mg(OH)_2$ nanosheets enabled us to construct a WCAMg water treatment reagent. It is demonstrated that WCAMg achieves an exceptional cadmium immobilization performance. Representatively, the best composite sample exhibits a maximum adsorption capacity of 351.1 mg g^{-1} and high removal efficiency (>99.9%) towards Cd^{2+} at concentrations $\leq$ 100 mg L^{-1}. The treated water utterly satisfies the Class I discharge limits of China's GB 8978-1996 standard. The removal mechanism involves dual pathways: (1) the $Mg(OH)_2$ component enables Cd^{2+} chemisorption through $Cd(OH)_2/CdCO_3$ precipitation, and (2) the three-dimensional carbon aerogel network effectively immobilizes the formed cadmium complexes. This unique architecture prevents the secondary sludge formation and facilitates straightforward separation. The synergistic design positions WCAMg as a sustainable water remediation agent that provides an economically viable solution for heavy metal contamination management.

Author Contributions: Conceptualization, Y.G.; data curation, Y.Y.; formal analysis, J.L.; investigation, R.A. and H.M.; supervision, Q.P. and S.L.; writing—original draft, R.A.; and writing—review and editing, Y.G., Q.P. and S.L. All authors have read and agreed to the published version of the manuscript.

Funding: This work is supported by the National Natural Science Foundation of China (22276046).

Data Availability Statement: The data presented in this study are available on request from the corresponding authors.

Acknowledgments: The Analyzing and Testing Center of Northeast Forestry University is acknowledged for the SEM tests.

Conflicts of Interest: The authors declare no conflicts of interest.

References

1. Piwowarska, D.; Kiedrzyńska, E.; Jaszczyszyn, K. A global perspective on the nature and fate of heavy metals polluting water ecosystems, and their impact and remediation. *Crit. Rev. Environ. Sci. Technol.* **2024**, *54*, 1436–1458. [CrossRef]
2. Liu, K.; Guan, X.; Li, C.; Zhao, K.; Yang, X.; Fu, R.; Li, Y.; Yu, F. Global perspectives and future research directions for the phytoremediation of heavy metal-contaminated soil: A knowledge mapping analysis from 2001 to 2020. *Front. Environ. Sci. Eng.* **2021**, *16*, 73. [CrossRef]
3. Rahman, M.M.; Karim, M.R.; Alharbi, H.F.; Aldokhayel, B.; Uzzaman, T.; Zahir, H. Cadmium Selenide Quantum Dots for Solar Cell Applications: A Review. *Chem. Asian. J.* **2021**, *16*, 902–921. [CrossRef] [PubMed]
4. Mavazzan, A.; Mendhe, A.C.; Bayannavar, P.K.; Sankapal, B.R.; Kamble, R.R.; Madar, S.F.; KM, M.P.; Bheemayya, L. Design of Metal Free Fluorescent Pyridine Dyes Anchored on Cadmium Sulfide Nanowires: Optical, Electrochemical and Photovoltaic Applications. *J. Fluoresc.* **2023**, *34*, 2405–2414. [CrossRef]
5. Cirovic, A.; Denic, A.; Clarke, B.L.; Vassallo, R.; Cirovic, A.; Landry, G.M. A hypoxia-driven occurrence of chronic kidney disease and osteoporosis in COPD individuals: New insights into environmental cadmium exposure. *Toxicology* **2022**, *482*, 153355–153364. [CrossRef]
6. Zieliński, J.; Huculak-Mączka, M.; Kaniewski, M.; Nieweś, D.; Hoffmann, K.; Hoffmann, J. Kinetic modelling of cadmium removal from wet phosphoric acid by precipitation method. *Hydrometallurgy* **2019**, *190*, 105157–105163. [CrossRef]
7. Liu, S.; Yang, S.; Zhang, X.; Luo, T.; Chen, Y. Removal of cadmium and arsenic from Cd-As-Pb-bearing dust based on peroxide leaching and coprecipitation. *Hydrometallurgy* **2022**, *209*, 105839–105851. [CrossRef]
8. Dabrowski, A.; Hubicki, Z.; Podkościelny, P.; Robens, E. Selective removal of the heavy metal ions from waters and industrial wastewaters by ion-exchange method. *Chemosphere* **2004**, *56*, 91–106. [CrossRef]
9. Elfeghe, S.; Anwar, S.; Zhang, Y. Adsorption and removal studies of cadmium ion onto sulphonic/phosphonic acid functionalization resins. *Can. J. Chem. Eng.* **2022**, *100*, 3006–3014. [CrossRef]

10. Xiang, H.; Min, X.; Tang, C.-J.; Sillanpää, M.; Zhao, F. Recent advances in membrane filtration for heavy metal removal from wastewater: A mini review. *J. Water Process Eng.* **2022**, *49*, 103023–103037. [CrossRef]
11. Maqbool, A.; Shahid, A.; Jahan, Z.; Bilal Khan Niazi, M.; Ali Inam, M.; Tawfeek, A.M.; Kamel, E.M.; Saeed Akhtar, M. Development of ZnO-GO-NiO membrane for removal of lead and cadmium heavy metal ions from wastewater. *Chemosphere* **2023**, *338*, 139622–139639. [CrossRef]
12. Ma, F.; Zhao, H.; Zheng, X.; Zhao, B.; Diao, J.; Jiang, Y. Enhanced adsorption of cadmium from aqueous solution by amino modification biochar and its adsorption mechanism insight. *J. Environ. Chem. Eng.* **2023**, *11*, 109747–109758. [CrossRef]
13. Liu, Z.; Li, X.; Zhan, P.; Hu, F.; Ye, X. Removal of cadmium and copper from water by a magnetic adsorbent of PFM: Adsorption performance and micro-structural morphology. *Sep. Purif. Technol.* **2018**, *206*, 199–207. [CrossRef]
14. Tan, I.A.W.; Chan, J.C.; Hameed, B.H.; Lim, L.L.P. Adsorption behavior of cadmium ions onto phosphoric acid-impregnated microwave-induced mesoporous activated carbon. *J. Water Process Eng.* **2016**, *14*, 60–70. [CrossRef]
15. Sufian, J.; Babaakbari Sari, M.; Marchelli, F.; Fiori, L.; Avanes, A.; Moradi, S. An Analysis of the Factors Influencing Cadmium Removal in Aquatic Environments by Chlorella vulgaris-Derived Solids. *J. Carbon Res.* **2023**, *10*, 2. [CrossRef]
16. Tolkou, A.K.; Maroulas, K.N.; Theologis, D.; Katsoyiannis, I.A.; Kyzas, G.Z. Comparison of Modified Peels: Natural Peels or Peels-Based Activated Carbons for the Removal of Several Pollutants Found in Wastewaters. *J. Carbon Res.* **2024**, *10*, 2. [CrossRef]
17. Mulla, B.; Ioannou, K.; Kotanidis, G.; Ioannidis, I.; Constantinides, G.; Baker, M.; Hinder, S.; Mitterer, C.; Pashalidis, I.; Kostoglou, N.; et al. Removal of Crystal Violet Dye from Aqueous Solutions through Adsorption onto Activated Carbon Fabrics. *J. Carbon Res.* **2024**, *10*, 19. [CrossRef]
18. Xu, H.; Ou, Z.; Li, W.; Hu, T.; Zhang, Y.; Xu, H.; Wang, J.; Li, Y. Cadmium(II) adsorption by recyclable Zeolite-Loaded Hydrogel: Extension to the removal of Cadmium(II) from contaminated soil. *Chem. Eng. J.* **2024**, *492*, 151842–151855. [CrossRef]
19. Perfecto Barragán, P.; Guadalupe Macedo, M.M.; Olguín, M.T. Cadmium sorption by sodium and thiourea-modified zeolite-rich tuffs. *J. Environ. Sci.* **2017**, *52*, 39–48. [CrossRef]
20. Xu, S.; Xing, Y.; Liu, S.; Luo, X.; Chen, W.; Huang, Q. Co-effect of minerals and Cd(II) promoted the formation of bacterial biofilm and consequently enhanced the sorption of Cd(II). *Environ. Pollut.* **2020**, *258*, 113774–113783. [CrossRef]
21. Otunola, B.O.; Ololade, O.O. A review on the application of clay minerals as heavy metal adsorbents for remediation purposes. *Environ. Technol. Innov.* **2020**, *18*, 100692–100706. [CrossRef]
22. Xu, C.; Wang, H.; Shang, Y.; Li, B.; Yu, D.; Wang, Y. Highly efficient Cd(II) removal using 3D N-doped carbon derived from MOFs: Performance and mechanisms. *J. Hazard. Mater.* **2022**, *436*, 129149–129161. [CrossRef] [PubMed]
23. Wei, S.; Huang, S.; Zhou, J.; Xiao, C.; Cao, J.; Xiao, J.; Xie, C. Magnetic Carbon Porous Polymer Prepared from a New Suspended Emulsion for the Absorption of Heavy Metal Ions. *Polymers* **2025**, *3*, 17–33. [CrossRef] [PubMed]
24. Dang, A.; Liu, X.; Wang, Y.; Liu, Y.; Cheng, T.; Zada, A.; Ye, F.; Deng, W.; Sun, Y.; Zhao, T.; et al. High-efficient adsorption for versatile adsorbates by elastic reduced graphene oxide/Fe_3O_4 magnetic aerogels mediated by carbon nanotubes. *J. Hazard. Mater.* **2023**, *457*, 131846–131857. [CrossRef]
25. Zhang, Y.; Wu, L.; Deng, H.; Qiao, N.; Zhang, D.; Lin, H.; Chen, Y. Modified graphene oxide composite aerogels for enhanced adsorption behavior to heavy metal ions. *J. Environ. Chem. Eng.* **2021**, *9*, 106008–106018. [CrossRef]
26. Li, S.C.; Hu, B.C.; Ding, Y.W.; Liang, H.W.; Li, C.; Yu, Z.Y.; Wu, Z.Y.; Chen, W.S.; Yu, S.H. Wood-Derived Ultrathin Carbon Nanofiber Aerogels. *Angew. Chem. Int. Ed.* **2018**, *57*, 7085–7090. [CrossRef]
27. Chen, Z.; Zhuo, H.; Hu, Y.; Lai, H.; Liu, L.; Zhong, L.; Peng, X. Wood-Derived Lightweight and Elastic Carbon Aerogel for Pressure Sensing and Energy Storage. *Adv. Funct. Mater.* **2020**, *30*, 1910292. [CrossRef]
28. Xiong, C.; Fan, X.; Xiong, Q.; Zhang, Y.; Su, Y. Double hydrogen bonding force improves the performance of composite phase change materials by improving the carbon skeleton of wood aerogel. *J. Energy Storage* **2025**, *114*, 115752–115761. [CrossRef]
29. Liang, C.; Qiu, H.; Song, P.; Shi, X.; Kong, J.; Gu, J. Ultra-light MXene aerogel/wood-derived porous carbon composites with wall-like "mortar/brick" structures for electromagnetic interference shielding. *Sci. Bull.* **2020**, *65*, 616–622. [CrossRef]
30. Xu, K.; Jiao, Y.; Li, J.; Xiao, H.; Fu, Q. FeP nanoparticle embedded in N,P-doped 3D porous wood-derived carbon aerogel for oxygen reduction reaction. *Carbon* **2024**, *228*, 119408–119416. [CrossRef]
31. Liu, Z.; Liu, T.; Dong, H.; Yang, B.; Li, X.; Li, X.; Wu, Y.; Xu, K. Diatom-Inspired Nanoscale Heterogeneous Assembly Strategy for Constructing Thermal Insulating Wood-Based Aerogels with Exceptional Strength, Resilience, Degradability, and Flame Retardancy. *ACS Nano* **2025**, *19*, 6826–6839. [CrossRef] [PubMed]
32. Yu, Y.; Wei, L.; Pang, Z.; Wu, J.; Dong, Y.; Pan, X.; Hu, J.; Qu, J.; Li, J.; Tian, D.; et al. Multifunctional Wood Composite Aerogel with Integrated Radiant Cooling and Fog–Water Harvesting for All-Day Building Energy Conservation. *Adv. Funct. Mater.* **2024**, *35*, 2414590–2414601. [CrossRef]
33. García-Zubiri, I.X.; González-Gaitano, G.; Isasi, J.R. Sorption models in cyclodextrin polymers: Langmuir, Freundlich, and a dual-mode approach. *J. Colloid Interface Sci.* **2009**, *337*, 11–18. [CrossRef]
34. Ryszko, U.; Rusek, P.; Kołodyńska, D. Transforming industrial processes: Effective removal of Cd(II) ions from wet phosphoric acid streams. *Chem. Eng. J.* **2025**, *505*, 159696–159713. [CrossRef]

35. Zhao, J.; Yang, S.; Bi, C.; Peng, C.; Wang, Y.; Li, Y.; Tao, E. Modulation of ciprofloxacin adsorption conjugation effect by specific conformation of manganese-nitrogen co-doped biochar and its mechanistic study. *Sep. Purif. Technol.* **2025**, 132304–132315. [CrossRef]
36. Dai, H.; Li, N.; Cui, Y.; Gao, W.; Peng, X.; Peng, W.; Si, H.; Mu, L.; Shi, Y.; Cheng, Z.; et al. Indispensable Synergy between C=C and C=O Sites in Biochar for Peroxomonosulfate Activation and Sulfamethoxazole Degradation. *ACS ES&T Eng.* **2024**, *4*, 2888–2897. [CrossRef]
37. Zhao, N.-D.; Wang, Y.; Zou, X.-H.; Yin, W.-M.; Wang, X.-Y.; Guo, Y.-R.; Pan, Q.-J. Fabrication of cellulose@$Mg(OH)_2$ composite filter via interfacial bonding and its trapping effect for heavy metal ions. *Chem. Eng. J.* **2021**, *426*, 130812–130825. [CrossRef]
38. Li, Y.; Zhou, M.; Waterhouse, G.I.N.; Sun, J.; Shi, W.; Ai, S. Efficient removal of cadmium ions from water by adsorption on a magnetic carbon aerogel. *Environ. Sci. Pollut. Res.* **2021**, *28*, 5149–5157. [CrossRef]
39. Jing, W.; Yang, C.; Lin, X.; Tang, M.; Lian, D.; Yu, Y.; Liu, D. $MnFe_2O_4$-loaded bamboo pulp carbon-based aerogel composite: Synthesis, characterization and adsorption behavior study for heavy metals removal. *RSC Adv.* **2024**, *14*, 39995–40005. [CrossRef]
40. Yuan, M.; Liu, D.; Liu, W.; Song, Z.; Shang, S.; Wang, Z.; Ren, J.; Cui, S. Graphene oxide/polydopamine modified montmorillonite/carboxymethyl chitosan composite aerogel for efficient removal of Pb^{2+}, Cu^{2+}, and Cd^{2+}: Adsorption behavior, mechanism and DFT study. *Sep. Purif. Technol.* **2024**, *339*, 126585. [CrossRef]
41. *GB 8978-1996*; Integrated Wastewater Discharge Standard. State Environmental Protection Administration (SEPA): Beijing, China, 1996.
42. Tran, H.N.; You, S.-J.; Hosseini-Bandegharaei, A.; Chao, H.-P. Mistakes and inconsistencies regarding adsorption of contaminants from aqueous solutions: A critical review. *Water Res.* **2017**, *120*, 88–116. [CrossRef] [PubMed]
43. Li, X.; Shuai, K.; Zhang, Y.; Jiao, G.; Zhou, H.; She, D. Removal of Cd^{2+} from wastewater to form a three-dimensional fiber network using Si-Mg doped industrial lignin-based carbon materials. *Int. J. Biol. Macromol.* **2023**, *229*, 62–69. [CrossRef] [PubMed]
44. Cai, Q.; Xu, M.; Ma, J.; Zhang, X.; Yang, G.; Long, L.; Chen, C.; Wu, J.; Song, C.; Xiao, Y. Improvement of cadmium immobilization in contaminated paddy soil by using ureolytic bacteria and rice straw. *Sci. Total Environ.* **2023**, *874*, 162594–162603. [CrossRef] [PubMed]
45. Singh, N.; Charan, S.; Patil, K.R.; Viswanath, A.K.; Khanna, P.K. Unusual formation of nano-particles of CdO and $Cd(OH)_2$ from the reaction of dimethyl cadmium with DMF. *Mater. Lett.* **2006**, *60*, 3492–3498. [CrossRef]
46. Li, J.; Chen, M.; Yang, X.; Zhang, L. Effect and mechanisms of soil functional groups in bacterial-enhanced cadmium contaminated soil phytoremediation. *Environ. Technol. Innov.* **2024**, *33*, 103531–103545. [CrossRef]
47. Wang, X.-Y.; Hao, Y.; Zhao, H.-B.; Guo, Y.-R.; Pan, Q.-J. 2D-layered $Mg(OH)_2$ material adsorbing cellobiose via interfacial chemical coupling and its applications in handling toxic Cd^{2+} and UO_2^{2+} ions. *Chemosphere* **2021**, *279*, 130617–130627. [CrossRef]

C Journal of Carbon Research

MDPI

Article

Innovative X-Ray Absorption Technology for Improved Monitoring of the Degradation and Oxidation of Granular Activated Carbon Filters Used in Hospital Water Treatment Systems

Jeamichel Puente Torres [1], Harold Crespo Sariol [2], Thayset Mariño Peacok [3], Tom Haeldermans [4], Guy Reggers [4], Jan Yperman [4,*], Peter Adriaensens [4], Robert Carleer [4] and Dries Vandamme [4]

1 Faculty of Electrical Engineering, University of Oriente, Santiago de Cuba 90500, Cuba; jeamichel.torres@meam.be or jeamichelp@gmail.com
2 Neurosciences, Signal and Images Processing, Laboratory of Applied Acoustics, University of Oriente, Santiago de Cuba 90500, Cuba; harold810922@gmail.com
3 Laboratory of Applied Acoustics, Faculty of Chemical Engineering, University of Oriente, Santiago de Cuba 90600, Cuba; tpeacok@gmail.com
4 Research Group of Analytical and Circular Chemistry, Hasselt University, Agoralaan Building D, 3590 Diepenbeek, Belgium; tomhaeldermans94@gmail.com (T.H.); guy.reggers@uhasselt.be (G.R.); peter.adriaensens@uhasselt.be (P.A.); robert.carleer@uhasselt.be (R.C.); dries.vandamme@uhasselt.be (D.V.)
* Correspondence: jan.yperman@uhasselt.be

Abstract: This study introduces a novel, non-invasive X-ray absorption analysis (XRA) method to evaluate the photonic absorption process of granular activated carbon (GAC) in hospital water purification systems. By leveraging digital radiographic images, this innovative technique monitors the deterioration and oxidation of the GAC filter, predicts its remaining lifetime, and estimates its water dechlorinating capacity. Analyzing the entire GAC filter and making a reuse possible, the new XRA method provides valuable insights into the filter's condition, enhancing water purification efficiency and costs without analyzing subsamples. Complementary analytical techniques on subsamples, taken at various depths, did not yield valuable additional information of the GAC filter exhaustion condition, nor additionally make a reuse impossible.

Keywords: XRA; activated carbon; photonic absorption; hospital water purification; digital image processing

1. Introduction

Activated carbon (AC) adsorption is a widely used technique for removing various pollutants due to its extensive surface area, high pore volume, porous structure, and specific surface functional groups. Available in both powder and granular forms (0.2–5 mm), granular activated carbon (GAC) has been extensively utilized for eliminating undesirable odours, colours, tastes, and organic and inorganic pollutants from domestic and industrial wastewaters [1–5]. In clinical laboratory water treatment systems, GAC is employed to remove free chlorine, commonly used in Cuba for water disinfection [6–12]. The disinfectant power of chlorine and its compounds relies on their oxidation capacity, with hypochlorous acid (HClO) formed through the hydrolysis of molecular chlorine. The primary mechanism for removing residual HClO by GAC involves an oxidation/reduction process, where GAC acts as a reducing agent, converting HClO into non-oxidative chloride ion (Cl^-) that can be efficiently removed by the RO unit [13].

Maintaining the required electrical conductivity (EC) for the outlet water (1–5 μS/cm) of the water treatment system necessitates an inlet water EC between 200 and 300 μS/cm. If the outlet water EC exceeds 5 μS/cm, the water pre-treatment system requires intervention to restore specifications. The performance of the GAC unit significantly impacts the lifespan of the most expensive components of the system, such as the following: the nanomembrane, the reverse osmosis (RO) membrane, and mixed-bed ion-exchange resin units (see Figure S1, Supplementary Material (S.M.)). This means that all filters, membranes, and ion-exchange cartridges of the water treatment system are tested as described in detail by the management flowchart of the hospital in the study of [14]. As this paper is now focused on the GAC unit, in the past, the management was executed by taking samples at different layers (as described in Section 2.1, Figure 1, XRA method 1)) and is discussed in detail in the studies of [12,13,15]. As a consequence of the sampling, the GAC filter is not suitable for its further use and must be discarded, exhausted or not, and replaced by an expensive and fresh GAC filter.

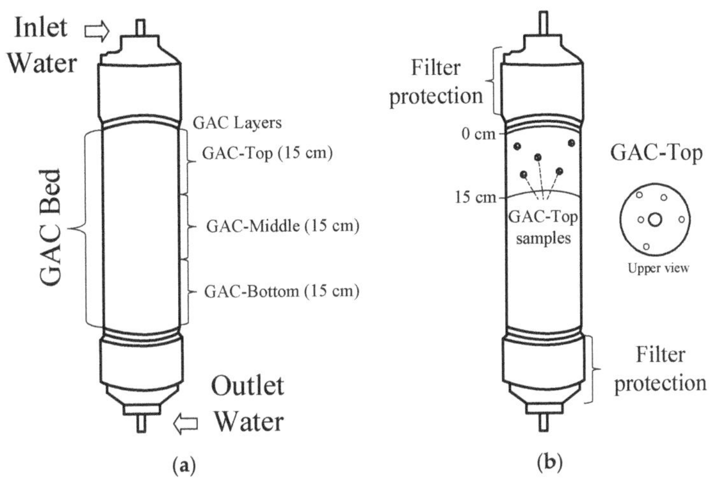

Figure 1. Schematic diagram of the water treatment filter (from Hyundai enterprise) and the sample locations at different layer regions in the GAC bed (**a**) and the sampling process (**b**). Adapted from the study of [13].

The novel XRA method 2 (fast, simple, and reliable), incorporating a new mathematical model, was developed to determine precisely and specifically when the entire cylindrical GAC filter should be replaced (or regenerated) because of complete deterioration/oxidation (making it ineffective to free chlorine removal). In this study, other analytical techniques are mentioned, addressing sample morphology (SEM), composition (TGA, EA, XRF), and porosity (BET). This approach allows us to evaluate the reliability of the proposed novel XRA method in describing the photonic absorption process after extended GAC oxidation, now with a possible reuse of the entire GAC filter during a hospital treatment.

In addition, the impact of parameters such as initial and final photonic intensity, the radius and apothem of the cylindrical filter, and the linear absorption coefficient of the entire GAC filter is investigated. Finally, data from breakthrough curves will be correlated with XRA data to predict the remaining lifetime of the filter and estimate its free chlorine removal capacity. As a result, the lifetime of the GAC filter can be extended in a cost-effective manner (which is the focus of this paper) and guarantees the required water quality (EC < 5 μS/cm).

2. Materials and Methods

Analytical procedures, experiments, and results on GAC samples using TGA, XRF, BET, SEM, and EA are described in Supplementary Material (S.M.) (Section S2).

2.1. GAC Samples and Sampling Method

Figure S1, in S.M., depicts a general set-up of the ultrapure water production system. A GAC filter (from Hyundai enterprise) was selected for the assessment of the photonic absorption process. Such a filter usually operates at a flow rate of 1.9 L/min, using pressure and temperature ranges between 1 and 3 kg/cm^2 and 34–100 °C, respectively. After six months of continuous exploitation, this filter was declared as follows: "out of service" only based on EC measurements and is labelled as GAC-Used.

Three different layer regions in the GAC bed were sampled using the procedure described in the study of [12]. These subsamples were collected within each 15 cm depth of the GAC filter as shown in Figure 1a [12,13]. Figure 1b depicts how the sampling process was performed. A collection of five samples per filter layer, radially distributed over the whole layer area, was gathered and mixed to obtain per layer a representative sample. In accordance with the GAC layer position (Figure 1a), samples were labelled as GAC-Top, GAC-Middle and GAC-Bottom.

Selected layers were characterized using different conventional techniques (TGA, EA, XRF, and SEM) and XRA method 1 (Section 2.3.1).

The filter was turned over 180° during exploitation, on regular basis, in order to extend its lifetime, having no negative effect on the water-cleaning efficiency and water quality. A sample of virgin GAC (GAC-Virgin) was also included and characterized for comparison using the same conventional techniques and both XRA methods.

2.2. Sample Preparation

Prior to analysis, the subsamples were pulverized and sieved using a 63 μm sieve in a WQS vibrating screen (0.3 mm/3000 min^{-1}) using XRA method 1 (Section 2.3.1). ASTM (Standard Test Method) D2867-04 for moisture determination [16] was used as reference for the drying process; all samples were kept in a silica-gel desiccator until analysis. As no radial dependency between the samples of each layer was found, the obtained subsamples at each layer were mixed and taken as one sample for each selected layer (GAC-Top, GAC-Middle and GAC-Bottom).

A sample of GAC-Virgin was subjected to an extended free chlorine exposure, in batch, in a dosage of 1 g of GAC to 10 mL using a solution of NaClO at a concentration of 4 mg/L during a period of 90 days. The initial concentration of NaClO, used for this experiment, was more than 20 times the real concentration that the filter would see in a real-life case scenario. The experiment was stopped when the concentration of NaClO in the solution remained the same during a period of a week. It resulted in an extremely oxidized GAC, which was used for calibration of the free chlorine removal capacity (Section 4.5). The sample was analyzed using only XRA method 1 (Section 2.3.1) and EA (Section 4.2), declared as exhausted, and labelled GAC-VExh.

The GAC-Used filter was also extra-treated with HClO, in a scaled column experiment, until exhaustion and labelled as GAC-UExh for comparison (Section 4.5).

2.3. XRA Experiments

The GAC-Used filter was analyzed by two XRA methods. First, the entire GAC filter was characterized using a new XRA method (XRA method 2) (Sections 2.3.2 and 3.1). Afterwards, the traditional XRA method (XRA method 1), as described in Section 2.3.1, was applied using the subsampling process on the same GAC filter.

2.3.1. XRA Method 1

Recent publications [12,13,15] demonstrated the use of digital radiographic images in order to detect differences in the X-ray absorption levels between GAC samples. Experiments for GAC samples were conducted using a TOSHIBA Mamorez-mgu 100d X-ray equipment used for mammography studies and very similar to the one depicted in Figure 2. Subsamples collected at various depths of the filter were placed under the X-ray equipment, following the methodology described in references [12,13]. Briefly, X-ray radiography experiments were in manual mode. An X-ray beam with energy of 22 keV-125 mAs and a sample thickness of 12 mm were used. A focal distance of 60 cm was applied for all the experiments.

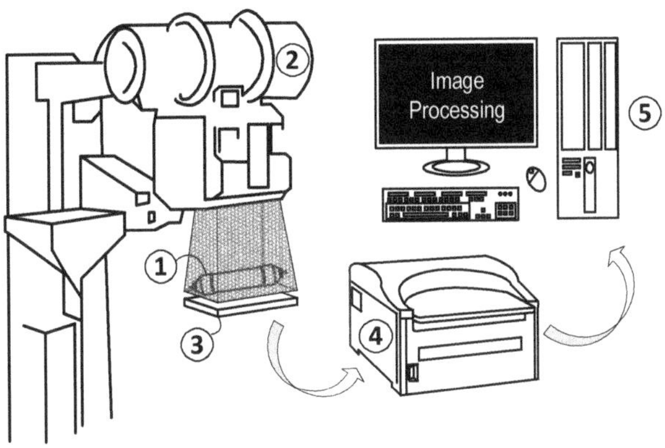

Figure 2. Scheme of experimental XRA set-up. Adapted from the study of [12]. Block 1: Cylindrical GAC filter from the water treatment system. Block 2: X-ray apparatus. Block 3: Photosensitive detector. Block 4: Digital conversion unit. Block 5: Computer.

The frequency-amplitude spectra of the X-ray images were determined using the discrete fast Fourier transform in two dimensions (2D-DFT) according to the studies of [12,13]. The mathematical analysis of the digital X-ray radiographic images was performed using dedicated software developed in MATLAB®15. XRA software version 1.0 specifically for this application.

2.3.2. XRA Method 2 (New Method)

XRA experiments on the entire GAC filter were conducted at eight different energies from 40 keV to 120 keV with 10 keV steps using a TOSHIBA KXO-36 s (clinical general radiography) X-ray equipment. Figure 2 depicts the experimental set-up applied for the new method.

The entire GAC filter (1) was placed under the X-ray apparatus (2) and was exposed at the high-energy X-ray radiation beam. The attenuated photons were registered by a photosensitive detector (3), and the registered intensity of attenuated photons was transformed in a grey-scale image using the conversion unit (4). Obtained grey-scale image was conveniently processed by PC (5) using dedicated software. The cylindrical shape of GAC filter caused different X-ray absorption levels in the same X-ray digital image and produced a grey-scale image with grey-scale intensity (GSI) values decreasing from the centre of the filter to the edges, independently of its exhaustion degree (virgin filters will show the same behaviour). However, a GAC filter with a higher photonic attenuation will always show a whiter image in comparison with a virgin GAC filter.

According to the original XRA images, each filter has an approximate radius of 220 pixels, equivalent with the 5 cm radius of the cylindrical filter; each GSI measurement was performed using 20 pixel steps (equivalent to approximately 0.46 cm). According to the proposed model (Section 3.1), apothem variations were measured in pixels using XRA images and dedicated software, and, afterwards, the conversion to cm was performed using the real filter dimensions.

The photons registered by the detector are the photons that cross through the material without any interaction, reaching the detector with the same initial energy. In contrast, the photons that interact with the GAC filter will produce photoelectrons and characteristic radiation that leave the GAC filter with lower energy [17]. Scattered photons from electron transitions are not relevant from a radiologic viewpoint and therefore neglected. This inconvenience is usually solved using anti-diffusion grids, which are installed in the X-ray apparatus eliminating about 95% of this radiation [12,17].

2.4. Breakthrough Curves for Free Chlorine Removal

First, a calibration curve for NaClO determination was obtained as described in S.M. (Figure S2). A 4 mg/L NaClO solution was used for the free chlorine removal experiments in scaled columns. To determine the free chlorine in samples collected after the experiments, a total sample volume of 3 mL was used. Specifically, 1 mL of the effluent was collected every 20 min and mixed with 0.5 mL of N, N-diethyl-p-phenylenediamine sulphate (DPD) solution (2 mg/mL). The mixture was then diluted with 1.5 mL of Milli-Q (MQ) water to reach a final volume of 3 mL. The free chlorine concentration was determined by photometry using a UV–visible spectrophotometer (Ultraspec-7000, Biochrom Ltd., Holliston, MA, USA) at a wavelength of 515 nm [18,19].

3. Theory and Calculation

3.1. XRA Method 2: Relationship Between Grey-Scale Intensity (GSI), Linear Absorption Coefficient (μ), Radius (r), and Apothem (a) for a Cylindrical GAC Filter

An X-ray photonic detector will record the number of photons that pass through a material of thickness x, the number of photons that reach the detector will be N, and the number of photons that interact with the material and are removed from the beam is n. The number of removed photons n is proportional to the number of initial photons present in the beam and the number of atoms in the material (proportional to x). Therefore, the relation between n, N, and x can be mathematically expressed as follows [17]:

$$n = \mu N x \tag{1}$$

with

n : Number of photons removed from the beam (dimensionless).

μ : Linear absorption coefficient (in mm^{-1}).

N : Number of photons that reach the detector (dimensionless).

x : Material thickness (in mm).

If ΔN represents the change in the number of photons in the incident X-ray beam that pass through x, then N is reduced by each interaction that occurs ($\Delta N = -n$), and Equation (1) adopts the following form:

$$\Delta N = -\mu N x \tag{2}$$

A sectional frontal view of the GAC filter is shown in Figure 3.

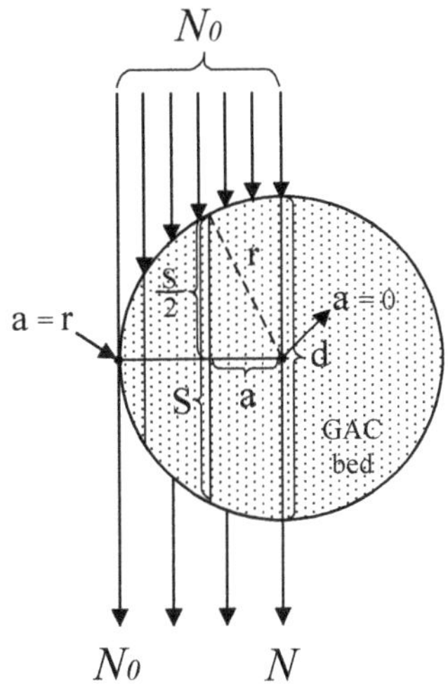

Figure 3. Frontal view of the Hyundai GAC filter for a hospital water purification system.

The photons that pass through the GAC filter without any interaction will be recorded by the photosensitive detector. For all experiments, the distance between the focus of the beam and the GAC sample should remain constant between experiments.

N_0: Initial photonic intensity (dimensionless).

N: Final photonic intensity (dimensionless).

a: Apothem of the circumference (in mm).

d: Diameter of the circumference (in mm).

r: Radius of the circumference (in mm).

S: Chord of the circumference (in mm).

For a constant distance between the focus of the beam and the GAC filter, the number of photons that reach the detector is only modified by the square of the distance, since the air attenuation is about 0.001% and therefore negligible. Considering the cylindrical shape of the GAC filter, the longitude of the circumference chords changes proportionally to the variation in the circumference apothems. Thus, S can be related with a applying the Pythagoras theorem:

$$r^2 = \left(\frac{S}{2}\right)^2 + a^2 \tag{3}$$

and

$$S = 2\sqrt{r^2 - a^2} \tag{4}$$

According to the shape of the filter, each circumference chord corresponds to a specific x:

$$S = x \tag{5}$$

Then,

$$x = 2\sqrt{r^2 - a^2} \tag{6}$$

Considering that the radius of the circumference (r) is constant, chord variations ($S = x$) are function of the apothem (a). Assuming the variations of Δx and a are small enough, expression (6) can be derived with respect to a:

$$\frac{dx}{da} = \left(2\sqrt{r^2 - a^2}\right)' \tag{7}$$

Then,

$$dx = \left(2\sqrt{r^2 - a^2}\right)' da \quad (8)$$

For small variations of N, Equation (2) can be written as follows:

$$dN = -\mu N dx \text{Thus, } \Delta N = \sum dN \text{ and also } x = \sum dx \quad (9)$$

Combining Equation (8) with Equation (9) gives the following:

$$dN = -\mu N\left(2\sqrt{r^2 - a^2}\right)' da \quad (10)$$

The differential equation shown in Equation (10) can be solved separating variables as follows:

$$\begin{gathered} \frac{dN}{N} = -\mu\left(2\sqrt{r^2 - a^2}\right)' da \rightarrow \int \frac{dN}{N} = -\mu \int \left(2\sqrt{r^2 - a^2}\right)' da \\ \ln N = -\mu\left(2\sqrt{r^2 - a^2}\right) + C \end{gathered} \quad (11)$$

where

C: Coefficient of the independent solution.

When photons pass through the cylindrical filter, the higher attenuation occurs at the centre of the filter ($S = x = 2r = d$ and $a = 0$). On the other hand, when $S = x = 0$ and $a = r$, the opposite phenomenon occurs (see Figure 3).

According to Figure 3, the boundary condition can be defined:

$$\text{For } a = r \rightarrow S = x = 0 \rightarrow N = N_o \rightarrow \ln N_o = C \text{ according to Equation (11)} \quad (12)$$

Replacing the independent solution C in Equation (11) gives the following:

$$\begin{gathered} \ln N = -\mu\left(2\sqrt{r^2 - a^2}\right) + \ln N_o \ and \ \ln N - \ln N_o = -\mu\left(2\sqrt{r^2 - a^2}\right) \\ \rightarrow \ln\left(\frac{N}{N_o}\right) = -\mu\left(2\sqrt{r^2 - a^2}\right) \end{gathered} \quad (13)$$

Then,

$$N = N_o e^{-\mu(2\sqrt{r^2 - a^2})} \quad (14)$$

When $a > r$, Equation (14) adopts a complex solution, meaning that the shape of the filter changed and corresponds with an ellipse, not with a circumference. In that case, the previous analysis is not valid; the parameter a is therefore restricted to the condition $a < r$.

According to recent publications [12,13], it is possible to establish a mathematical relationship between attenuated photons (N) and GSI using Equation (15).

$$\text{GSI} = f(N) \quad (15)$$

From Equation (15), GSI can be expressed as an equivalent exponential function of the photonic intensity (N). Therefore, considering the filter shape, it is possible to express the photonic absorption process using the GSI values as given by Equation (16):

$$\text{GSI} = \text{GSI}_0 \cdot e^{\mu(2\sqrt{r^2 - a^2})} \quad (16)$$

The grey-scale intensity of incident photons (GSI_0) was calculated using the information at the edge of the filter (without GAC sample) according to the study of [12]. The GSI value at 40 mAs and 40–120 keV was approximately 0.0039 for all the measurements. At filter edge, GSI = GSI_0, and $r = a$, $\mu = 0$ (Equation (16)). According to the proposed new mathematical model, GSI values in the cylindrical filter change as a function of the initial photonic intensity, the linear absorption coefficient of the material, and apothem variations (reflected in chord variations).

Considering this model, the GSI value will be more intense for $a = 0$ (at the centre of the cylindrical filter) by direct influence of the filter shape, regardless of the exhaustion degree of the GAC, and it will decrease from that point up to $a = r$, at the edge of the filter, to its lowest level (GSI = GSI_0).

3.2. Parameters for the Breakthrough Curves for Free Chlorine Removal

The breakthrough curves were plotted using the ratio between the effluent free chlorine and the initial free chlorine concentrations (C_e/C_o) versus time (in min). The dechlorinating capacity of the GAC was calculated and quantified taking into account the removed free chlorine concentration from the inlet water by the GAC filter and the time duration that the filter is capable to maintain its removal capacity with respect to the virgin material.

The breakthrough time (t_p) was determined as the time for reaching the equilibrium at which the derivative of the ratio (C_e/C_o) to the time reaches its maximum, and the breakthrough curve shows an inflexion point. At this point, the removal capacity of the GAC towards the adsorbates (free chlorine and hardness ions) can be estimated using Equation (17):

$$CA = \frac{\int_0^{t_{p(i)}} \left[1 - \left(\frac{1}{C_o} \cdot C_{ei}(x)\right)\right] dx}{\int_0^{t_{p(virgin)}} \left[1 - \left(\frac{1}{C_o} \cdot C_{e(virgin)}(x)\right)\right] dx} \tag{17}$$

where

CA: Removal capacity towards the adsorbate (dimensionless).

$t_{p_{(i)}}$: Breakthrough time of the sample i for adsorbate removal (in min).

$t_{p_{(virgin)}}$: Breakthrough time of GAC-Virgin sample for adsorbate removal (in min).

C_{ei}: Adsorbate concentration of the effluent for sample i (in mg/L).

$C_{e(virgin)}$: Adsorbate concentration of the effluent for GAC-Virgin sample (in mg/L).

C_o: Adsorbate concentration at the inlet of the column (in mg/L).

Equation (17) calculates the ratio between the areas under the breakthrough curves for each analyzed GAC sample and the GAC-Virgin sample. The result is expressed in percent with respect to the adsorbate removal capacity of the virgin GAC sample (CA (%)). Equation (17) utilizes the complete shape of the breakthrough curve (changes in the effluent concentration as well as the breakthrough time) for the estimation of the GAC removal capacity. The calculations for the numerical estimation of the area under the curve for each analyzed GAC sample were performed using the software packages Origin® 6.1 and MATLAB® 15 [14,20].

4. Results and Discussion

4.1. XRF Analysis

Only the contribution of inorganic cations and anions is monitored when measuring the EC of water from the hospital water treatment system. According to XRF results (see S.M., Section S3.2, Table S1), no significant differences in composition are found between GAC-Virgin and GAC-Used samples, except for the Cl, Ca, and Mg content, which differs as expected due to the HClO treatment. Additionally, XRF results do not show any evidence of precipitation or adsorption of inorganic components that could cause filter exhaustion.

4.2. Elemental Analysis (EA) Results

N, C, H and O contents for GAC layer samples and GAC-Virgin are depicted in Table 1a.

Table 1. (**a**): N, C, H, and O content of the GAC samples (in wt. %). (**b**): N, C, H, and O content of the GAC-Virgin sample after extended free chlorine exposure (in wt. %).

(a)								
	N	SD	C	SD	H	SD	O	SD
GAC-Virgin	0.44	0.01	93.54	0.62	0.81	0.07	2.22	0.57
GAC-Bottom	0.57	0.02	76.41	0.46	2.14	0.07	17.89	0.41
GAC-Middle	0.54	0.01	75.92	0.74	2.23	0.11	18.31	0.76
GAC-Top	0.56	0.02	76.63	0.01	1.04	0.02	18.78	0.25
(b)								
	N	SD	C	SD	H	SD	O	SD
GAC-VExh	0.46	0.03	86.44	0.03	0.91	0.01	9.21	0.04

With SD: standard deviation.

In the hospital water purification system, when the EC starts to fall out of range, trained technicians reorient the filter by turning it around. This procedure is feasible due to the filter's small size and easy-to-manipulate mechanical characteristics, thereby extending its safe use. As a result of this practice, no significant differences in element concentrations between the GAC layers of the used filter are observed (Table 1a) [21,22].

Based on EA results, a clear increase in the oxygen content of the used GAC is observed, indicating significant oxidation. This increase in oxygen content is related to the deterioration and oxidation of the GAC as confirmed by an artificial long-term free chlorine-treated GAC-Virgin (GAC-VExh, Section 2.2) and is reported in Table 1b, demonstrating a 7% increase in oxygen content compared to GAC-Virgin and consistent with the EA results presented in Table 1a [14,20].

However, this does not mean that GAC-Used is exhausted. On the contrary, it can still effectively reduce chlorine and chloramines, providing high-quality water. This suggests that EC measurements alone do not accurately describe the exhaustion stage of the filter.

A drawback of the EA is that it renders the filter unusable for further patient treatment since sample collection was required.

4.3. XRA Method 1

Figure 4 depicts XRA images of three sampled GAC-Used layers and GAC-Virgin as well as the image histograms and 2D-DFT frequency spectra for each analyzed image.

The digital images obtained using XRA method 1 show a grey-scale image where the GAC-Used layer samples with higher photonic absorption are closer to one in the normalized grey scale (whiter zones) and vice versa.

In Figure 4b, a quite similar pattern in the histograms from radiographic images of the different GAC-Used layer samples is observed, revealing a broad distribution of the number of pixels along the normalized grey scale. Image histograms from GAC-Top, GAC-Middle, and GAC-Bottom depict two characteristic peaks: the first displaced to the left in the normalized grey scale (GSI $\approx$ 0.08, closer to zero) and the second one closer to the right side of the normalized grey scale (GSI $\approx$ 0.99, closer to one), which is an indication of a higher photonic absorption process in comparison with GAC-Virgin [14,20].

GAC-Virgin depicts one single peak displaced to the left of the normalized grey scale (GSI $\approx$ 0.04) without the presence of high photonic absorption zones in the rest of the image histogram.

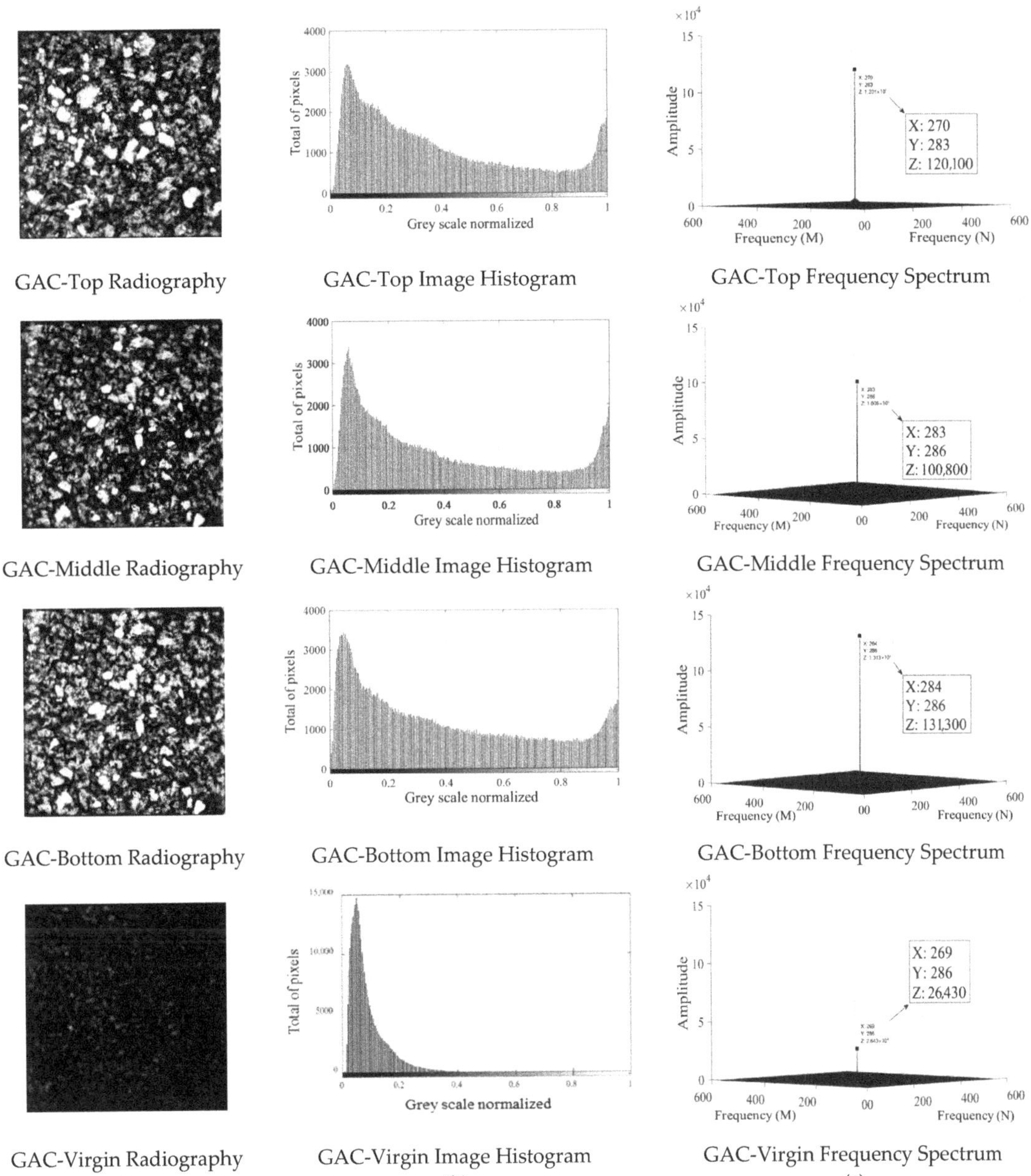

Figure 4. Digital radiographic images (**a**), image histograms (**b**), and image frequency spectra (**c**) of three GAC-Used layer samples and GAC-Virgin (XRA method 1).

As GAC-Virgin has not yet been used, its photonic absorption is lower in comparison with the GAC-Used layer samples, thus producing a darker image. When analyzing the frequency spectra of obtained radiographic images (Figure 4c), similar patterns between the direct components of the 2D-DFT in samples from the used filter can be noticed. On the other hand, significant differences in amplitude are found between GAC-Virgin and the GAC-Used layer samples as the result of oxidation. However, obtained differences in the direct components of the 2D-DFT between GAC-Used layer samples from XRA method 1 are not related with differences in the exhaustion degree or oxidation degree between layers or differences in the filter last orientation as confirmed by TGA (S.M., Section S3.1), EA (Table 1a), XRF (S.M., Section S3.2), and SEM (S.M., Section S3.4) results. In addition, XRA method 1 requires sampling of the GAC-Used filter, which makes it unusable in the water treatment system.

4.4. XRA Method 2

Figure 5 depicts XRA images from the entire GAC-Used filter (right side of each image) and an entire filter filled with GAC-Virgin (left side of each image) at different irradiation energies.

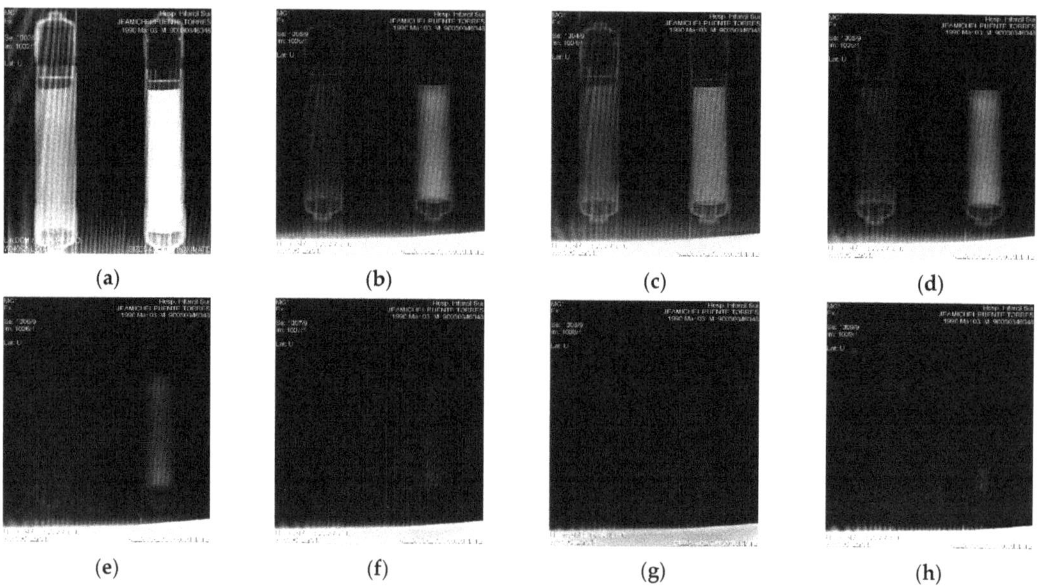

Figure 5. XRA images of the filter filled with GAC-Virgin (left in each image) and GAC-Used (right in each image) at different irradiation energies by XRA method 2, (**a**) 40 keV, (**b**) 50 keV, (**c**) 60 keV, (**d**) 70 keV, (**e**) 80 keV, (**f**) 90 keV, (**g**) 100 keV, and (**h**) 120 keV.

According to the proposed model (Equation (16)), the shape of the filter has a significant influence on the photonic absorption process, regardless of the exhaustion differences between filters. From Figure 5, significant differences in brightness (reflected in GSI levels) between GAC-Used and GAC-Virgin can be noticed. Again, no significant difference can be observed between layers in the used filter at lower energies, confirming the results obtained by TGA (S.M, Section S3.1), EA (Table 1), XRF (S.M., Section S3.2), SEM (S.M., Section S3.4), and XRA method 1.

Obtained differences in grey-scale intensity at different energies at the edges of the used filter (Figure 5) could be related with slight modifications in initial photonic intensity at the extremes of the filter (top and bottom). The same phenomena can also be observed in the GAC-Virgin filter between top and bottom layers in the energy range between 60 and 80 keV (not visible for GAC-Used).

In Figure 5a–h, the GAC-Used filter (right) presents more brightness in comparison with the GAC-Virgin (left) filter, indicating significant differences in photonic absorption between both cylindrical filters, which is related to a drastic change in elemental composition (Table 1), especially an increased oxygen content in GAC-Used.

The increased O content will change the effective Z value of the GAC matrix and therefore its photonic absorption. An increased O content can be considered as a measure of the oxidation degree of the GAC by the feed water. In addition, BET measurements (S.M., Section S3.3) point to a small decrease in the surface area of GAC-Used compared to GAC-Virgin.

SEM analysis (S.M., Section S3.4) revealed a smooth surface with small and medium cavities for GAC-Virgin. In contrast, deterioration of the GAC-Used layer samples is observed, confirming clear differences in porosity (Figure S4) as the result of oxidation by the feed water. According to Figure 5, a decrease in the brightness of GAC-Used and GAC-Virgin filters is noticed when the energy increases: an increase in energy causes a decrease in the grey-scale level obtained, due to a proportional increase in the penetration power of the incident photonic beam, making the differentiation by contrast very difficult (darker images) [14,20]. Some images (b–h) in Figure 5 appear to be rather identical, but this is partly misleading and certainly subjective. The difference between the different images (Figure 5) becomes only clear from a numerical point of view when analyzing the images using the dedicated XRA software (see Section 2.3.2). Hereafter, a correct conclusion can be made

Obtained GSI values from XRA method 2 were correlated with apothem (a) variations in both GAC-Used and GAC-Virgin filters, specifically for irradiation energies within the range of 50 keV to 80 keV. For all selected energies, an initial photonic intensity of 40 mAs was applied in all experiments. Table 2 presents the GSI (a), (S) values (see Figure 3), and the energy applied for each filter. At lower energies (40 keV), the penetrating power decreases, and only a few photons reach the detector, resulting in XRA images with very high brightness values (see Figure 5a). Below 40KeV and from 80 keV, the produced XRA images exhibit very low brightness values for both filters (see Figure 5f–g). Consequently, the difference in GSI values between the two filters is not significant at these low and high energy levels, and, thus, the results are omitted in Table 2.

Table 2. S, a and GSI parameters found at different irradiation energies (50–80 keV range).

		50 keV		60 keV		70 keV		80 keV	
a	S	$GSI_{(Used)}$	$GSI_{(Virg)}$	$GSI_{(Used)}$	$GSI_{(Virg)}$	$GSI_{(Used)}$	$GSI_{(Virg)}$	$GSI_{(Used)}$	$GSI_{(Virg)}$
5.00	0.00	0.11	0.07	0.12	0.13	0.11	0.07	0.10	0.07
4.54	4.19	0.15	0.09	0.25	0.15	0.13	0.08	0.11	0.07
4.09	5.75	0.19	0.10	0.28	0.17	0.21	0.09	0.13	0.07
3.64	6.85	0.25	0.10	0.34	0.17	0.24	0.11	0.15	0.08
3.18	7.72	0.34	0.11	0.35	0.19	0.26	0.12	0.16	0.08
2.73	8.34	0.35	0.13	0.35	0.21	0.32	0.14	0.19	0.08
2.27	8.91	0.40	0.13	0.37	0.21	0.39	0.15	0.20	0.08
1.82	9.32	0.41	0.15	0.39	0.21	0.42	0.16	0.21	0.08
1.36	9.62	0.42	0.12	0.41	0.22	0.45	0.18	0.22	0.08
0.91	9.83	0.43	0.15	0.42	0.25	0.46	0.19	0.23	0.08
0.45	9.96	0.52	0.12	0.44	0.27	0.51	0.21	0.25	0.10
0.00	10.0	0.53	0.18	0.57	0.33	0.54	0.25	0.32	0.12

With the following: $GSI_{(Used)}$: grey-scale intensity for the GAC-Used filter; $GSI_{(Virg)}$: grey-scale intensity for the GAC-Virgin filter; and a and S in cm.

According to Table 2, GSI values depict a decreasing trend from the centre of both filters to the extreme edge.

In order to discuss the results of Table 2 in more detail and more clearly, Figure S5 is included in S.M., depicting the correlation between the GSI values (from XRA method 2) and the apothem (a) variations for both GAC filters.

The correlation parameters between GSI values obtained applying XRA method 2 and apothem variations in the GAC-Used filter (GSI_{Used}) and the GAC-Virgin (GSI_{Virgin}) filter are presented in Table 3.

Table 3. Fitting parameters found for apothem (a) variations and GSI values (XRA) for the GAC-Used filter and the GAC-Virgin filter.

50 keV-40 mAs					
Evaluated Parameter	GSI_0	$e(GSI_0)$	m	e(m)	R^2
GSI_{Virgin}	0.160	0.008	−0.010	0.002	0.86
GSI_{Used}	0.550	0.006	−0.080	0.002	0.98
60 keV-40 mAs					
Evaluated Parameter	GSI_0	$e(GSI_0)$	m	e(m)	R^2
GSI_{Virgin}	0.290	0.009	−0.030	0.003	0.95
GSI_{Used}	0.510	0.020	−0.060	0.008	0.93
70 keV-40 mAs					
Evaluated Parameter	GSI_0	$e(GSI_0)$	m	e(m)	R^2
GSI_{Virgin}	0.230	0.004	−0.030	0.001	0.99
GSI_{Used}	0.560	0.001	−0.080	0.003	0.99
80 keV-40 mAs					
Evaluated Parameter	GSI_0	$e(GSI_0)$	m	e(m)	R^2
GSI_{Virgin}	0.090	0.004	−0.007	0.001	0.79
GSI_{Used}	0.280	0.008	−0.040	0.003	0.97

$GSI = ma + GSI_0$ (with e(i) the error on parameter i).

According to Table 3, strong correlation coefficients, reaching a 99% goodness of fit, can be found between GSI and (a) when an optimal energy of 70 keV (40 mAs) for the incident photonic beam is used. Acceptable correlation coefficients between GSI and (a) are observed at 50 keV, 60 keV, and 80 keV for both the GAC-Used filter and the GAC-Virgin filter (except for the GAC-Virgin filter at 80 keV).

Figure 6 shows the differences in grey-scale intensity (DGS) values calculated for each irradiation energy to determine the maximum DGS value between extreme samples to optimize the irradiation energy and sample thickness, achieving the best possible sensitivity for XRA method 2.

From Figure 6, the maximum DGS value around 0.3 was obtained at 70 keV-40 mAs. It can be concluded based on Equation (16) that the higher the diameter of the cylindrical filter to be analyzed, the higher the energy needed to achieve proper sensitivity values using XRA method 2.

In order to evaluate the effectiveness of the XRA method 2 for the description of the photonic absorption process, correlation plots between GSI and S are displayed in Figure 7 for the GAC-Used and the GAC-Virgin filters using the optimal experimental conditions (70 keV-40 mAs).

Table 4 shows the fitting correlation parameters between GSI and S found for both filters at an irradiation energy of 70 keV-40 mAs with very good regression coefficients (R^2) of approximately 99%.

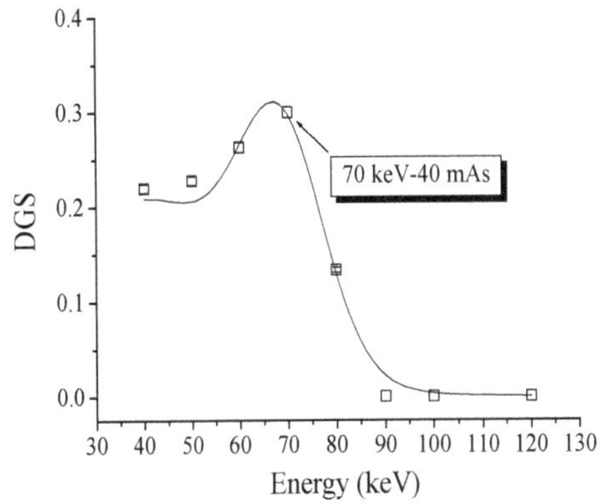

Figure 6. Differences in grey-scale intensities (DGS) obtained for GAC-Used and the GAC-Virgin filters at different irradiation energies.

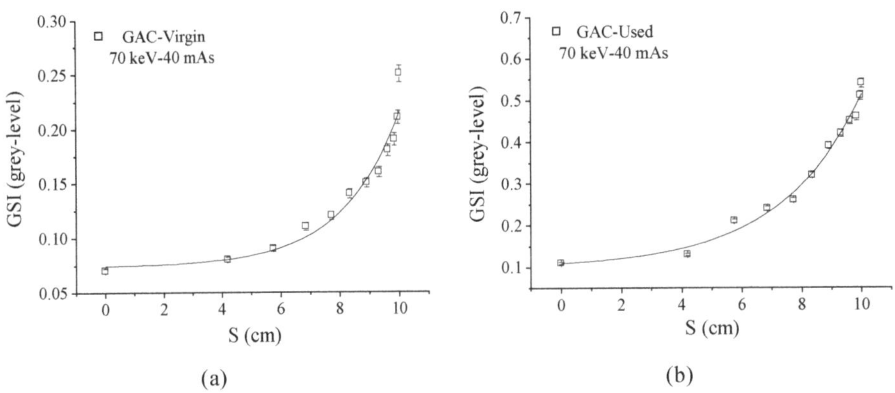

Figure 7. Correlation graphs between GSI and *S* for GAC-Virgin (**a**) and GAC-Used (**b**) filters at 70 keV-40 mAs.

Table 4. Fitting correlation parameters between GSI and *S* for both filters at optimized conditions (70 keV-40 mAs).

Evaluated Parameter	GSI_0	$e(GSI_0)$	m	e(m)	t	e(t)	R^2
GSI_{Virgin}	$7 \cdot 10^{-2}$	$1 \cdot 10^{-3}$	$7 \cdot 10^{-4}$	$1 \cdot 10^{-5}$	2.0	$4 \cdot 10^{-1}$	0.99
GSI_{Used}	$1 \cdot 10^{-1}$	$1 \cdot 10^{-3}$	$1 \cdot 10^{-2}$	$5 \cdot 10^{-3}$	3.0	$3 \cdot 10^{-1}$	0.99

$GSI = me^{(S/t)} + GSI_0$ (with e(i) the error on parameter i).

Figure 7a,b demonstrate that almost all the photons are absorbed at the centre of the filter when the *S* parameter maximizes, producing higher GSI values in the normalized grey scale. Figure 8 depicts XRA images, the image histogram, and frequency spectra from the entire GAC-Used and GAC-Virgin filter when applying XRA method 2.

From Figure 8, significant differences can be noticed between image histograms and frequency spectra from the GAC-Used filter and GAC-Virgin filter. The image histogram from the GAC-Used filter (Figure 8b) depicts a broad peak of a relative low number of pixels around 0.45 in the normalized grey scale, consistent with a higher photonic absorption process in this filter and therefore a higher GSI value in comparison with the GAC-Virgin filter.

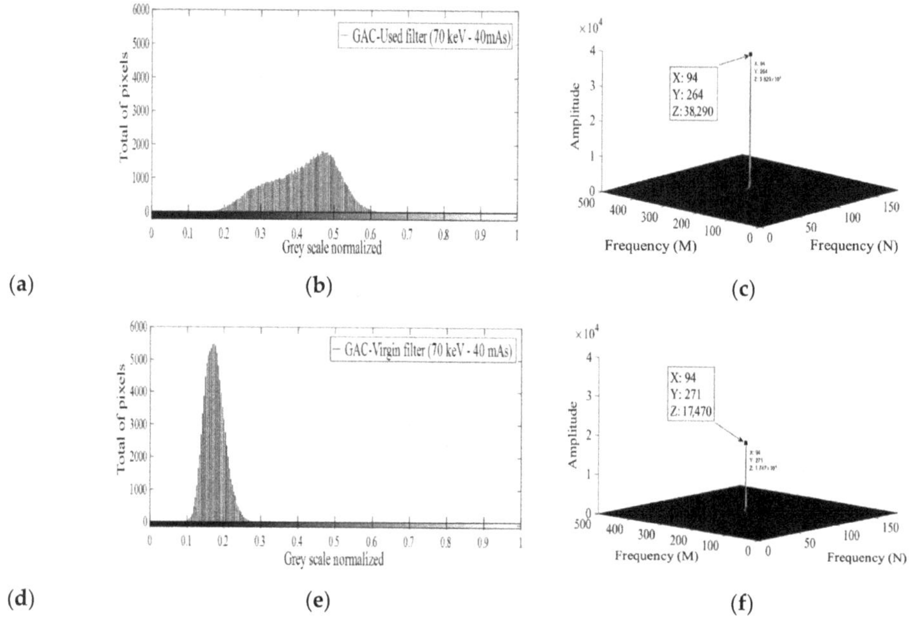

Figure 8. Digital radiographic image, image histogram, and frequency spectrum of GAC-Used filter (**a**–**c**); digital radiographic image, image histogram, and frequency spectrum GAC-Virgin filter (**d**–**f**) (XRA method 2).

On the other hand, the image histogram from the GAC-Virgin filter (Figure 8e) shows a narrow peak with a maximum number of pixels around 0.18 in the normalized grey scale consistent with a lower photonic absorption process for this filter, showing relatively lower GSI values (closer to ¨0¨). The direct component of the frequency spectra (2D-DFT) of both filters reinforces the obtained results from image histograms, showing a significant higher peak for the GAC-Used filter in comparison with the GAC-Virgin filter.

While XRA method 1 yields similar results and conclusions (see Figure 4), the novel non-invasive XRA method 2 was developed to provide more detailed information about the oxidation state of the entire filter without the need for time-consuming and destructive sampling, unlike XRA method 1, which requires extensive damage to the filter for sampling, making it unusable in the water treatment system.

In addition, XRA method 2 preserves the integrity of the filter, allowing it to remain in use. This significant advantage enables comprehensive insights into the filter's condition, as demonstrated by the distinct GSI values between GAC-Used and GAC-Virgin filters, correlating with their oxidation and deterioration levels.

Therefore, the development of XRA method 2 addresses the limitations of XRA method 1 by eliminating the need for invasive sampling, which can be labour-intensive and disruptive. This new method allows for continuous, real-time monitoring of the GAC filter's performance throughout its operational life in a hospital-grade water treatment unit. By accurately determining the oxidation and deterioration degree of GACs, in other words, its exhaustion degree, XRA method 2 provides a reliable and efficient alternative for ensuring optimal filter performance and longevity.

4.5. Relation Between XRA Method 2 and the GAC Free Chlorine Removal Capacity

Breakthrough curves from the GAC-Used filter were analyzed to gather information about the free chlorine removal capacity of the GAC sample using scaled columns. GAC-

VExh provides a controlled reference point for comparison with GAC-UExh from the hospital setting (Section 2.2). This comparison is crucial for validating the effectiveness of XRA method 2, correlating the obtained data from breakthrough curves with the actual oxidation and deterioration states of the GAC filters, and estimating their remaining lifespan [21,22].

Figure 9a,b depict obtained breakthrough curves as well as the breakthrough times for the GAC-Virgin filter, GAC-Used filter, and GAC-UExh filter after the free chlorine removal experiments until saturation of GAC-Used.

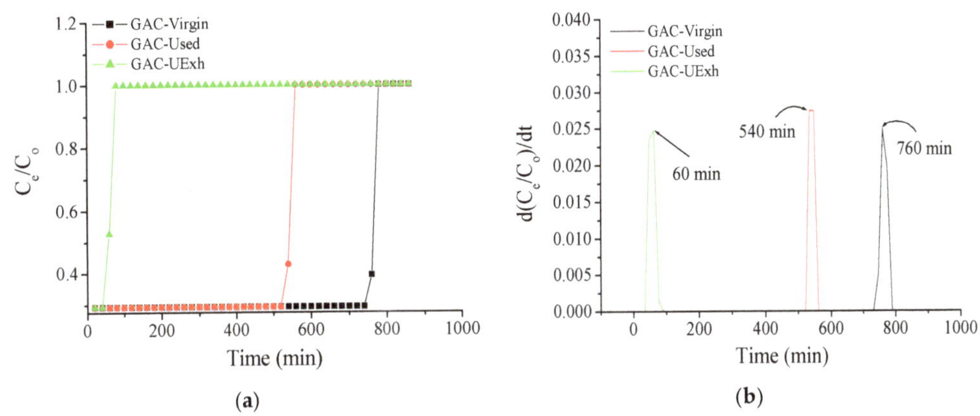

Figure 9. Breakthrough curves for GAC-Virgin, GAC-Used, and GAC-UExh filter (**a**), derivative function of the breakthrough curves with breakthrough times (**b**).

The area under the breakthrough curves was used as a numeric indicator of the GAC filter free chlorine removal capacity using Equation (17). The breakthrough time used as reference for determining the exhaustion level for the GAC-Virgin filter was approximately 760 min. From Figure 9a,b, significant differences in free chlorine removal capacity between analyzed GAC filters can be noticed with the following ranking: GAC-Virgin > GAC-Used > GAC-UExh. For the GAC-Used filter, it must be pointed out that the exhaustion level was only determined based on EC measurements in the hospital water purification system being not the optimal method since it is still effective in free chlorine removal.

Table 5 reports the calculated values of the area under the breakthrough curves and the free chlorine removal capacity according to Equation (17).

Table 5. Area under breakthrough curves (A_C) and free chlorine removal capacity ($CA_{(HClO/ClO^-)}$) (%).

GAC-Sample	A_C	$CA_{(HClO/ClO^-)}$
GAC-Virgin	522.0	1.00
GAC-Used	338.0	0.65
GAC-UExh	30.52	0.06

The GSI levels for the GAC-Virgin filter and GAC-Used filter were 0.1725 and 0.4667, respectively, utilizing XRA method 2. The GSI value of GAC-UExh is close to ¨1¨ in the normalized grey scale [13,14]. Figure 10 depicts the free chlorine removal capacity obtained from breakthrough curves plotted against the GSI values derived from the application of XRA method 2 for the GAC-Virgin, GAC-Used, and GAC-UExh.

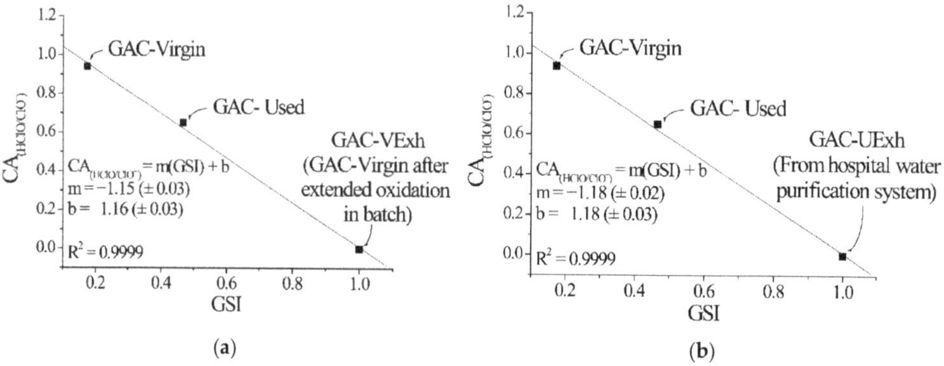

Figure 10. Correlation graphs between $CA_{(HClO/ClO^-)}$ and GSI for a GAC-Virgin filter, GAC-Used filter, and GAC-VExh. (**a**) Additionally, a $CA_{(HClO/ClO^-)}$ value of 0.97 is obtained for the GAC-UExh (**b**).

The CA $_{(HClO/ClO^-)}$ capacity (0.94) and GSI value (0.99) of GAC-VExh was calculated according to Equation (S1) (see S.M., Section S2.5) and XRA method 1. Some differences in exhausted characteristics between GAC-VExh and GAC-UExh can be expected due to its different history treatment, and, thus, some small differences in the GSI value can be obtained. Therefore, an average value of both linear equations could be used. This comparison was conducted to assess the performance of the novel XRA method 2 for the entire filter from different filtration histories.

This demonstrates that XRA method 2 can predict the free chlorine removal capacity during the operation of the water purification system, using GSI values and expressing $CA_{(HClO/ClO^-)}$ (%) as

$$CA_{(HClO/ClO^-)}(\%) = [m\cdot(GSI) + b]\cdot 100 \tag{18}$$

for GSI > 0.1645.

Equation (18) provides an indication of the dechlorinating capacity (in %) of the entire GAC filter during operation, relative to the GAC-Virgin sample, using GSI values. A GSI value below 0.1645 indicates a 100% free chlorine removal capacity, while GSI values around 0.99 signify a complete loss of this capacity [14,20].

Therefore, XRA method 2 introduces a more accurate and reliable criterion for replacing the GAC filter compared to the current EC measurements. Additionally, the entire GAC-Used filter is analyzed, rendered it further available in the treatment. Another significant outcome of this research is the more precise determination of the free chlorine removal capacity using XRA method 2, which will extend the GAC filter's lifespan and be cost-effective (see S.M, Section S1), directly indicating their oxidation/deterioration degree and monitoring their removal capacity for adsorbates.

5. Conclusions

A novel X-ray absorption analysis (XRA) method has been proposed and optimized to deliver detailed information about the oxidation and deterioration degree of GAC filters used in hospital water purification systems. This method also assists in predicting the remaining lifetime of the filter in a cost-effective manner. Applicable to entire cylindrical GAC filters, the method is based on digital radiographic images, grey-scale intensity levels, and a modified mathematical model. The grey-scale intensity is influenced by the relationship between the filter's radius and apothem, the initial photonic beam intensity, and the linear absorption coefficient of the analyzed filter. Optimal experimental conditions are 70 keV-40 mAs. Notably, this novel XRA method does not require subsamples, preserving the filter for continued use.

The proposed XRA method stands out as a non-invasive technique that maximizes the filter's usability after each XRA monitoring session, unlike previously described XRA methods and other conventional analytical techniques. This innovative method is a valuable tool for monitoring the performance of cylindrical GAC filters, providing a direct indication of their oxidation and deterioration degree. Additionally, it aids in predicting the filter's remaining lifespan and estimating its dechlorinating capacity.

Supplementary Materials: The following supporting information can be downloaded at https://www.mdpi.com/article/10.3390/c11020030/s1, Figure S1: General set-up of an ultrapure water production system (min. pressure and water flow: 1–3 kg/cm^2 and 1.9 L/min). Block 1: Sediment filter. Block 2: GAC filter (from Hyundai enterprise). Block 3: Nano membrane. Block 4: Reverse Osmosis (RO) membrane. Block 5: Mixed-bed ion exchange cartridges.; Figure S2: Calibration curve of absorbance (A) as a function of different NaClO concentrations (in mg free Cl_2/L).; Figure S3: Mass loss in wt. % of GAC-Top, GAC-Bottom and GAC-Virgin samples as a function of temperature, (a) 25 °C to 600 °C (N_2 atmosphere) and (b) 25°C to 900 °C, with 600 °C to 900 °C (O_2) atmosphere).; Figure S4: (a) SEM image GAC-Top, (b) GAC-Bottom and (c) GAC-Virgin.; Figure S5: Correlation graphs between GSI and *a* variations for the GAC-Virgin filter and GAC-Used filter between 50 keV and 80 keV.; Table S1: Semi-quantitative analysis of the inorganic fraction of the different layers in the GAC filter (on dry basis). Refs. [21,22] are cited.

Author Contributions: Conceptualization, validation, software, formal analysis, data curation, writing original draft: J.P.T.; methodology: J.P.T., T.H. and T.M.P.; investigation: J.P.T. and G.R.; resources, project management, supervison: J.Y., P.A. and D.V.; writing—review and editing: H.C.S., J.Y., R.C., P.A. and D.V. All authors have read and agreed to the published version of the manuscript.

Funding: This research was funded by: UHasselt BOF funding program: grant number BOF23KV01 and VLIR-IUC-UOS funding program: grant number: Universidad de Oriente CU2019IUC030A105-77143.

Data Availability Statement: Data can be received by mailing the corresponding author.

Acknowledgments: The authors would like to thank Hasselt University BOF funding, especially the BOF-Incoming mobility -BOF23KV01 scholarship for Jeamichel Puente Torres. The authors would like to thank the Programme VLIR-IUC-UOS—Universidad de Oriente CU2019IUC030A105-77143 between Belgium and Cuba for providing funding and granting the support of the current study and future studies.

Conflicts of Interest: The authors declare no conflicts of interest.

References

1. Bansal, R.C.; Goyal, M. *Activated Carbon Adsorption*; CRC Press: Boca Raton, FL, USA, 2005.
2. Hsieh, C.-T.; Teng, H. Influence of mesopore volume and adsorbate size on adsorption capacities of activated carbons in aqueous solutions. *Carbon* **2000**, *38*, 863–869. [CrossRef]
3. Ying, W. *Proceedings of the 44th Purdue Industrial Waste Conference, West Lafayette, IN, USA, 9–11 May 1989*; Lewis Publishers: Chelsea, MI, USA, 1989.
4. DeSilva, F.J. *Activated Carbon Filtration*; Water Quality Products: Arlington Heights, IL, USA, 2000; p. 16.
5. Tran, H.N.; You, S.J.; Hosseini-Bandegharaei, A.; Chao, H.P. Mistakes and inconsistencies regarding adsorption of contaminants from aqueous solutions: A critical review. *Water Res.* **2017**, *120*, 88–116. [CrossRef]
6. Lim, M.; Delehomme, C.; Capece, J. *Removal of Residual Chlorine from Drinking-Water by Solar Radiation (UV) and Activated Carbon Filtration*; Intelligentsia International, Inc.: LaBelle, FL, USA, 2008.
7. Magee, V. The application of granular active carbon for dechlorination of water supplies. In Proceedings of the Society of Water Treatment Examination, Chorleywood, UK, 12–14 June 1956.
8. Abdul-Ahad, M.Y. Design Criteria of an Activated Carbon Bed for Dechlorination of Water. *Iraqi J. Chem. Pet. Eng.* **2008**, *9*, 41–49. [CrossRef]

9. Adachi, A.; Kimata, S.; Noguchi, M.; Okano, T. Decomposition of monochloramine from water with rice bran. *J. Health Sci.* **2002**, *48*, 126–129. [CrossRef]
10. Soice, N.P.; Pellegrino, J.; Kavanagh, J.; Riley, R.L. Oxidative degradation of polyamide reverse osmosis membranes: Studies of molecular model compounds and selected membranes. *J. Appl. Polym. Sci.* **2003**, *90*, 1173–1184. [CrossRef]
11. Adachi, A.; Okano, T. Generation of cyanide ion by the reaction of hexamine and losartan potassium with sodium hypochlorite. *J. Health Sci.* **2008**, *54*, 581–583. [CrossRef]
12. Puente Torres, J.; Crespo Sariol, H.; Yperman, J.; Brito Sauvanell, Á.; Carleer, R.; Navarro Campa, J. A novel X-ray radiography approach for the characterization of granular activated carbons used in the rum production. *J. Anal. Sci. Technol.* **2018**, *9*, 1–15. [CrossRef]
13. Puente Torres, J.; Crespo Sariol, H.; Yperman, J.; Adriaensens, P.; Carleer, R.; Peacok, T.M.; Brito Sauvanell, Á.; Thijssen, E.; Reggers, G.; Haeldermans, T.; et al. X-ray absorption as an alternative method to determine the exhausting degree of activated carbon layers in water treatment system for medical services. *Talanta* **2019**, *205*, 120058. [CrossRef] [PubMed]
14. Puente Torres, J.; Crespo Sariol, H.; Mariño Peacok, T.; Brito Sauvanell, Á.; Cuyvers, G.; Reggers, G.; Samyn, P.; Vandamme, D.; Yperman, J.; Adriaensens, P.; et al. Chemical Regeneration of Activated Carbon Used in A Water Treatment System for Medical Services. *Adv. Environ. Eng. Res.* **2023**, *4*, 1–28. [CrossRef]
15. Puente Torres, J.; Crespo Sariol, H.; Mariño Peacok, T.; Yperman, J.; Adriaensens, P.; Carleer, R.; Brito Sauvanell, Á. X-Ray absorption (XRA): A new Technique for the characterization of Granular Activated Carbons. *Materials* **2021**, *14*, 91. [CrossRef] [PubMed]
16. *D 2867-04*; ASTM, Standard Test Methods for Moisture in Activated Carbon. ASTM International: West Conshohocken, PA, USA, 2011.
17. Johns, H.E.; Cunningham, J.R. *The Physics of Radiology*; Charles C. Thomas: Springfield, IL, USA, 1983.
18. Hossain, S.; Cook, D.; Chow, C.W.K.; Hewa, G.A. Development of an optical method to monitor nitrification in drinking water. *Sensors* **2021**, *21*, 7525. [CrossRef] [PubMed]
19. Danial, L.H. *Current Technology of Chlorine Analysis for Water and Wastewater*; Technical Information Series-Booklet No. 17; Hach Company: Loveland, CO, USA, 1995.
20. Mariño Peacok, T.; Crespo Sariol, H.; Sánchez Roca, Á.; Puente Torres, J.; Gryglewicz, G.; Yperman, J. Efficiency evaluation of thermally and chemically regenerated activated carbons used in a water cleaning system by acoustic emission analysis. *J. Porous Mater.* **2021**, *28*, 451–469. [CrossRef]
21. Turlapati, V.Y.; Kumar, S.; Singh, R.K.; Sharma, M. Analysis of micropore size distribution using Dubinin's theory of volume filling-Effect of particle size on pore characterization of organic-rich Indian shales. *J. Nat. Gas Sci. Eng.* **2022**, *106*, 104746. [CrossRef]
22. Harp, D.L.; Hach, C.C. *Current Technology of Chlorine Analysis for Water and Wastewater*; Technical Information Series-Booklet No. 22; Hach Company: Loveland, CO, USA, 2003.

C Journal of Carbon Research

MDPI

Article

Reuse of Activated Carbons from Filters for Water Treatment Derived from the Steam Cycle of a Nuclear Power Plant

Beatriz Ledesma Cano [1], Eva M. Rodríguez [2], Juan Félix González González [1] and Sergio Nogales-Delgado [1,*]

[1] Department of Applied Physics, University of Extremadura, Avda. De Elvas, 06006 Badajoz, Spain; beatrizlc@unex.es (B.L.C.); jfelixgg@unex.es (J.F.G.G.)
[2] Department of Chemical Engineering and Physical-Chemistry, University of Extremadura, Avda. De Elvas, 06006 Badajoz, Spain
* Correspondence: senogalesd@unex.es

Abstract: Nuclear energy has a great impact on the global energy mix. In Spain, it supplies over 20% of current energy requirements, demonstrating the relevance of nuclear power plants. These plants generate different types of waste (apart from radioactive) that should be managed. For instance, the activated carbon included in filters (which neutralize isotopes in a possible radioactive leakage) should be periodically replaced. Nevertheless, these activated carbons might present long service lives, as they have not undergone any adsorption processes. Consequently, a considerable amount of activated carbon can be reused in alternative processes, even in the same nuclear power plant. The aim of this work was to assess the use of activated carbons (previously included in filters to prevent possible radioactive releases in primary circuits) for water treatment derived from the steam cycle of a nuclear power plant. A regeneration process (boron removal) was carried out (with differences between untreated carbon and after treatments, from S_{BET} = 684 m^2 g^{-1} up to 934 m^2 g^{-1}), measuring the adsorption efficiency for ethanolamine and triton X-100. There were no significative results that support the adsorption effectiveness of the activated carbon tested for ethanolamine adsorption, whereas a high adsorption capacity was found for triton X-100 (q_{L1} = 281 mg·g^{-1}), proving that factors such as porosity play an important role in the specific usage of activated carbons.

Keywords: ethanolamine; activated carbon; boron; triton X-100; water treatment; nuclear power plant

1. Introduction

1.1. Nuclear Power Plants: Current Status

Nuclear energy plays an important role in the energy mix of different countries and is extremely controversial in political and social spheres. The development of demanding environmental policies has taken place, requiring lower waste generation from industrial processes (including nuclear programs) and stricter control of these residues. Initially, much effort was made that focused on minimizing the effect of pollutants. However, these measures were not entirely effective, as pollution can be transferred to other environments or can accumulate, increasing environmental risks.

As a result, the search for alternatives to reducing waste generation might have a more positive impact. Many different industries (including nuclear power plants, where different kinds of waste are generated) tend to select the right process to avoid the production of residues with difficult environmental management (or, if residues are produced, to enhance them through valorization), with their discharge to the environment as the last option.

In this sense, terms such as circular economy have gained attention, in which different products and by-products can be reused or recycled, which is highly related to green processes [1,2]. This way, green policies promote the implementation of technologies to protect the environment.

In this regard, nuclear energy plays an important role in the energy mix of different countries, with much controversy about the sustainability and energy independence of this energy source [3]. Indeed, scientific interest in this topic has recently increased, as observed in the number of publications about this subject [4].

Without a doubt, many different industries are interested in the transition towards this new scenario, including nuclear power plants (NPPs). Specifically, NPPs produce nuclear waste that is considered problematic, including products that are in contact with radioactive materials, which can be managed by following green or circular economic practices [5]. For instance, in the case of Europe, after all Europe's reactors are decommissioned, the total amount of waste is estimated to be 6.6 million m^3, with France as the greatest producer of low- and intermediate-level waste (30%) [6]. Spain produces 4% of these wastes at the European level [7,8]. Specifically, in the case study of this work, Almaraz NPP (with a yearly average generation of 115 m^3), total low- and intermediate-level waste amounts of 1447 and 5278 m^3 were generated from its implementation to 2019; compare this to the Spanish generation (with 31,319 and 41,116 m^3, respectively) [9].

The European context could be explained by the consolidated presence of NPPs in this continent, which can be extrapolated to other regions like Asia or America. Indeed, there are currently more than 400 nuclear power plants operating around the world, distributed according to Figure 1. The number of NPPs can provide an idea about nuclear activity and the subsequent residues that can be generated and must be properly managed. As observed in this figure, there is a relatively homogeneous distribution of nuclear power plants in the northern hemisphere, where most plants are located, with the United States (exceeding 90 plants) and France and China (with over 50 plants each) as the most representative countries. This figure shows how even though there is concern about nuclear energy, the number of operating plants, along with projected ones, is considerable and essential for the energy mix of many countries and regions.

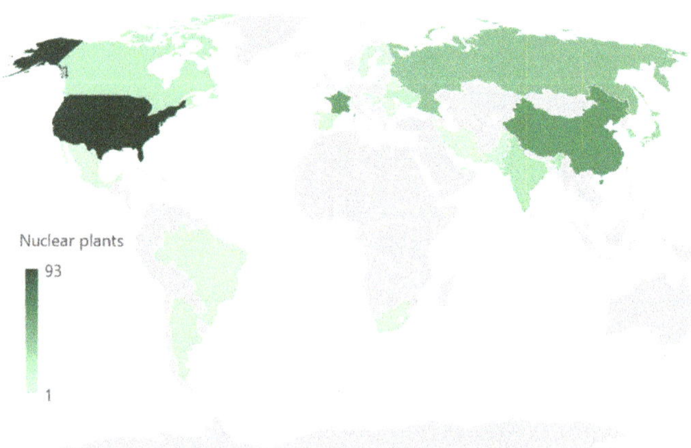

Figure 1. Distribution of nuclear power plants around the world [10].

Regarding Spain, there are seven NPPs in total, including eight operating reactors [11]. In this sense, although Spain presents great potential with renewable sources like solar or wind energy, the energy produced in nuclear power plants represents around 20% of the total energy generation in the national electricity system. The NPP selected for this study is located in Almaraz (Extremadura, Spain), with two reactors.

1.2. Use of Activated Carbons in Npps

As expected, a wide range of materials are used in NPPs, like activated carbons. These are carbonaceous and porous materials obtained through carbonization, which take part in reactions with gas during or after carbonization (optionally with the previous addition of chemical products) to improve their adsorption. Activated carbons present a disorderly and porous microcrystalline structure, providing an amorphous structure. The space existing between these microcrystals is the cause of the porosity in activated carbons. The main requirements of an activated carbon are the following:

- They should have high adsorption capacity and selectivity towards the studied ionic or molecular species.
- They should present suitable kinetic characteristics for the specific adsorption process, considering the particle size of the activated carbon, adsorbate diffusion, and structure of the bed.
- From an industrial point of view, concerning filter systems, it is important that activated carbons present long retention times (that is, the adsorbate should be kept in the activated carbon for long periods without undergoing desorption) to reduce the frequency of replacement of these filters.
- They should have a good cost/efficiency ratio, meaning they are easily prepared and supplied.

During the operating stage of reactors in NPPs or other facilities, diverse radioactive waste is generated. Due to their potential environmental and health hazards, it is important that these substances be removed through different safety mechanisms [5].

Activated carbons are suitable for effective use in filters to retain radioactive material produced in different stages in NPPs [12–14], with retention efficiencies of radioactive methyl iodide of 100% in many cases. Table 1 shows the minimum requirements for the NPP located in Almaraz (Spain), the subject of study in this work. As can be inferred from this table, demanding requirements for these filters assure safety in all the stages where activated carbons are used, with the subsequent regular replacement of these materials once their service life has expired. In this sense, processes such as the aging effect could decrease the effectiveness of activated carbons, especially in this case study, as adsorption of radioiodine could provoke this negative effect. Nevertheless, unless a leakage has taken place, these activated carbons are not normally used as adsorbents, and they present an interesting alternative for other applications in the same nuclear power plant.

Table 1. Different systems in Almaraz NPP where activated carbons are used, and lower limits for their retention efficiency.

System	Minimum Retention Efficiency, %
Fuel building	85.0
Hydrogen purging	95.0
Safeguard building	97.5
Control room	99.5
Auxiliary building	95.0
Vacuum system of condenser	99.3
Purging building	95.0
Pre-accession area	95.0
Purge in the area	95.0

Figure 2 shows the specific use of the activated carbon included in this study. Specifically, the filters are in the vacuum system of the condenser in the Almaraz Nuclear Power Plant, and the objective is the improvement of the thermodynamic yield in the nuclear cycle. For this purpose, a vacuum at the low-pressure turbine outlet is created by removing air and non-condensed gas from steam. Radiation monitors continuously control the flue gas with the aim of detecting radioactive isotopes (in case of breakage of the steam generator pipes). In this case, flue gas goes through the filter, retaining methyl iodide with an efficiency of 99.27%.

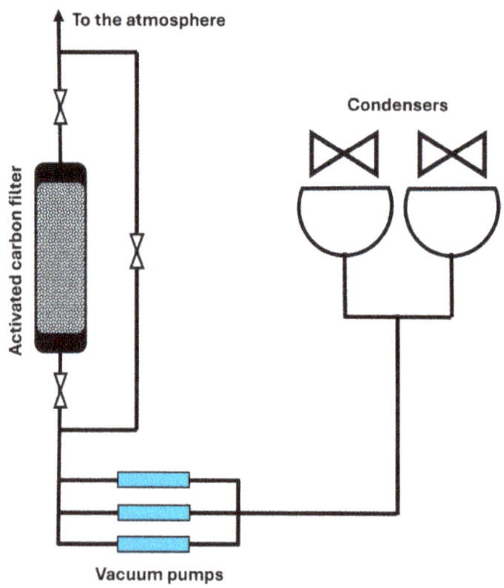

Figure 2. Example of the role of activated carbon in a nuclear power plant.

Activated carbons used in these filters are usually microporous, with a homogeneous pore size distribution and high specific surface. These characteristics are highly dependent on different factors, such as the raw material or precursor or the activation process used to prepare the final product, which are decisive in physical adsorption.

On the other hand, regarding the surface specificity of a solid towards a certain adsorbate, the reactivity of the surface can be modified by adding a suitable impregnation agent.

Among the most popular impregnation agents that are used in activated carbons, there is a wide range of compounds, such as iodides (KI, BaI, or PbI), $AgNO_3$, and amino compounds such as triethylenediamine (TEDA) and hexamethylenetetramine (HMTA). Among them, KI and TEDA are the most studied at the industrial level to remove elemental iodine and $^{131}ICH_3$ [15–17]. The ASTM D3803-91/2022 standard was established for carbon testing for $^{31}ICH_3$ adsorption, specifically for activated carbons that are impregnated with KI or TEDA [18].

The main drawback related to activated carbons in this context is their exposure to moisture during storage or their service life, along with a decrease in retention efficiency of $^{131}ICH_3$, implying an aging effect and, consequently, reducing their service life [19,20]. This effect is related to surface oxidation of activated carbons, increasing their hydrophilicity and the subsequent steam adsorption. Some authors point out that the generation of oxygenates reduces the retention efficiency of $^{131}ICH_3$ due to the increase in blocked pores due to steam adsorption [21]. In this sense, the reuse of these spent activated carbons could be an interesting contribution to the circular economy in this context.

On the other hand, thanks to its high capability of neutron absorption, boron (specifically B-10) is used as a component in neutron shields to control the safe operation and detect neutrons (included in specific devices) in nuclear power plants. Consequently, B-10 can act as a neutron controller in reactors, where neutrons can be dispersed in different directions. In this sense, its presence in filters in NPPs is common, explaining its presence in the activated carbons used in this research.

1.3. Ethanolamines in NPPs

Apart from different materials, considerable amounts of waste (apart from nuclear waste) can be generated in NPPs due to routine activities. For instance, ethanolamines could present an environmental problem in this context. Thus, thermoelectric plants convert the heat generated in combustion or nuclear fission into steam, which is used in turbines to operate a generator to convert mechanical energy into electricity. After the turbine, a condenser is installed to condense steam and obtain water, which is placed in the corresponding feeding container to restart the process in the steam boiler. In parallel, the coolant water flows into the condenser, removing heat through a heat exchanger. Nuclear power plants with pressurized water reactors have an additional water circuit, the so-called primary circuit.

All thermoelectric plants use water for this purpose. In nuclear plants, water also moderates neutrons derived from fission, controlling this nuclear reaction. A thorough chemical analysis of water ensures the safety and efficiency of these facilities. Around 50% of unexpected outages are due to the presence of pollutants or the chemical composition of water in water–vapor systems. In many cases, corrosion is a real challenge, and a weak base such as ethanolamine is added to extend the service life of these facilities [22]. Hence, ethanolamine (EA) is an organic compound that is a primary amine and, at the same time, a primary alcohol. This compound is normally used as a corrosion inhibitor and presents a challenge in environmental management due to its low biodegradability, requiring different oxidation processes [23]. Along with other industrial uses, wastewater with ethanolamine could increase chemical oxygen demand and total nitrogen, which is a considerable environmental problem (recommended human exposure to this compound is below 0.2 ppm, not exceeding 0.6 ppm in drinking water). Consequently, suitable management of this waste is required, mainly including oxidation processes [24] or adsorption methods with several solids, including activated carbon [25].

1.4. Triton X-100 and Its Environmental Concerns

Equally, the application of different surfactants as detergents for household use is recurrent, with widespread use throughout the world. For instance, triton X-100 is a non-ionic surfactant with multiple applications, which is generally present in laundry detergents, detergent sanitizers, and pesticide products, with a specific use for enhancing bioremediation of PAH. However, the environmental management of these products is a challenge, as there are concerns about the fate of surfactants once released into the environment, especially in terms of water and soil pollution. For this reason, triton X-100 removal is essential, with adsorption and biodegradation as the main treatments, especially when anaerobic and aerobic municipal wastewater sludges are used [26].

1.5. Aim of This Work

Considering the above, the goal of this study was the regeneration of activated carbons from filters devoted to avoiding leaks or spills in primary circuits in a nuclear power plant located in Spain. For this purpose, cooling water from the same NPP was used, and different methods were used to remove B and organic compounds from the activated carbons to reduce costs. Afterwards, the resulting activated carbon was used for the adsorption of two pollutants included in wastewater, ethanolamine and triton X-100. Accordingly, the innovation of this work lies in the proposal of a possible reuse of a waste generated in

NPPs, offering different alternatives for the regeneration of activated carbons for their use in environmental management, mainly in water treatment, as to the best of our knowledge, there are no practical studies with this specific application.

2. Materials and Methods

The filters selected for this study were provided by Almaraz NPP, located in Extremadura, Spain (39°48′29″ N, 5°41′49″ W). This plant has two reactors, with an open-circuit refrigeration system to a reservoir 1 km away (Arrocampo-Almaraz, 39°49′0″ N, 5°42′0″ W). Its activity started in 1983, and the gross energy production of this plant is 15,500 GWh. This NPP has a considerable amount of activated carbon that is used in filters to prevent spills in the primary circuit, which should be replaced due to standards.

The activated carbon collected from this NPP was labeled as ACA. The activated carbon was directly extracted without being exposed to radioactive leakages in ventilation filters after its service life. The treatments (washing, oxidation, and pyrolysis) that were carried out to pre-treat these activated carbons are described in the following subsections; they were characterized and used for adsorption of ethanolamines and triton X-100. Specifically, non-treated samples were labeled as $ACA_{control}$.

2.1. Boron Removal Through Washing Treatment

As the filters used in this work contained activated carbon with boron for neutron capture in NPPs, boron removal as a pre-treatment was required for activated carbon reutilization. For this purpose, two methods were used. First, an acrylic column with circulation of HCl solution was used. Second, an immersion method was carried out.

Regarding the first case (Figure 3), activated carbon was sieved and a specific amount (34.5 g) was introduced and packed in the acrylic column. Afterwards, the activated carbon was washed with distilled water to remove the air included in interstices. To fill the volume of the column, glass spheres were added above and below the activated carbon bed, and these phases were separated with metal mesh.

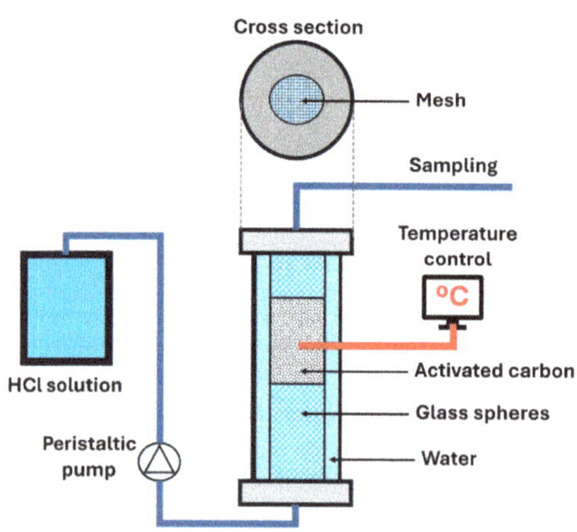

Figure 3. Scheme for boron removal in a column.

To start the process, the first step was to pump the specific flow through the column, which was kept constant during the whole experiment at 5 mL·min^{-1}. Afterwards, different samples were taken to carry out boron determination (explained in further subsections).

Concerning the immersion method, it consisted of adding a certain amount of solid in a fixed volume of HCl solution (with a specific concentration), then stirring the mixture until a stationary regime was reached. Due to desorption of B from the adsorbent, changes in the solution took place, allowing the quantification of desorption in the activated carbon. For these tests, different quantities of adsorbent (0.05 and 0.1 mg) were added in test tubes; we added 10 mL of HCl solution with a certain HCl concentration, which is specified in further subsections. The samples obtained were labeled as $ACA_{HCl\ washing}$.

The choice of HCl to remove boron from activated carbon is based on its well-known properties to facilitate desorption of species adsorbing on activated carbon surfaces. As a strong acid, HCl can effectively break interactions between boron compounds, such as boric acid, and the activated carbon surface, which often involve weak electrostatic forces, hydrogen bonds, or the formation of complexes. In an acidic environment, these interactions are disrupted, allowing boron desorption. HCl has been extensively used in previous experiments to regenerate activated carbon and desorb various adsorbed species, including metals and other inorganic compounds. In addition, HCl is an inexpensive and readily available acid, making it a practical option for large-scale applications.

2.2. Oxidation of Activated Carbon

For this test, activated carbon samples (6 g), after the washing treatment with HCl solutions, were oxidized by varying the oxidation time between 1.5 and 9.0 h with a nitric acid solution (7 mol·L^{-1}), using 60 mL of this solution per gram of activated carbon. The mixture was placed in a Soxhlet apparatus at the boiling point of the solution during the corresponding oxidation times. Additionally, the activated carbon was oxidized by using different nitric acid concentrations (from 4 to 7 mol·L^{-1}) at a constant oxidation time of 1.5 h following the abovementioned procedure. Finally, the activated carbon was washed with distilled water to remove nitrate content, which was assured by measuring the pH in water until pH = 5.5 was reached. The resulting samples were labeled as $ACA_{oxidation}$.

2.3. Pyrolysis of Activated Carbon After Washing Treatment with HCl Solution

The experimental facility for ACA pyrolysis is shown in Figure 4. As observed, a cylindrical reactor (length = 700 mm; external diameter = 42 mm; inner diameter = 36 mm) was used, which charged the sample at the top of the reactor and placed it in a basket. Different gas-phase compounds, such as nitrogen (99.999%, Linde, Dublin, Ireland) or steam were introduced in the reactor, whose temperature was kept by using a PID controller, allowing different heating rates. The flue gas was cooled down in a condensation system (two flasks immersed in water with ice) to collect the condensable fraction. The resulting samples were labeled as $ACA_{pyrolysis}$.

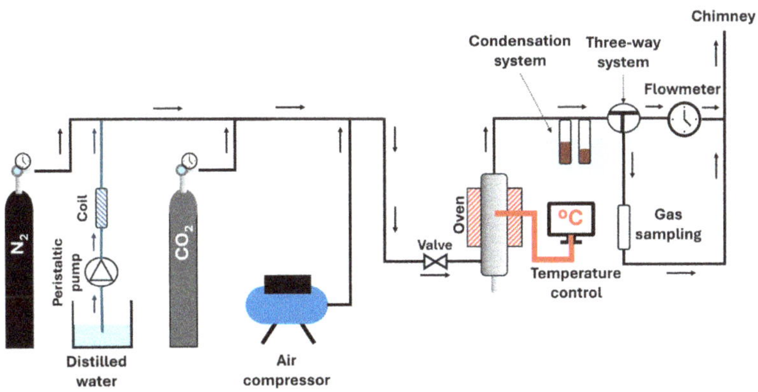

Figure 4. Experimental scheme for pyrolysis of activated carbon.

2.4. Characterization of Activated Carbons

In this subsection, the main textural and chemical characterizations of the activated carbons obtained in previous processes are described.

2.4.1. Textural Characterization

Nitrogen gas adsorption–desorption isotherms at 77 K were determined by using a semi-automatic piece of equipment (Autosorb-1, Anton Paar, Graz, Austria). The sample (1 g) was previously degasified at 120 °C for 12 h. Experimental data of adsorbed nitrogen for each relative pressure P/P_0 were analyzed, using the Brunauer–Emmet–Teller method to calculate BET surface area (S_{BET}). Also, the Dubinin–Radushkevich method was used to determine micropore volume (V_{mi}), whereas the Gregg and Sing procedure was selected to determine mesopore volume (V_{me}). Finally, the Sing method was used to determine the external surface (S_{ext}), using as a non-porous material a product prepared from olive stone. The inner surface (S_{in}) was determined by difference (S_{BET}–S_{ext}).

SEM analysis was carried out to observe surface characteristics in treated activated carbon. A high-resolution electron microscope (Quanta 3D FEG, FEI company, Hillsboro, OR, USA) was used. The working mode was high vacuum, with accelerating voltage from 200 V to 30 kV. The detectors were Everhardt–Thornley SED and gaseous analytical BSED. For comparison, magnifications of 200×, 1500×, and 2500× were used.

2.4.2. Chemical Characterization

To characterize the surface chemistry of ACA, the point-of-zero-charge (PZC) method was used. Thus, the method proposed by Carrot et al. [27] was followed: 0.5 g of activated carbon was weighed, then 10 mL of $NaNO_3$ solution (0.1 M) was added. The solutions were introduced in a thermostatic bath with a stirring system for 48 h at 25 °C. Afterwards, the samples were filtered with Whatman #1 filter paper, and the pH of the filtered solution was measured with a pH meter (pH meter basic 20+, Crison, Barcelona, Spain).

For further chemical analysis, Fourier transform infrared (FTIR) spectroscopy was employed to examine the surface functional groups of the activated carbon. The FTIR spectra were recorded on a PerkinElmer Spectrum 100 FTIR spectrometer (PerkinElmer Inc., Waltham, MA, USA) in the range of 4000–400 cm^{-1} using KBr pellets. This technique provided valuable information on the surface functional groups and bonding types, which are crucial for understanding the material's reactivity and interaction with adsorbates.

2.5. Boron Determination

Once the activated carbon was treated according to the abovementioned methods, the amount of B included in ACA was determined by spectrophotometry (λ = 410 nm) with a previous reaction with azomethine, included in previous standards [28]. The samples were analyzed as follows: 5 mL of each sample was taken in a test tube, adding 4 mL of buffer solution and 2 mL of azomethine. The solution was homogenized, and after 2 h, the absorbance at 410 nm was measured in a spectrophotometer (UV-2005, J. P. Selecta, Barcelona, Spain). A calibration line was obtained in order to compare experimental data and determine boron content. Thus, 5 mL of solutions with different B concentrations in deionized water (0, 0.4, 0.8, 1.2, 1.6, 2.0, 2.4, 2.8, and 3.8 mg·L^{-1}) underwent the same procedure previously explained, showing the calibration curve and the main results in Section 3.2.

2.6. Ethanolamine Adsorption Tests in Activated Carbon

Concerning activated carbons that underwent different purification treatments, a final adsorption test with ethanolamine was carried out with the aim of assessing their suitability for this purpose. At this point, a treatment to modify the surface chemistry of ACA was carried out to improve the selectivity towards the pollutant of interest (in this case, ethanolamine).

The selected experimental method was immersion, consisting of keeping in contact both adsorbate and ACA in the same solution with constant stirring rates and temperature to reach the adsorption–desorption equilibrium. Thus, changes in adsorbate concentration were found, which were quantified by the difference between the initial and stationary states. Using this method, if different adsorbate solutions (with different concentrations) are tested with the same amount of activated carbon (or vice versa), the corresponding adsorption isotherm can be determined by representing the adsorbed amount (q, $mg \cdot g^{-1}$) versus the concentration of adsorbate in a stationary state (C_e, $mg \cdot L^{-1}$).

In these tests, 0.01 (or 0.05 g) of activated carbon was added to test tubes, followed by the addition of an ethanolamine solution (5, 10, or 20 ppm). These experiments were carried out in a thermostatic bath (Unitronic, Selecta, Barcelona, Spain) with constant stirring at 25.0 ± 0.1 °C. Once the contact time finished, the activated carbon was separated from the solution through centrifugation. Ethanolamine concentration was determined through ionic chromatography (Metrohm 850 Professional IC, Metrohm, Herisau, Switzerland) with a Metrosep C6 column. The tests were performed in duplicate with the same starting solution in all cases.

The initial ethanolamine concentrations selected for this experiment were based on the real ethanolamine content in purge water found in Almaraz Nuclear Power Plant, which was below 5 ppm. However, the concentration range was higher and more restrictive to assess the effectiveness of activated carbons in this case.

2.7. Triton X-100 Adsorption in Activated Carbon

As a possible alternative for ethanolamine adsorption, the adsorption of triton X-100 was also evaluated, which could contribute to the environmental management of this surfactant in soil and water [26].

For this purpose, equilibrium adsorption isotherms for TX-100 were carried out following the same method as for ethanolamine, but in this case, a fixed amount (0.1 g) of each adsorbent was added to 15 mL of an organic solution of TX-100 with initial concentrations (C_0) ranging from 10 to 100 $mg \cdot L^{-1}$.

After filtering the adsorbents, the concentration in the supernatant solutions (C_e) was analyzed using UV–Vis spectrophotometry (UNICAM Helios-λ spectrophotometer, Scribd, San Francisco, CA, USA) at a wavelength of 277 nm. This wavelength was selected after prior spectral scanning tests, which showed signal stability regardless of possible pH variations. The absorbances of a series of standard solutions with different concentrations were measured in triplicate, confirming the validity of Beer's law with regression coefficients very close to one in the concentration range used. From the C_e values, the equilibrium solid-phase concentration (q_e) was obtained.

Kinetic studies were performed by adding 100 mg of each activated carbon (AC) to 25 mL of TX-100 solution with an initial concentration of 20 $mg \cdot L^{-1}$. The flasks were placed in a thermostatic bath at 20 °C with stirring, and samples were taken at appropriate time intervals. The slurries were filtered through 0.45 μm membranes, and the concentrations of TX-100 in solution (c) and the corresponding solid-phase concentrations (q) were obtained. Tests were carried out at least in duplicate.

3. Results and Discussion

The main characteristics of ACA are described in the following subsections, before and after undergoing the different treatments explained in Section 2. The treatments devoted to surface chemical modification were applied once ACA was washed with HCl solution.

3.1. Characterization of Activated Carbons

Table 2 shows C, H, N, and O content in different ACA samples derived from ultimate analysis. As can be inferred from this table, there was a high amount of fixed carbon, which is common in activated carbon obtained from biomass waste. It should be noted that the ACA preparation procedure is not known. Nevertheless, according to the literature, the activation of carbons from coconut shell (among other wastes) is generally carried out with different activation agents, such as steam and carbon dioxide [29,30]. From an economic point of view, steam would be a more advantageous option due to its low cost.

Table 2. Ultimate analysis of different activated carbons.

Activated Carbon	C, %	H, %	N, %	O, %
$ACA_{control}$	78.17	0.36	0.75	20.72
$ACA_{HCl\ washing}$	77.84	0.16	0.64	21.36
$ACA_{oxidation}$	76.67	0.01	1.02	22.30
$ACA_{pyrolysis}$	85.21	0.06	0.77	13.96

Furthermore, the treatments applied to ACA implied some changes in ultimate analysis. Regarding $ACA_{HCl\ washing}$ and $ACA_{oxidation}$, there was a decrease in fixed carbon, whereas for $ACA_{pyrolysis}$ this amount was increased. This fact could be due to the removal of oxygenated groups due to the thermal treatment at high temperatures. On the other hand, the introduction of oxygenated groups in oxidation processes could imply an increase in oxygen and nitrogen at the expense of fixed carbon.

Thus, the effect of the modification of surface functional groups on the adsorption of activated carbons plays an important role currently, and due to the more restrictive policies related to water pollution, there has been increasing interest in activated carbons with certain functional groups for pollutant removal. In this sense, textural analysis can provide interesting information about this subject, as observed in Figure 5 for nitrogen gas adsorption–desorption isotherms. As observed in this figure, these curves correspond to type I, according to the IUPAC classification [31], related to microporous materials. Adsorption mainly takes place at low P/P_0 ratios, with a relatively constant adsorbed quantity once the micropores are filled with the adsorbate. Consequently, there is a well-defined plateau corresponding to monolayer adsorption where the adsorbed quantity is increased with pressure until an upper limit is reached, corresponding to the full covering of the monolayer. It is interesting that after regeneration processes of ACA, microporosity is recovered, which implies that boron was included in the small pores. Subsequently, cleaning the original activated carbon is important to improve its adsorption capacity.

Once the different models (BET, Dubinin–Radushkevich, and Greeg and Sing) were applied, the following values related to surface characteristics were obtained. Also, the chemical characterization was carried out showing PZC (see Table 3). There were considerable differences between the control sample and the treated ones. The specific surface of treated samples increased, as well as the rest of the textural characteristics, which could make these activated carbon samples suitable for certain applications. It is interesting to note that, as previously explained, the increase in pore volume was mainly due to the presence of micropores. Regarding PZC, it should be noted that after the corresponding treatments, two activated carbons presented acidic properties ($ACA_{HCl\ washing}$ and $ACA_{oxidation}$), whereas $ACA_{pyrolysis}$ presented basic properties as in the case of control samples.

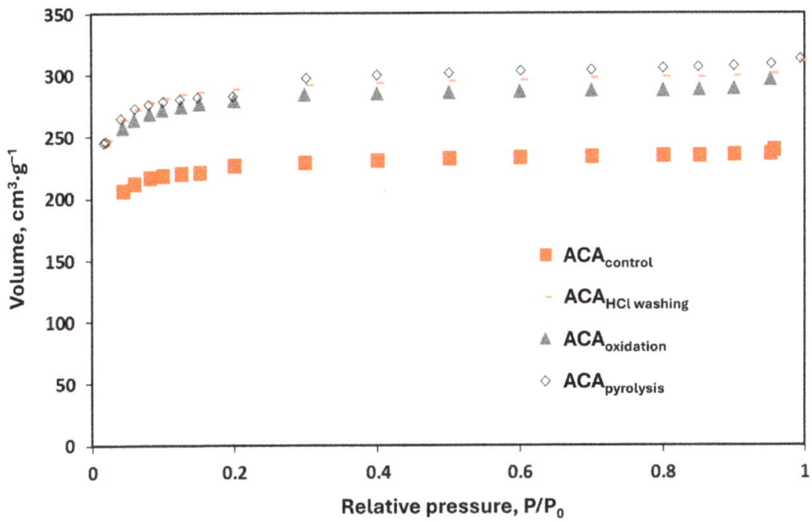

Figure 5. Nitrogen gas adsorption–desorption isotherms for different activated carbons.

Table 3. Textural and chemical analysis of different activated carbons. N_2 adsorption at 77 K and point of zero charge.

Activated Carbon	SBET, $m^2 \cdot g^{-1}$	Vmi, $cm^3 \cdot g^{-1}$	Vme, $cm^3 \cdot g^{-1}$	PZC
$ACA_{control}$	684	0.361	0.005	9.04
$ACA_{HCl\ washing}$	921	0.475	0.011	4.52
$ACA_{oxidation}$	893	0.464	0.002	2.60
$ACA_{pyrolysis}$	934	0.485	0.009	10.24

The SEM images (Figure 6a–c) show the morphology of $ACA_{pyrolysis}$ activated carbon, highlighting a highly porous structure with well-defined cavities. Irregular surfaces with a rough texture and interconnected macropores are observed, suggesting high accessibility to adsorption sites. This porous structure is consistent with the measured BET surface area, indicating a high adsorption capacity and significant porosity development.

The FT-IR absorption spectrum of the $ACA_{pyrolysis}$ sample (see Figure 7) was recorded using a Perkin-Elmer 1720 spectrometer operating within a wavenumber range of 450–4000 cm^{-1}. Prior to analysis, KBr was used as a dispersing and binding agent to prepare pellets. The sample (previously dried at 110 °C for 12 h) was mixed with KBr to obtain a total mass of 250 mg, maintaining a sample-to-KBr ratio of 1:450. The mixture was then compressed using a hydraulic press (Perkin-Elmer) under a load of 10 $t \cdot cm^{-2}$ for 10 min. Each analysis consisted of 40 scans with a resolution of 2 cm^{-1}. The blank spectrum (a KBr-only pellet prepared under the same conditions) was subtracted from the sample spectrum.

Band assignments were based on vibrational frequencies reported in the literature for activated carbons. As shown in Figure 7, a broad band located around 3300–3600 cm^{-1} is associated with ν(O-H) stretching vibrations in alcohols, specifically OH functional groups (intra- and intermolecular hydrogen bonding). This band is also linked to the presence of moisture in the sample.

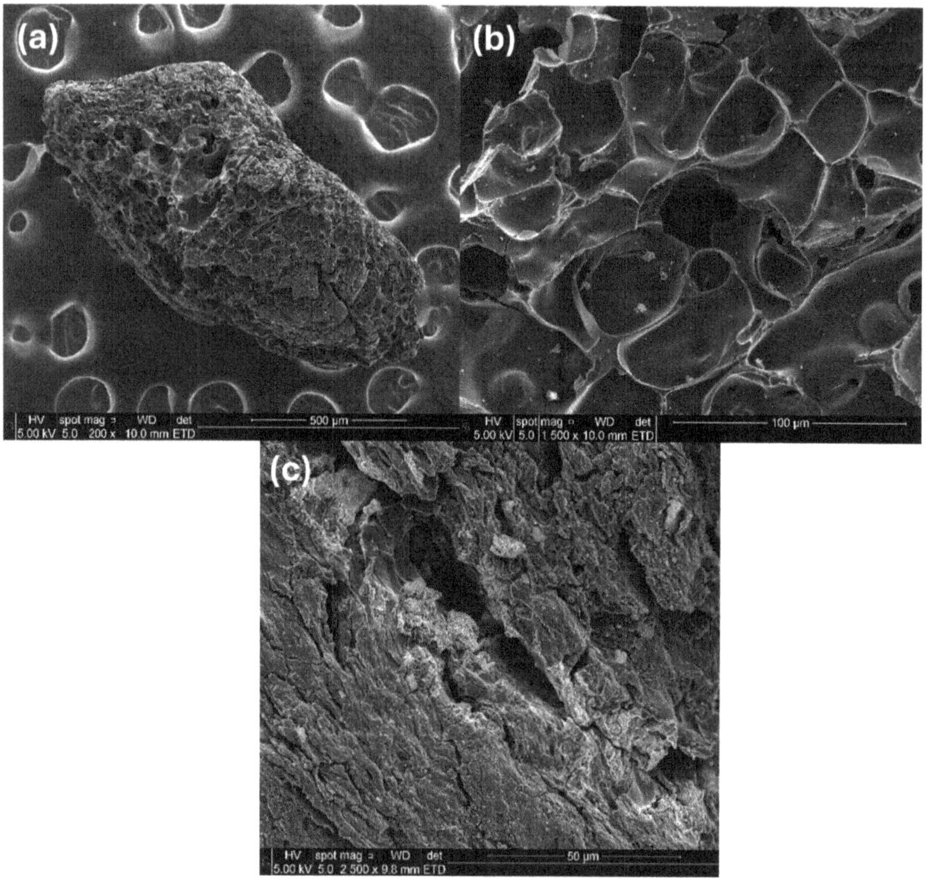

Figure 6. SEM images for ACApyrolysis activated carbon at different magnification levels: (**a**) 200 times; (**b**) 1500 times; (**c**) 2500 times.

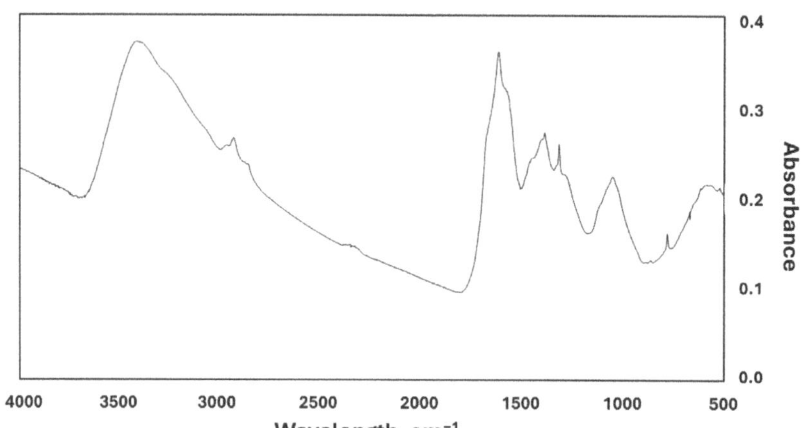

Figure 7. FT-IR absorption bands for ACApyrolysis.

The band around 1700–1600 cm^{-1} is attributed to C=O stretching vibrations, which may correspond to lactone, quinone, or carboxylic acid groups. Bands closer to 1600 cm^{-1} are associated with aromatic rings or C=C stretching vibrations, indicating the presence of alkene groups. C-H bonds in the 1150–1000 cm^{-1} region, along with C-H vibrations observed in the range of 900–650 cm^{-1}, suggest the presence of aromatic compounds.

3.2. Boron Removal

Firstly, to quantify B removal from ACA, its determination in the liquid phase was carried out, with the corresponding calibration curve for this purpose, as explained in Section 2 (the curve is shown in Figure 8). As shown, an increase in absorbance with B concentration was observed, showing a good adjustment to a linear trend with a coefficient of determination of 0.9997. As a result, this procedure and calibration line were used for the quantification of B removal through the two different methods covered in this study.

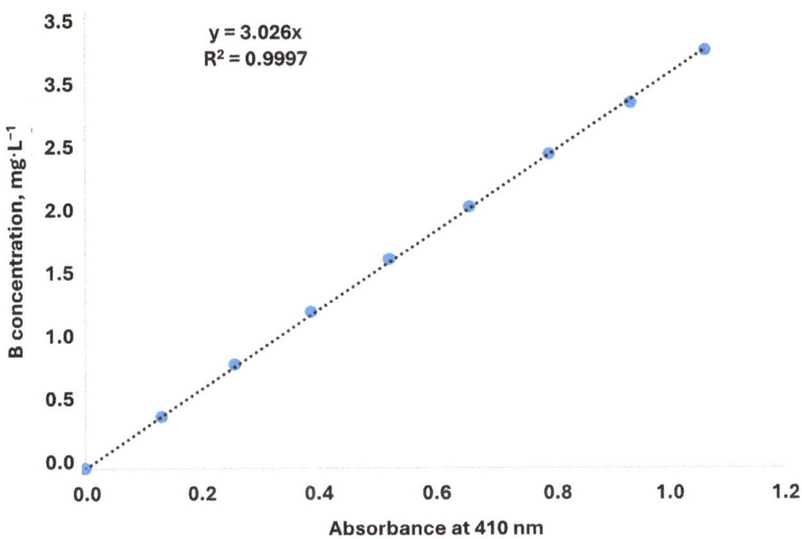

Figure 8. Calibration curve for B determination.

3.2.1. Boron Removal in Column

In this case, Figure 9 shows the main findings about B removal from this method, showing B content in eluted samples from the column. It can be seen that boron is highly desorbed from the activated carbon at an initial stage (during the first 10 h), at the beginning of the contact between the activated carbon and the HCl solution. Comparing the two different concentrations used for this experiment, it should be noted that the lowest concentration used (that is, 0.05 M) would be enough for boron removal from the activated carbon, as the final result in both cases is relatively similar (although slightly higher amounts were removed by using 0.1 M solution, that is, 0.58 mg B per g of ACA). Using this method, a stationary stage took place after 10 h, with constant B concentration in the samples analyzed. From then on, HCl solution did not extract B from the activated carbon bed.

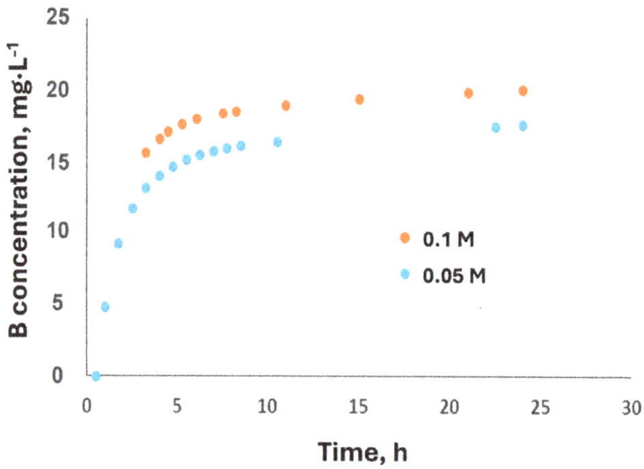

Figure 9. Comparison between different concentrations for B removal.

According to these data, different kinetic models were considered. Regarding the pseudo-first-order model, Lagergren's first-order equation is probably the first known equation applied to the description of adsorption or desorption systems in the liquid phase, being highly used currently [32,33] (see Equation (1)):

$$\frac{dq}{dt} = k_1(q_e - q) \tag{1}$$

where q and q_1 are the amounts of adsorbed solute per mass unit of activated carbon (at a certain time t and at stationary state, respectively), and k_1 is the mass transfer constant. By integrating this equation, for the initial conditions t = 0 and q = 0, Equation (2) is obtained:

$$q = q_e\left(1 - e^{-k_1 t}\right) \tag{2}$$

In order to obtain adjustment parameters, several authors [34–36] linearize Equation (2), obtaining Equation (3):

$$ln(q_e - q) = -k_1 t + ln(q_e) \tag{3}$$

Thus, for the right use of this expression, it is necessary to know the adsorbed amount in the stationary state (q_e). Concerning the pseudo-second-order model, it was proposed by Ho and McKay [37], being expressed as follows (Equation (4)):

$$\frac{dq}{dt} = k_2(q_e - q)^2 \tag{4}$$

Integrating with the initial condition (t = 0 and q = 0), Equation (5) is obtained:

$$q = \frac{k_2 t q_e^2}{q_e k_2 t + 1} \tag{5}$$

Different linearization procedures have been described for a pseudo-second-order equation, with Equation (6) being one of the most recurring ones:

$$\frac{t}{q} = \frac{1}{k_2 q_e^2} + \frac{1}{q_e}t \tag{6}$$

In this case, linearized expressions of pseudo-first- and pseudo-second-order reactions (Equations (3) and (6), respectively) were used. In most cases, experimental data are better adjusted to a pseudo-second-order reaction. A detailed analysis of the procedure used for linearization can explain this fact. It should be considered that the linear adjustment in the pseudo-first order requires knowledge of the q value in the stationary stage (q_e).

In general, the maximum experimental value obtained can be assumed as q_e. The problem related to the mathematical adjustment to the pseudo-first-order model is the logarithmic function, $\ln(q_e\text{-}q)$, which tends to infinity with time as it approaches q_e. Therefore, this model does not adjust accurately over long periods of time. In conclusion, linearization of this model is frequently used but not always used properly.

Considering the above, 0-dimensional models were applied to boron desorption in the original activated carbon by using linear adjustment (Figure 10a,b). The kinetic parameters obtained for the pseudo-first and pseudo-second orders are included in Table 4. In light of these figures, the theoretical curves obtained are significantly close to experimental data, which are supported by the high determination coefficient in both cases (especially concerning the pseudo-second order, with R^2 above 0.9945 and 0.9997 at 0.05 and 0.10 M, respectively).

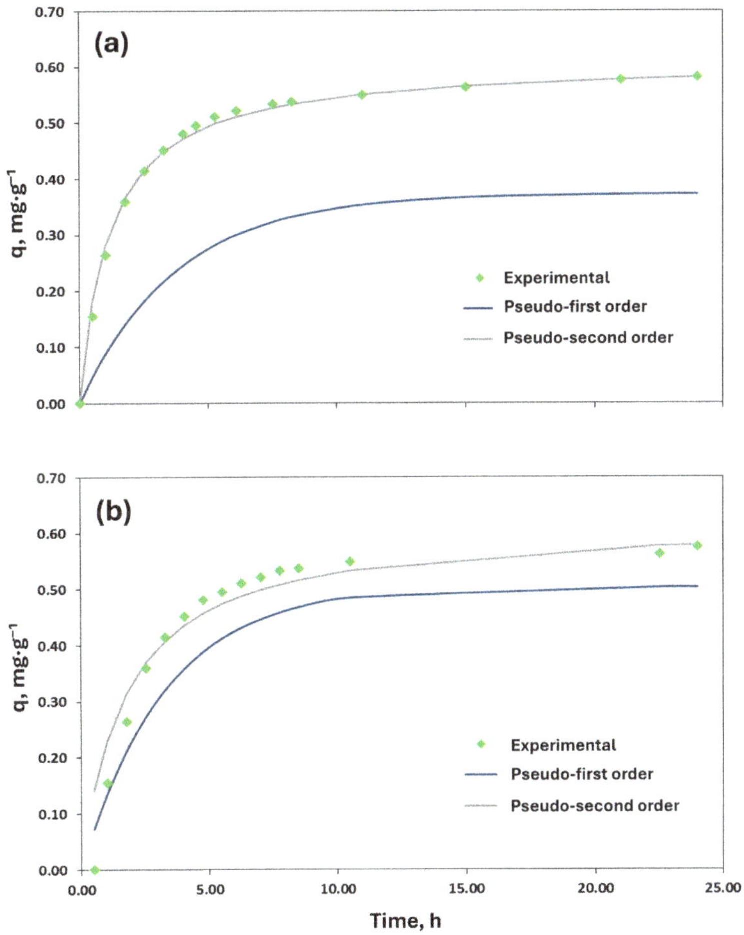

Figure 10. Amount of desorbed B over time with different HCl concentrations: (**a**) 0.1 M; (**b**) 0.05 M. Adjustment to pseudo-first- and pseudo-second-order models.

Table 4. Parameters obtained for pseudo-first- and pseudo-second-order models in boron desorption in original activated carbon.

Model	Concentration, M	k_1, s^{-1}	q, $mg \cdot g^{-1}$	R^2
Pseudo-first order	0.10	$7.47 \cdot 10^{-5}$	0.37	0.9217
	0.05	$8.54 \cdot 10^{-5}$	0.50	0.9722
Pseudo-second order	0.10	$3.89 \cdot 10^{-4}$	0.61	0.9997
	0.05	$2.64 \cdot 10^{-4}$	0.62	0.9945

Concerning kinetic parameters (see Table 4), it should be noted that both models present different kinetic constants and qualities of adjustment.

3.2.2. Boron Removal Through Immersion Method

Table 5 shows the main results related to this removal method, where different initial concentrations, amounts of ACA, and times were used. A constant amount of B per gram of activated carbon was extracted regardless of the immersion time.

Table 5. B removal over time by using immersion method.

Initial Concentration, M	Amount of ACA, g	Time, h	B, mg	$mg_B \cdot g_{ACA}{}^{-1}$
0.05	0.05	5	0.0402	0.80
		10	0.0375	0.75
		15	0.0413	0.83
		20	0.0401	0.80
		25	0.0397	0.79
	0.01	5	0.0099	0.99
		10	0.0094	0.94
		15	0.0110	1.10
		20	0.0100	1.00
		25	0.0095	0.95
0.1	0.05	5	0.0352	0.70
		10	0.0358	0.72
		15	0.0348	0.70
		20	0.0393	0.79
		25	0.0358	0.72
	0.01	5	0.0100	1.00
		10	0.0102	1.02
		15	0.0093	0.93
		20	0.0092	0.92
		25	0.0120	1.20

Therefore, long immersion periods are not necessary, as 5 h seems to be enough to accomplish the maximum boron extraction. Also, low HCl concentrations were enough to remove boron from the activated carbon, as the results obtained in both cases are relatively similar.

When both methods are compared, the immersion of activated carbon removed higher amounts of boron from the samples. However, a complete desorption of B from the surface of activated carbon could not take place, as an amount of B kept on the surface could take part in the equilibrium with desorbed B in HCl solution.

3.3. *Ethanolamine Adsorption Tests*

Table 6 shows the main results related to the adsorption of ethanolamine on the resulting activated carbon after HCl washing treatment (through the immersion method).

Different amounts of activated carbon were put in contact with 10 mL of ethanolamine at different concentrations (from 5 to 20 ppm) for 24 h. It should be noted that the final EA concentration is presented as a concentration range. No significant results were observed that support the adsorption effectiveness of the activated carbons tested in this experiment. The lack of significant adsorption could be attributed to factors such as the interactions between the adsorbate and the carbon surface, as well as the distribution and accessibility of the pores. These factors may have limited the efficiency of the adsorption process. Also, the surface characteristics of the activated carbon did not play an important role in the adsorption process. In this regard, future studies could be focused on modifying activated carbons with functional groups such as amino or carboxyl groups, which have been shown to enhance adsorption of nitrogen-containing compounds like amines [38].

Table 6. Ethanolamine (EA) content in different kinds and quantities of activated carbons.

Activated Carbon	ACA Amount	Initial EA Concentration	Final EA Concentration
	g		
$ACA_{HCl\ washing}$	0.01	4.86 10.48 18.72	3.15–4.27 10.42–11.06 16.54–17.11
	0.05	4.86 10.48 18.72	5.74–5.83 12.26–12.44 19.68–20.20
$ACA_{Oxidation}$	0.01	4.86 10.48 18.72	4.75–4.92 10.37–11.15 16.19–17.06
	0.05	4.86 10.48 18.72	4.44–5.04 10.69–11.25 19.27–20.15
$ACA_{pyrolysis}$	0.01	4.86 10.48 18.72	4.18–5.25 10.14–11.36 16.45–17.85
	0.05	4.86 10.48 18.72	4.44–5.23 9.69–11.86 18.55–19.84

Furthermore, the use of alternative adsorbents, such as ion-exchange resins, could present an interesting option. These products have shown superior adsorption capacities for amines compared to traditional activated carbons, constituting an interesting research line in further works. Equally, the degradation of ethanolamine with UV–chlorine process [39] or oxidation by sodium N-bromobenzenesulfonamide [40] could be interesting alternatives.

In addition to the above, adsorption tests were carried out using TX-100, a non-ionic surfactant that is considered a representative model pollutant in many industrial processes. It is important to underline that this study contributes to the understanding of the behavior of NAC-activated carbons in the adsorption of different types of pollutants and suggests that these activated carbons could have useful applications in industrial water treatment.

3.4. *Triton X-100 Adsorption Tests*

3.4.1. Equilibrium Adsorption Isotherms

As a first step, the adsorption process of triton X-100 under equilibrium conditions was studied. Figure 11 shows the equilibrium adsorption isotherms (q_e vs. c_e) obtained for the pyrolyzed activated carbon ($ACA_{pyrolysis}$). These isotherms describe the relationship between the adsorbate concentration and the amount adsorbed onto an adsorbent at a given temperature. The phenomena responsible include adsorption and desorption, surface coverage, equilibrium, adsorption capacity, and the observed isotherm type.

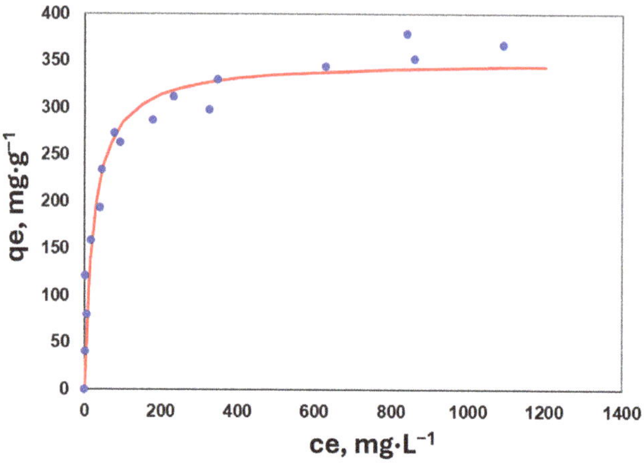

Figure 11. Adsorption isotherm of TX-100 on ACApyrolysis.

An initial, nearly vertical section is observed up to ce 20 mg·L^{-1} and qe 160 mg·g^{-1}. Then, as ce increases, the isotherm slope decreases until a plateau is reached at ce = 346 mg·L^{-1} and qe = 330 mg·g^{-1}. According to Giles's classification [41], the isotherm can be classified as type L. The isotherms were fitted to the Langmuir (Equation (7)) and Double-Langmuir (Equation (8)) models, which accurately describe class L isotherms [42]. The equations for these models are

$$q = q_1 \frac{b \cdot c}{1 + b \cdot c} \quad (7)$$

$$q = q_{L,1} \frac{b_1 \cdot c}{1 + b_1 \cdot c} + q_{L,2} \frac{b_2 \cdot c}{1 + b_2 \cdot c} \quad (8)$$

where q is the amount of adsorbate adsorbed on the surface (mg·g^{-1}), qL is the Langmuir maximum adsorption capacity (mg·g^{-1}), b is the Langmuir constant (L·mg^{-1}), c is the equilibrium concentration of the adsorbate in the liquid (mg·L^{-1}), q_{L1} and q_{L2} are the maximum adsorption capacities associated with the first and second adsorption layers in the Double-Langmuir model, and b_1 and b_2 are the equilibrium constants associated with the first and second adsorption layers. The results obtained are summarized in Table 7. The regression coefficients (R^2) indicate an excellent fit for both models. The high capacity observed in the first adsorption layer (q_{L1} = 281 mg·g^{-1}) indicates a strong interaction between TX-100 and the active sites of the carbon. This can be explained by hydrophobic interactions between the alkyl chains of TX-100 and the carbonaceous surface of the adsorbent, as well as possible specific interactions between functional groups of the adsorbent and the hydrophilic groups of the surfactant. In contrast, the second adsorption layer (q_{L2} = 435.8 mg·g^{-1}) may be dominated by weaker interactions, such as Van der Waals forces, consistent with the lower b_2 value in the Double-Langmuir model.

Table 7. Main results for equilibrium adsorption isotherms according to Equations (7) and (8).

Langmuir			Double-Langmuir				
q_{max}, (mg·g^{-1})	b, (L·mg^{-1})	R^2	$qmax_1$, (mg·g^{-1})	b_1, (L·mg^{-1})	$qmax_2$, (mg·g^{-1})	b_2, (L·mg^{-1})	R^2
350.3	0.0433	0.9319	281.0	0.0948	435.8	$2.8 \cdot 10^{-4}$	0.9518

The results show that the adsorption of Triton X-100 on activated carbon follows a characteristic two-step behavior. At low concentrations, adsorption is rapid and dominated by hydrophobic interactions between the alkyl chains of the surfactant and the carbonaceous surface of the adsorbent, suggesting a direct adsorption mechanism onto the active sites of the carbon. However, as the concentration of Triton X-100 approaches the CMC, the slope of the isotherm decreases and the formation of a second adsorption layer is observed, indicating that surfactant–surfactant interactions start to play a relevant role in the retention of the compound. This behavior is consistent with the Double-Langmuir model, which describes two distinct phases: a first stage of monolayer adsorption on the carbon surface and a second stage in which the newly adsorbed molecules are organized through cooperative interactions with the previously adsorbed molecules. In addition, the accessibility of Triton X-100 to the carbon micropores may influence the adsorption capacity since the structure of the adsorbent determines the availability of active sites. These results are in agreement with those reported in the literature [42], a similar behavior in the adsorption of non-ionic surfactants on activated carbon was observed, where the isotherm presents an inflection in the CMC region and an initial phase dominated by adsorbate–adsorbent interactions, and a second stage governed by interactions between the adsorbed molecules can be distinguished.

3.4.2. Kinetics of TX-100 Adsorption

Figure 12 shows the kinetic curves of triton X-100 concentration in the liquid phase (c) and the adsorbed amount in the solid phase (q) for $ACA_{pyrolysis}$. The mass transfer rate of triton X-100 from the solution to the particle decreases substantially at 90 min, reaching equilibrium at approximately 350 min (horizontal plateau). This fact suggests that the adsorption process is primarily controlled by intraparticle diffusion after an initial rapid adsorption on the adsorbent surface. This behavior is consistent with adsorption systems controlled by mass transport mechanisms in heterogeneous systems.

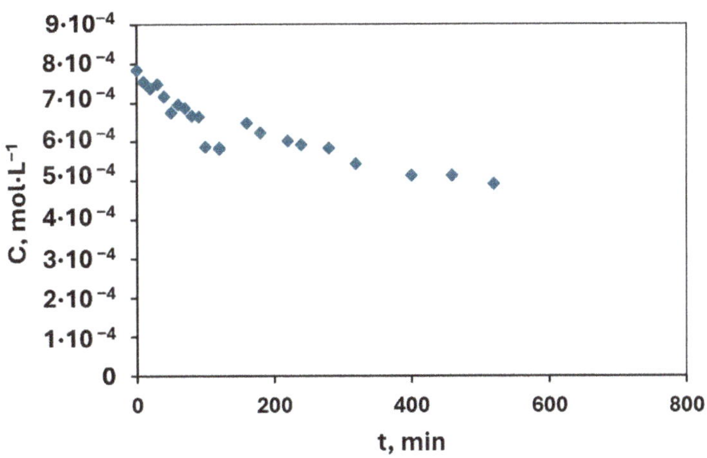

Figure 12. Kinetic curve of TX-100 adsorption from ACApyrolysis.

4. Conclusions

In this work, a study about the possible reuse of activated carbons used as filters in steam cycle treatment in a nuclear power plant was carried out. Once the activated carbon underwent washing treatment (including B removal), its textural characteristics improved, which is interesting for specific applications after its expiration date. Specifically, the adsorption capacity of this activated carbon was improved after washing treatment.

Regarding boron removal through an acrylic column, the most effective stage was at the beginning of the cycle, that is, when HCl solution came into direct contact with the activated carbon, while the effectiveness was reduced to a constant value afterwards. When the immersion method was used, shorter times were required (5 h) compared to the use of the acrylic column (around 24 h), with a higher removal of B per gram of activated carbon.

Even though the activated carbons did not show an effective adsorption of ethanolamine under the current experimental conditions, these results provided valuable insights into the characteristics of the recovered activated carbons from the corresponding filters. Nevertheless, an alternative usage of this activated carbon was proposed, showing a high effectiveness in triton X-100 adsorption.

Consequently, further studies should be focused on modifying activated carbons, for example, by introducing functional groups such as amino or carboxyl groups, which could improve the adsorption capacity for ethanolamine capture. Additionally, the use of alternative adsorbents, such as ion-exchange resins, could be an interesting research point for the future, as they have demonstrated higher effectiveness in adsorbing amines compared to traditional activated carbons. Optimization of experimental parameters, such as pH and temperature, could also play a key role in improving adsorption efficiency. Finally, economic studies to assess the feasibility of activated carbon reuse in this context would be another interesting contribution to further research. These approaches could be the starting point for the development of more efficient materials devoted to the removal of nitrogen-containing pollutants from aqueous solutions.

Author Contributions: Conceptualization, J.F.G.G. and B.L.C.; methodology, J.F.G.G. and B.L.C.; validation, J.F.G.G. and B.L.C.; formal analysis, B.L.C. and E.M.R.; investigation, J.F.G.G., B.L.C. and E.M.R.; resources, J.F.G.G.; data curation, B.L.C.; writing—original draft preparation, S.N.-D.; writing—review and editing, B.L.C., E.M.R. and S.N.-D.; visualization, B.L.C. and S.N.-D.; supervision, J.F.G.G. and B.L.C.; project administration, J.F.G.G.; funding acquisition, J.F.G.G. All authors have read and agreed to the published version of the manuscript.

Funding: Project "Reutilización de los carbones activados presentes en los filtros de la Central Nuclear de Almaraz para el tratamiento del agua del circuito de refrigeración". Funders: Centrales Nucleares Almaraz-Trillo (CNAT) and University of Extremadura (UEx). Funding number: 038/12.

Data Availability Statement: The original contributions presented in the study are included in the article; further inquiries can be directed to the corresponding author.

Acknowledgments: We would like to thank the members of Almaraz Nuclear Power Plant for their support and the supply of materials for the development of this work.

Conflicts of Interest: The authors declare no conflicts of interest.

References

1. Liu, Y.; Lyu, Y.; Tian, J.; Zhao, J.; Ye, N.; Zhang, Y.; Chen, L. Review of Waste Biorefinery Development towards a Circular Economy: From the Perspective of a Life Cycle Assessment. *Renew. Sustain. Energy Rev.* **2021**, *139*, 110716. [CrossRef]
2. Sharma, N.K.; Govindan, K.; Lai, K.K.; Chen, W.K.; Kumar, V. The Transition from Linear Economy to Circular Economy for Sustainability among SMEs: A Study on Prospects, Impediments, and Prerequisites. *Bus. Strategy Environ.* **2021**, *30*, 1803–1822. [CrossRef]
3. Gralla, F.; Abson, D.J.; Møller, A.P.; Lang, D.J.; von Wehrden, H. Energy Transitions and National Development Indicators: A Global Review of Nuclear Energy Production. *Renew. Sustain. Energy Rev.* **2017**, *70*, 1251–1265. [CrossRef]
4. Obregón, L.; Orozco, C.; Camargo, J.; Duarte, J.; Valencia, G. Research Trend on Nuclear Energy from 2008 to 2018: A Bibliometric Analysis. *Int. J. Energy Econ. Policy* **2019**, *9*, 542–551. [CrossRef]
5. Alwaeli, M.; Mannheim, V. Investigation into the Current State of Nuclear Energy and Nuclear Waste Management—A State-of-the-Art Review. *Energies* **2022**, *15*, 4275. [CrossRef]

6. World Nuclear Waste Report. *The World Nuclear Waste Report 2019*; The World Nuclear Waste Report (WNWR): Brussels, Belgium, 2019.
7. Ministerio para la Transición Ecológica y el Reto Demográfico. *7° Plan General de Residuos Radiactivos*; Ministerio para la Transición Ecológica y el Reto Demográfico: Madrid, Spain, 2023.
8. Consejo de Seguridad Nuclear. *Residuos Radiactivos*; Consejo de Seguridad Nuclear: Madrid, Spain, 2003. Available online: https://www.csn.es/residuos-radiactivos (accessed on 25 February 2025).
9. Centrales Nucleares Almaraz-Trillo. *Informe Semestral Primer Semestre 2019*; Centrales Nucleares Almaraz-Trillo: Almaraz, Spain, 2019.
10. World Nuclear Energy Status Report. World Nuclear Power Reactors 1951–2024; World Nuclear Energy Status Report. Available online: https://www.worldnuclearreport.org/ (accessed on 25 February 2025).
11. Ministerio para la Transición Ecológica y el Reto Demográfico Centrales. *Nucleares en España*; Ministerio para la Transición Ecológica y el Reto Demográfico Centrales: Madrid, Spain, 2023. Available online: https://www.miteco.gob.es/es/energia/nuclear/centrales/espana.html (accessed on 25 February 2025).
12. Magomedbekov, E.P.; Merkushkin, A.O.; Obruchikov, A.V.; Pokalchuk, V.S. Argon, Krypton and Xenon Adsorption Coefficients on Various Activated Carbons under Dynamic Conditions. *J. Radioanal. Nucl. Chem.* **2022**, *331*, 1091–1100. [CrossRef]
13. Matsuo, T.; Nishi, T. Activated Carbon Filter Treatment of Laundry Waste Water in Nuclear Power Plants and Filter Recovery by Heating in Vacuum. *Carbon* **2000**, *38*, 709–714. [CrossRef]
14. So, J.Y.; Cho, H.R. Thermal Characteristics of Spent Activated Carbon Generated from Air Cleaning Units in Korean Nuclear Power Plants. *Nucl. Eng. Technol.* **2017**, *49*, 873–880. [CrossRef]
15. Seon Choi, B.; Park, G.I.L.; Hyung Kim, J.; Wook Lee, J.; Kon Ryu, S. Adsorption Equilibrium and Dynamics of Methyl Iodide in a Silver Ion-Exchanged Zeolite Column at High Temperatures. *Adsorption* **2001**, *7*, 91–103. [CrossRef]
16. Won Park, S.; Kook Lee, W.; Moon, H. Adsorption and Desorption of Gaseous Methyl Iodide in a Triethylenediamine-Impregnated Activated Carbon Bed. *Sep. Technol.* **1993**, *3*, 133–142. [CrossRef]
17. Deitz, V.R. Interaction of Radioactive Iodine Gaseous Species with Nuclear-Grade Activated Carbons. *Carbon* **1987**, *25*, 31–38. [CrossRef]
18. ASTM International. *ASTM International Standard Test Method for Nuclear-Grade Activated Carbon*; ASTM International: West Conshohocken, PA, USA, 2022.
19. Ho, K.; Chun, H.; Lee, H.C.; Lee, Y.; Lee, S.; Jung, H.; Han, B.; Lee, C.H. Design of Highly Efficient Adsorbents for Removal of Gaseous Methyl Iodide Using Tertiary Amine-Impregnated Activated Carbon: Integrated Experimental and First-Principles Approach. *Chem. Eng. J.* **2019**, *373*, 1003–1011. [CrossRef]
20. Ho, K.; Moon, S.; Lee, H.C.; Hwang, Y.K.; Lee, C.H. Adsorptive Removal of Gaseous Methyl Iodide by Triethylenediamine (TEDA)-Metal Impregnated Activated Carbons under Humid Conditions. *J. Hazard. Mater.* **2019**, *368*, 550–559. [CrossRef] [PubMed]
21. Nacapricha, D.; Taylor, C.G. Quality Control of Nuclear Charcoals: Particle Size Effect and Trapping Mechanism. *Carbon*, 1996; 34, 155–163.
22. Shutikov, A.V.; Ivanov, V.N.; Tyapkov, V.F.; Yerpylyova, S.F.; Bykova, V.V. Introduction of Water Chemistry Regimes with Ethanolamine Metering at Nuclear Power Plants Equipped with VVER-Type Reactors. *Therm. Eng.* **2008**, *55*, 402–408. [CrossRef]
23. Lee, S.D.; Mallampati, S.R.; Lee, B.H. Enhanced Removal of Ethanolamine from Secondary System of Nuclear Power Plant Wastewater by Novel Hybrid Nano Zero-Valent Iron and Pressurized Ozone Initiated Oxidation Process. *Environ. Sci. Pollut. Res.* **2017**, *24*, 17769–17778. [CrossRef]
24. Khan, A.; Mahmood, H.; Yasin, S.; Moniruzzaman, M.; Iqbal, T. A Comprehensive Overview of Advanced Oxidation Process Assisted Mono-Ethanolamine Degradation in Aqueous Phase: Current Advances and Future Challenges. *J. Environ. Chem. Eng.* **2022**, *10*, 108078. [CrossRef]
25. Razali, M.N.; Yunus, R.M.; Jemaat, Z.; Alias, S. Monoethanolamine Wastewater Treatment via Adsorption Method: A Study on Comparison of Chitosan, Activated Carbon, Alum and Zeolite. *J. Appl. Sci.* **2010**, *10*, 2544–2550. [CrossRef]
26. Abu-Ghunmi, L.; Badawi, M.; Fayyad, M. Fate of Triton X-100 Applications on Water and Soil Environments: A Review. *J. Surfactants Deterg.* **2014**, *17*, 833–838. [CrossRef]
27. Carrott, P.J.M.; Nabais, J.M.V.; Ribeiro Carrott, M.M.L.; Menéndez, J.A. Thermal Treatments of Activated Carbon Fibres Using a Microwave Furnace. *Microporous Mesoporous Mater.* **2001**, *47*, 243–252. [CrossRef]
28. *BOE-A-1982-1323*; Orden de 1 de Diciembre de 1981 Por La Que Se Establecen Métodos Oficiales de Análisis de Aguas, Aceites y Grasas, Carne y Productos Carnícos, Fertilizantes, Productos Fitosanitarios, Leche y Productos Lácteos, Productos Orgánicos, Fertilizantes, Suelos y Productos Derivados de La Uva y Similares. Agencia Estatal Boletín Oficial del Estado: Madrid, Spain, 1982.
29. Sing, K.S.W. Characterization of Porous Materials: Past, Present and Future. *Colloids Surf. A Physicochem. Eng. Asp.* **2004**, *241*, 3–7. [CrossRef]

30. Rodríguez-Reinoso, F.; Molina-Sabio, M. Activated Carbons from Lignocellulosic Materials by Chemical and/or Physical Activation: An Overview. *Carbon* **1992**, *30*, 1111–1118. [CrossRef]
31. IUPAC. *Recommendations for the Characterization of Porous Solids*; International Union of Pure and Applied Chemistry: Research Triangle Park, NC, USA, 2009.
32. Tseng, R.L.; Wu, F.C.; Juang, R.S. Characteristics and Applications of the Lagergren's First-Order Equation for Adsorption Kinetics. *J. Taiwan Inst. Chem. Eng.* **2010**, *41*, 661–669. [CrossRef]
33. Plazinski, W.; Rudzinski, W.; Plazinska, A. Theoretical Models of Sorption Kinetics Including a Surface Reaction Mechanism: A Review. *Adv. Colloid. Interface Sci.* **2009**, *152*, 2–13. [CrossRef]
34. Hameed, B.H.; Rahman, A.A. Removal of Phenol from Aqueous Solutions by Adsorption onto Activated Carbon Prepared from Biomass Material. *J. Hazard. Mater.* **2008**, *160*, 576–581. [CrossRef] [PubMed]
35. Kumar, K.V.; Sivanesan, S. Pseudo Second Order Kinetics and Pseudo Isotherms for Malachite Green onto Activated Carbon: Comparison of Linear and Non-Linear Regression Methods. *J. Hazard. Mater.* **2006**, *136*, 721–726. [CrossRef] [PubMed]
36. Tan, I.A.W.; Ahmad, A.L.; Hameed, B.H. Adsorption of Basic Dye on High-Surface-Area Activated Carbon Prepared from Coconut Husk: Equilibrium, Kinetic and Thermodynamic Studies. *J. Hazard. Mater.* **2008**, *154*, 337–346. [CrossRef]
37. Ho, Y.S.; Mckay, G. Pseudo-Second Order Model for Sorption Processes. *Process Biochem.* **1999**, *34*, 451–465. [CrossRef]
38. Das, D.; Meikap, B.C. Comparison of Adsorption Capacity of Mono-Ethanolamine and Di-Ethanolamine Impregnated Activated Carbon in a Multi-Staged Fluidized Bed Reactor for Carbon-Dioxide Capture. *Fuel* **2018**, *224*, 47–56. [CrossRef]
39. Ra, J.; Lee, Y. Efficient Degradation of Ethanolamine by UV/Chlorine Process via Organic Chloramine Photolysis: Kinetics, Products, and Implications for Ethanolamine Wastewater Treatment. *Chem. Eng. J.* **2021**, *412*, 128631. [CrossRef]
40. Puttaswamy; Vaz, N.; Made Gowda, N.M. Oxidation of Ethanolamines by Sodium N-Bromobenzenesulfonamide in Alkaline Buffer Medium: A Kinetic and Mechanistic Study. *Int. J. Chem. Kinet.* **2001**, *33*, 480–490. [CrossRef]
41. Giles, C.H.; Smith, D.; Huitson, A. A General Treatment and Classification of the Solute Adsorption Isotherm. I. Theoretical. *J. Colloid. Interface Sci.* **1974**, *47*, 755–765. [CrossRef]
42. Azizian, S.; Eris, S. Adsorption Isotherms and Kinetics. *Interface Sci. Technol.* **2021**, *33*, 445–509.

C Journal of Carbon Research

MDPI

Article

A New Monohydrogen Phosphate-Selective Carbon Composite Membrane Electrode for Soil Water Samples

Ozlem Tavukcuoglu [1,*], Vildan Erci [2], Fatih Ciftci [3,4,*], Ibrahim Isildak [5] and Muhammed Zahid Kasapoglu [6]

1 Department of Biochemistry, Faculty of Hamidiye Pharmacy, University of Health Sciences, Istanbul 34668, Turkey
2 Department of Soil Science and Nutrition, Faculty of Agriculture, Selcuk University, Konya 42130, Turkey; vildanerci@selcuk.edu.tr
3 Department of Biomedical Engineering, Fatih Sultan Mehmet Vakıf University, Istanbul 34015, Turkey
4 Department of Technology Transfer Office, Fatih Sultan Mehmet Vakıf University, Istanbul 34015, Turkey
5 Department of Bioengineering, Faculty of Chemical and Metallurgical Engineering, Yildiz Technical University, Istanbul 34220, Turkey; isildak@yildiz.edu.tr
6 Institute of Nanotechnology and Biotechnology, İstanbul University-Cerrahpaşa, Istanbul 34320, Turkey; muhammed.kasapoglu@iuc.edu.tr
* Correspondence: ozlemoztolan@gmail.com (O.T.); faciftcii@gmail.com or fciftci@fsm.edu.tr (F.C.)

Abstract: This study focused on developing a novel composite phosphate-selective electrode for on-site and real-time applications using a silver polyglutaraldehyde phosphate and carbon nanotube (CNT) matrix. CNT-silver polyglutaraldehyde phosphate compound was synthesized and characterized using Fourier-transform infrared spectroscopy (FTIR), scanning electron microscopy (SEM), X-ray diffraction (XRD), and thermogravimetric analysis (TGA). The potentiometric performance of the composite phosphate-selective electrode was then investigated. The results demonstrated that the composite phosphate-selective electrode exhibited good sensitivity, with a linear response in the concentration range of 1.0×10^{-4} to 1.0×10^{-2} M for phosphate ions. The electrode also showed high selectivity towards phosphate ions compared to other anions, such as chloride and nitrate. Additionally, the electrode displayed a quick response time of less than 15 s, making it suitable for real-time measurements. The electrode was applied to surface and soil water samples. The results obtained from the water samples showed a strong correlation with those obtained from the preferred spectrophotometry method, highlighting the potential of the developed electrode for on-site and continuous monitoring of phosphate and offering an efficient and practical solution for various fields that require phosphate detection.

Keywords: composite matrix; phosphate-selective electrode; solid contact; soil analysis

1. Introduction

The development of phosphate-selective electrodes has gained significant attention due to the increasing need for the efficient, reliable, and cost-effective detection of phosphate ions across various applications, including in environmental monitoring, agriculture, and biomedical analysis [1–3]. Phosphates are essential for all living organisms, existing in both inorganic and organic forms, with orthophosphates (HPO_4^{2-} and $H_2PO_4^{-}$) being the most prevalent in natural environments [4]. They play critical roles in biological functions such as energy metabolism, genetic information storage, and cell signaling [4–6].

While phosphates are vital for human health and agricultural productivity, their overuse, particularly through phosphate-based fertilizers, poses serious environmental challenges. Phosphate pollution significantly degrades groundwater and surface water quality, leading to eutrophication, characterized by excessive algal blooms that disrupt aquatic ecosystems. The leaching of phosphates into water bodies due to improper fertilizer application necessitates substantial investments in wastewater treatment systems to mitigate these impacts. Therefore, monitoring phosphate levels in natural water sources and agricultural soils is crucial for maintaining water quality and optimizing agricultural outputs, which can help prevent economic losses for farmers [4,7,8].

Electrochemical sensors have emerged as a promising solution for phosphate monitoring, offering a high sensitivity for detecting trace levels of phosphates and other nutrients in soil and water samples [9–17]. These sensors provide rapid results, enabling real-time monitoring essential for dynamic environmental assessments. Their cost-effectiveness and compact design make them suitable for large-scale agricultural applications, allowing for in-field measurements without the need for extensive laboratory setups. Moreover, electrochemical sensors can be customized for specific ions, achieving high selectivity even in complex soil matrices. This selectivity is crucial for accurate monitoring in environments where multiple ions may be present. Compared to traditional chemical analysis methods, electrochemical techniques typically require minimal sample preparation, reducing analysis time and the risk of contamination. Additionally, these methods are environmentally friendly, as they generally consume fewer reagents and generate less waste, aligning with sustainable practices in environmental monitoring [18].

The growing need for the rapid and on-site analysis of environmental samples has driven the development of ion-selective electrode (ISE) technology for phosphate measurement [19–22]. ISE technology enables fast, selective, and quantitative measurement of phosphate levels [18]. Researchers have utilized various nanomaterials to enhance the sensitivity and stability of ion-selective electrodes for phosphate determination, with carbon nanotubes (CNTs) being widely employed due to their large surface area, high electrical conductivity, and chemical stability [23–27].

This study focuses on developing a composite matrix potentiometric electrode that utilizes carbon nanotube (CNT) and silver polyglutaraldehyde phosphate (Ag-pGAP) for the precise detection of monohydrogen phosphate ions. The synthesis and characterization of this composite material are outlined, alongside an assessment of the potentiometric performance of the phosphate-selective electrode and an exploration of its commercial viability and scalability. A thorough examination of the composite membrane matrix, which includes Ag-pGAP and CNT, will be conducted using various analytical techniques, including Fourier-transform infrared spectroscopy (FTIR), scanning electron microscopy (SEM), X-ray diffraction (XRD), thermogravimetric analysis (TG-DTG), and differential thermal analysis (DTA). Furthermore, the potentiometric performance attributes of the optimized electrode, such as selectivity and sensitivity, will be rigorously evaluated. The practical applicability of the developed phosphate-selective electrode will be illustrated through the measurement of phosphate concentrations in wastewater and soil samples [1]. Consequently, this planned study aims to provide an in-depth understanding of how a novel composite material, comprising Ag-pGAP and CNT, can improve the functionality of phosphate-selective electrodes. The findings will pave the way for the development of practical and scalable sensors for detecting phosphate ions, effectively addressing significant challenges related to sensitivity, selectivity, and stability.

2. Materials and Methods

2.1. Materials and Solutions

Thiourea dioxide, tetrahydrofuran (THF), silver (I) sulfide, Tris buffer solution (pH 7.0), copper (I) sulfide, and Kelowna extractant solution containing 0.25 mol/L CH_3COOH and 0.015 mol/L NH_4F were purchased from Sigma-Aldrich (St. Louis, MO, USA); glutaraldehyde and graphite were purchased from Fluka (Buchs, Switzerland); CNT was purchased from www.us-nano.com (Houston, TX, USA); silver nitrate was purchased from Carlo Erba (Milan, Italy); disodium monohydrogen phosphate was purchased from Merck (Darmstadt, Germany); hardener was purchased from Bayer (Leverkusen, Germany); and epoxy was purchased from Araldite (Basel, Switzerland). All chemicals and sodium salts of the anions used were of analytical grade. Throughout the experimental studies, double-distilled deionized water was used. A stock disodium monohydrogen phosphate solution (1.0×10^{-1} M) was prepared. Dilute solutions (1.0×10^{-1}–1.0×10^{-5} M) of disodium monohydrogen phosphate were then prepared by appropriate dilution of the stock solution with deionized water.

2.2. Synthesis of Silver Polyglutaraldehyde Phosphate (Ag-pGAP)

In the synthesis process, 7.4 g of thiourea dioxide and 21 g of 50% glutaraldehyde by weight were mixed in a flask. The mixture was then heated in an oil bath at 50 °C for 2 h, forming a light yellow polyglutaraldehyde dissolved in water. Subsequently, 230 mL of a 0.2 M solution of silver nitrate ($AgNO_3$) was added to the mixture, forming a silver complex. To precipitate Ag-pGAP, a dropwise addition of a 10^{-1} M Na_2HPO_4 solution was performed into the resulting complex solution. The precipitate was then washed with water and dried.

2.3. Characterization

The morphology of the silver polyglutaraldehyde phosphate, CNT, and CNT/Ag-pGAP used in the composite matrix was observed and examined using a Zeiss EVO/LS10 scanning electron microscope (SEM) (Carl Zeiss AG, Oberkochen, Germany). X-ray diffraction (XRD) patterns were recorded at ambient temperatures on an X'Pert XRD Philips Panalytical X'Pert Pro diffractometer (Malvern Panalytical, Almelo, The Netherlands) using Ni-filtered CuKα radiation (λ = 1.54059 Å; 45 kV and 40 mA). Fourier-transform infrared spectroscopy (FTIR) spectra of the samples were collected in the 4000–400 cm^{-1} region using a Shimadzu Corp IrPrestige-21 spectrophotometer (Shimadzu Corporation, Kyoto, Japan) with a resolution of 4 cm^{-1}. Thermal analysis curves (TG/DTG, DTA) were monitored using a TA Instruments SDT Q600 apparatus (TA Instruments, New Castle, DE, USA) in a dynamic air atmosphere with a heating rate of 10 °C min^{-1}, using platinum crucibles with a mass of approximately 10 mg, and a temperature range of 0–800 °C. Calcinated α-alumina was taken as the reference.

2.4. Fabrication of the Electrode

The composite membrane electrodes were prepared using the synthesized Ag-pGAP complex. The membrane compositions used with silver sulfide, copper sulfide, and epoxy are also listed in Table 1. The specific composition of the electrode, as shown in Table 1, was selected to optimize performance characteristics, including stability, sensitivity, and selectivity. The chosen materials and their ratios could reflect a carefully balanced compromise that enhanced the electrode's performance in detecting phosphate ions, while minimizing interference and ensuring long-term functionality. Each membrane mixture was homogeneously stirred with the appropriate amount of THF on a watch glass. Using a high-speed mechanical drill system with inserted bit for mixing, carbon nanotubes (CNTs) were well dispersed throughout the membrane matrix. Subsequently, the mixtures were compressed into plastic pipes and dried at room temperature.

Table 1. Ratios of composite phosphate electrodes prepared with Ag-pGAP.

No.	CNT (mg)	Ag-pGAP (mg)	Silver Sulfide (Ag_2S) (mg)	Copper Sulfide (Cu_2S) (mg)	Epoxy (mg)
1	100	10	80	10	800
2	100	10	80	10	600
3	100	10	80	10	400
4	100	10	80	10	200
5	–	11	55	–	34

Thin sections were extracted from the drying tubes and affixed to the end of a copper wire of appropriate length (15 cm, 0.5 mm thickness) using a solid contact material. This solid contact material was created by thoroughly dissolving a mixture consisting of 50% (*w*/*w*) graphite, 35% (*w*/*w*) epoxy, and 15% (*w*/*w*) hardener in a suitable amount of THF. After allowing it to dry for 5 h, the electrodes were insulated with a strong adhesive similar to "Patex" to prepare them for measurement.

2.5. Potentiometric Measurements

Potentiometric measurements were carried out using a multi-channel potentiometer developed by Medisen Medical Technologies Research and Development Ind. Co., Ltd., in Istanbul, Turkey. An Elit solid-state Ag/AgCl dual-junction reference electrode with a lithium acetate internal reference solution was employed for these measurements.

2.6. Statistical Analysis

In this study, the Bland–Altman method and regression analysis were utilized for 30 soil samples to plot the differences against the averages of the measurements obtained from the two distinct methods, allowing for the examination of any relationship between the differences and the averages to compare the outcomes for the same samples. If the differences exhibited a normal distribution, they were expected to be randomly distributed around zero, with 95% falling between the limits of agreement. The Bland–Altman method was implemented using the Origin package program, while the regression analysis was conducted using the MINITAB package program. Additionally, this study utilized the concept of "agreement boundaries" within the Bland–Altman method. If the differences followed a normal distribution, the differences were expected to be randomly distributed around zero, with 95% falling between "$\overline{d} - 1.96$ s and $\overline{d} + 1.96$ s". In this method, the range "$\overline{d} \pm 1.96$ s" refers to "limits of agreement". Furthermore, this study considered the use of regression analysis for calibrating one measurement against another or detecting bias between two measurement methods.

The practical applicability of the composite phosphate-selective electrode was further assessed for analyzing soil surface water samples obtained from three different agricultural areas near Istanbul. To validate the correlation between the two methods, statistical tests such as Student's *t*-test and Fisher's F-test were employed, with calculated values provided.

3. Results and Discussion

3.1. FTIR Analysis

Figure 1a shows the FTIR spectrum for Ag-pGAP, CNT, and the CNT/Ag-pGAP composite. In the FTIR spectrum of silver polyglutaraldehyde phosphate (Ag-pGAP), we observed a significant absorption band between 1740 and 1725 cm^{-1}, associated with C=O stretching [28]. Additionally, the peaks at 2850 and 2750 cm^{-1} indicate C-H stretching related to the aldehyde (-CHO) group. A peak at 2920 cm^{-1} suggests the presence of silver metal centers. Notably, the intensity of the hydroxyl (OH^-) species was prominently observed around 3450 and 3655 cm^{-1} [29]. When examining the FTIR spectra of CNT and CNT/Ag-pGAP, only minor differences were noted. This is likely due to the low weight percentage (<10%) of Ag-pGAP in the composite. The spectrum is primarily dominated by intense and broad absorption bands from the CNT, which may have obscured the weaker signals from the functional groups of Ag-pGAP. However, the weak split peaks observed at 900 and 950 cm^{-1} in the CNT/Ag-pGAP spectrum indicate the presence of P-O and P-C bonds. This suggests successful bonding between the carbon and phosphorus atoms in CNT and Ag-pGAP. Although these peaks are subtle, they are characteristic of phosphoryl and phosphonate groups, confirming the presence of Ag-pGAP even at low concentrations.

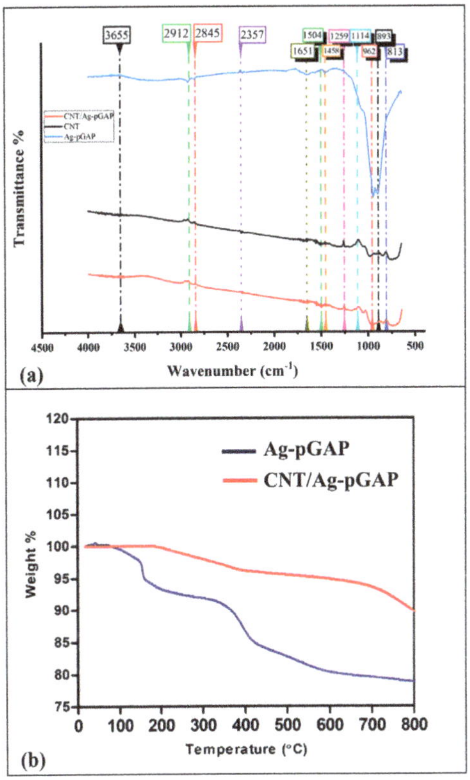

Figure 1. (**a**) FTIR spectrum of Ag-pGAP, CNT, and CNT-Ag-pGAP. (**b**) TGA graphs of Ag-pGAP and CNT-Ag-pGAP.

3.2. Thermogravimetric Analysis

The thermogravimetric behavior of Ag-pGAP and CNT/Ag-pGAP was investigated. The thermogravimetric curves depicting their behavior are shown in Figure 1b. The graphs show that Ag-pGAP experienced weight loss after 120 °C, while CNT/Ag-pGAP compound exhibited weight loss after 200 °C. Complex structures containing metals are known to be stable up to 150 °C, but tend to break down into smaller fragments beyond this temperature. This indicates that CNT/Ag-pGAP behaves like a stable complex. Weight losses between 150 and 600 °C are believed to arise from the degradation of the polyglutaraldehyde phosphate structure. In contrast, CNT/Ag-pGAP shows limited weight loss in this temperature range, suggesting the mixture's high stability. At the final temperature (800 °C), the residues are predicted to have originated from the "silver" content in the

Ag-pGAP structure (77%) and both the "silver and CNT" content in the CNT/Ag-pGAP structure (88%).

3.3. SEM Analysis

Figure 2 presents SEM images of the silver polyglutaraldehyde phosphate (Ag-pGAP), carbon nanotube (CNT), and CNT/Ag-pGAP compounds at various magnifications. The micrographs indicate that Ag-pGAP has an amorphous structure, with silver appearing as tiny particles (Figure 2a). In Figure 2b, the CNT structures are clearly visible with smooth surfaces, showcasing their pristine morphology. In contrast, the SEM image of the CNT/Ag-pGAP composite (Figure 2c) retains the CNT structures but shows a clouded and less defined surface compared to Figure 2b.

Figure 2. SEM images of (**a**) Ag-pGAP, (**b**) CNT, and (**c**) CNT/Ag-pGAP.

This clouded appearance results from a thin, uniform layer of Ag-pGAP coating the CNTs, which slightly blurs the contours of the nanotubes. Since the Ag-pGAP layer is extremely thin—likely at the nanoscale—its overall morphology remains largely unchanged, but its surface texture experiences subtle alterations. The differences between Figure 2b,c can be challenging to discern due to the thin and homogeneous nature of the Ag-pGAP coating, which does not significantly impact its tubular structure but does affect surface roughness and texture. Additionally, the clouded surface and slight increase in agglomeration seen in Figure 2c suggest enhanced interactions between the tubes, implying that Ag-pGAP acts as a binding agent.

Given the nanoscale thickness of the Ag-pGAP layer, SEM imaging may have caused a struggle to distinguish the coating from the CNT substrate. This limitation arose from the resolution constraints of SEM in detecting ultrathin, amorphous layers. Nevertheless, the observed clouded surface and slight textural changes provide indirect evidence of the coating, aligning with findings from previous studies, which showed similar challenges in visualizing thin polymeric coatings using SEM [30,31].

3.4. XRD Analysis

The X-ray diffraction spectra of silver polyglutaraldehyde phosphate, CNT, and CNT/Ag-pGAP are shown in Figure 3. The obtained spectra exhibit strong intensity diffraction peaks. In the spectra of the CNT and CNT/Ag-pGAP samples, the peaks observed at approximately 26, 45, 52, and 73 can be attributed to CNT reflections. The broad peaks observed in the spectra indicate that the carbon nanotubes had small dimensions. In the spectrum of CNT/Ag-pGAP, the peaks observed at 31, 33, and 380, along with significant reductions in peak intensities compared to the CNT spectrum, indicate that Ag-pGAP was homogeneously adhered to and surrounded by the CNT surface.

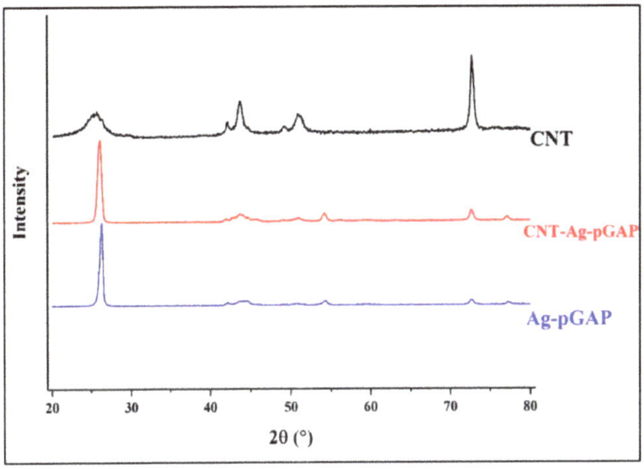

Figure 3. XRD spectra of Ag-pGAP, CNT, and CNT/Ag-pGAP.

3.5. Potentiometric Performance of the Electrode

The electrode membranes, containing Ag-pGAP as an ionophore, demonstrated sensitive responses to HPO_4^{2-} concentrations ranging from 10^{-5} to 10^{-1} mol/L in Tris buffer at pH 8.0. Interestingly, the electrodes exhibited minimal sensitivity to phosphate when Kelowna soil extractant was used as the base solution. In the Kelowna solution, the electrode displayed a sensitive response to HPO_4^{2-} concentrations spanning from 10^{-5} to 10^{-1} mol/L for total phosphate concentration, with a detection limit of 10^{-5} mol/L. This detection range encompasses the typical soil phosphorus concentration range in soil water samples. The electrode's selectivity was deemed satisfactory for measuring phosphates in the presence of six interfering ions, including Cl^-, CH_3COO^-, Cu^{2+}, SO_4^{2-}, HCO_3^-, and NO_3^-, with the electrodes exhibiting significantly higher responsiveness to phosphate compared to the interfering ions. To mitigate any pH influence, the pH levels of the tested solutions were standardized across various tested phosphate concentrations, maintaining the Kelowna solution at pH 8.0, with pH adjustments monitored using an Orion pH meter.

The Kelowna extraction for available and total dissolved P was conducted following the protocols outlined by Ebuele, involving mixing ten grams of soil with 100 mL of Kelowna solution at pH 8.0 in a 250 mL Erlenmeyer flask [32].

Potentiometric measurements were conducted using standard phosphate-selective electrodes with concentrations ranging from 1.0×10^{-1} to 1.0×10^{-5} M and $HPO_4{}^{2-}$ solutions prepared according to the ratios specified in Table 1. The pH effect on the electrode performance was investigated and the best response for $HPO_4{}^{2-}$ species was obtained at pH 7.0 with good reproducibility. It differed slightly from the pH value (8.5) reported in the literature [33]. The slight change in the optimal pH for the electrode likely arose from a combination of factors related to differences in electrode composition, phosphate species distribution, surface chemistry, and experimental conditions. The modified electrode materials in our study may simply have had a different affinity for phosphate ions at a lower pH than the electrode used in previous studies. No variations were observed in the measurements obtained, especially with Composition 2 of these electrodes. Figure 4 illustrates the relationship between potential and concentration for the composite phosphate-selective electrode prepared with Composition 2. This specific composition was selected for performance evaluation due to its optimal potentiometric response and selectivity towards phosphate ions. The favorable performance of Composition 2 is attributed to its balanced epoxy content, which offers both mechanical stability and consistent ion exchange properties, ensuring reliable signal transduction.

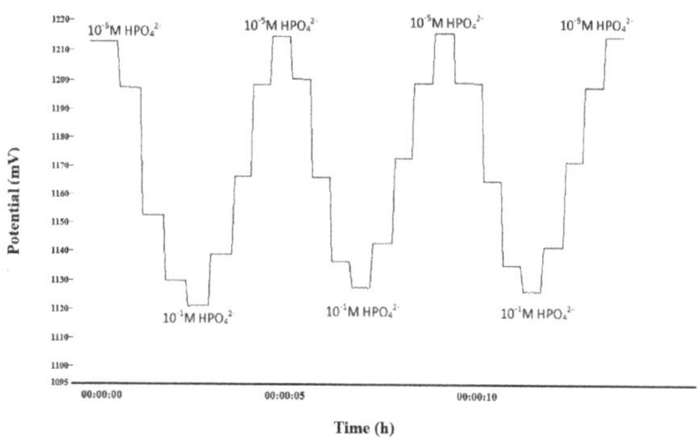

Figure 4. Potentiometric behavior of composite phosphate-selective electrode prepared with Ag-pGAP of composition.

The preparations of Compositions 1 and 3–5 were designed to systematically investigate how varying epoxy content influences potentiometric performance. The effects on membrane adhesion, electrical conductivity, and ion selectivity were examined by adjusting the epoxy ratios while keeping the amounts of CNT, Ag-pGAP, Ag_2S, and Cu_2S constant. Although not all compositions were evaluated in detail, preliminary testing showed that Composition 2 provided the best balance between selectivity, sensitivity, and stability. These findings led to the conclusion that Composition 2 was the most promising candidate for in-depth potentiometric analysis, resulting in the Nernstian response and high selectivity observed in this study.

Figure 5b presents the potentiometric response of the composite membrane phosphate-selective electrode prepared using Composition 2 listed in Table 1 against standard $HPO_4{}^{2-}$ ions and other ions in the concentration range of 1.0×10^{-1} to 1.0×10^{-5} M.

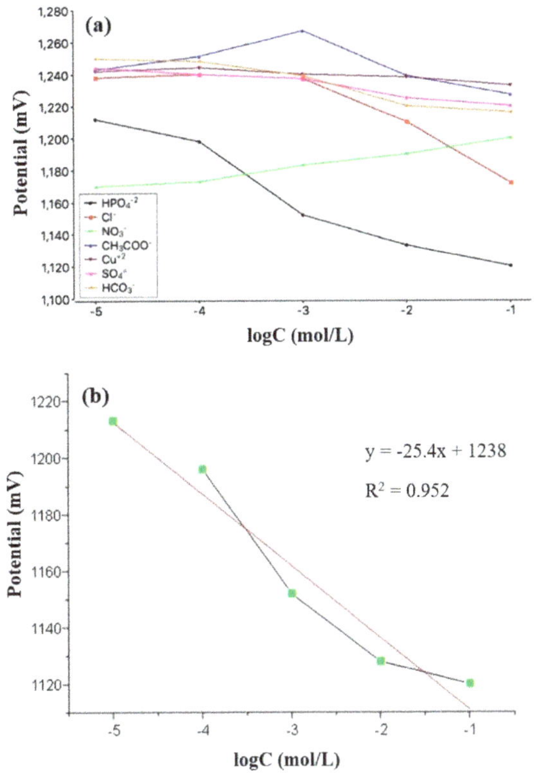

Figure 5. (**a**) Potentiometric behavior of the composite phosphate-selective electrode with Composition 2 towards ions HPO_4^{2-}, Cl^-, CH_3COO^-, Cu^{2+}, SO_4^{2-}, HCO_3^-, and NO_3^- (**b**) Calibration graph employed for determining the selectivity coefficient of the composite phosphate-selective electrode of Composition 2.

As shown in Figure 5a, the composite phosphate-selective electrode prepared using Composition 2 demonstrated specific sensitivity to HPO_4^{2-} ions, as well as to Cl^-, CH_3COO^-, Cu^{2+}, SO_4^{2-}, HCO_3^-, and NO_3^- ions. To better understand the selective behavior of the composite phosphate-selective electrode, selectivity coefficients were calculated. For this purpose, standard $HPO4^{2-}$ solutions ranging from 1.0×10^{-1} to 1.0×10^{-5} M were prepared. The potential changes exhibited by the composite phosphate-selective electrode in response to these HPO_4^{2-} solutions were analyzed, and calibration curves were plotted using the obtained data. From the calibration curve, a linear equation was derived. Subsequently, the same electrode was employed to measure the potential changes in different standard anion solutions, including sodium and potassium, at concentrations ranging from 1.0×10^{-1} to 1.0×10^{-5} M. Selectivity coefficients were calculated based on the potential changes in each anion and the linear equation derived from the calibration curve. The calibration curve used for calculating selectivity coefficients is illustrated in Figure 5b.

Potentiometric measurements were conducted using a composite phosphate-selective electrode (Composition 2), which was prepared with silver polyglutaraldehyde phosphate. The electrode demonstrated a linear response to standard HPO_4^{2-} ions within a concentration range of 1.0×10^{-4} to 1.0×10^{-2} M at a pH of 7.0. The linear equation representing the potential–concentration relationship in the linear working range was determined as the following: $E = -25.4[HPO_4^{2-}] + 1238$ $R^2 = 0.952$. It was observed that within the linear working range, there was an approximate 30 ± 1 mV potential difference for every 10-fold concentration change. The detection limit of the electrode with Composition 2 was determined to be 4.7×10^{-5} M. The results collectively demonstrate that the phosphate-selective composite electrode with Composition 2 not only provided a linear response but also showcased excellent sensitivity and a low detection limit for phosphate ions. In terms of repeatability, the electrode's performance was evaluated using standard HPO_4^{2-} solutions at concentrations of 1.0×10^{-2}, 1.0×10^{-3}, and 1.0×10^{-4} M. The measurements were conducted in a controlled environment, ensuring that the electrode surface was thoroughly washed with deionized water between different solutions to prevent cross-contamination. The potentiometric response behavior of the electrode is graphically presented in Figure 6 and numerically in Table 2, along with the standard deviations, to demonstrate the repeatability of the results.

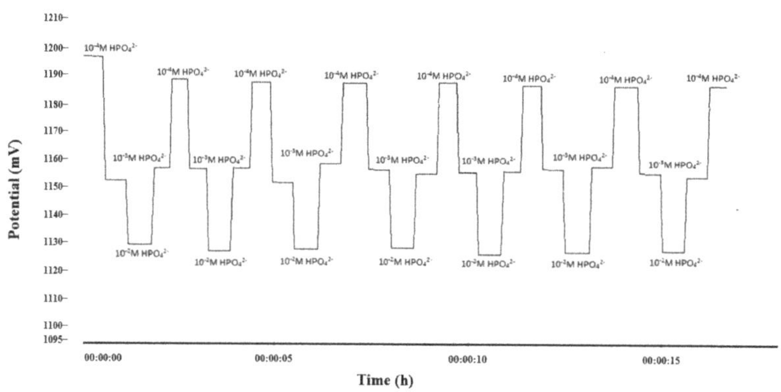

Figure 6. Reproducibility assessment of the composite phosphate-selective electrode with Composition 2, fabricated using Ag-pGAP.

Table 2. Average and standard deviation values showing the repeatability of the composite phosphate-selective electrode of Composition 2.

	Measurement Results (mV)							
HPO_4^{2-}	1	2	3	4	5	6	7	$X \pm S$ *
1×10^{-2} M	1130	1128	1128	1129	1127	1128	1128	1128.28 ± 0.95
1×10^{-3} M	1158	1158	1158	1153	1158	1157	1157	1157 ± 1.82
1×10^{-4} M	1189	1188	1188	1188	1187	1188	1188	1188 ± 0.58

* Average and standard deviation values are given for $n = 7$.

To measure the response time of the composite phosphate-selective electrode prepared with Ag-pGAP for Composition 2, the transition times to reach the equilibrium potential were examined in standard HPO_4^{2-} solutions within the concentration range of 1.0×10^{-1} to 1.0×10^{-5} M. The response time graph of the composite phosphate-selective electrode is shown in Figure 7. It was observed that the response time of the electrode, which exhibited repeatable and linear responses to standard HPO_4^{2-} ions in the concentration range of 1.0×10^{-2} to 1.0×10^{-4} M, was particularly less than 15 s, especially as the concentration increased towards higher concentrations.

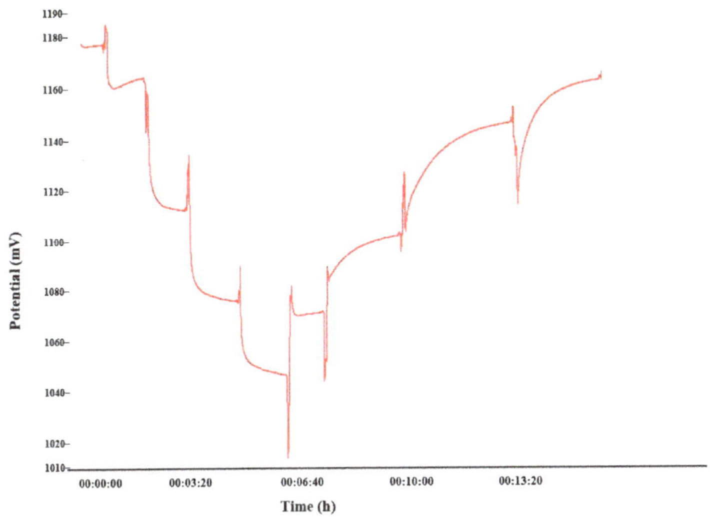

Figure 7. Response time of the composite phosphate-selective electrode prepared with Ag-pGAP, utilizing Composition 2.

3.6. Applications

The application section of this study involved the utilization of spectrophotometric [34] and potentiometric methods to extract separate PO_4^{3-} values from 30 soil samples collected from diverse locations in Istanbul. Extraction was carried out using Kelowna solution in Tris buffer at pH 8.0. Following extraction, the samples underwent filtration and were stored at room temperature for subsequent analysis using a solid-state phosphate-selective electrode. The acquired values underwent statistical assessment by employing the Bland–Altman method and regression technique [35]. Table 3 details descriptive statistics for the data obtained from the 30 soils, encompassing the mean, standard deviation, minimum, and maximum values for both the spectrophotometric (P_2O_5) and potentiometric (HPO_4^{2-}) methods.

Table 3. The descriptive statistics for the phosphate values obtained from the soil.

		n	Mean (ppm)	S.D.	Min.	Max.
Spectrophotometric	**P_2O_5**	30	2.35557	2.08279	0.28047	9.54652
Potentiometric	**HPO_4^{2-}**	30	2.29233	1.98941	0.3	9

In Table 3, the mean and standard deviation values of the data obtained through both measurement methods are very close to each other. The relevant data sets were initially subjected to the Bland–Altman method using the Origin program, and graphs depicting the deviations of the observed values from the mean were obtained, as shown in Figure 8.

Upon reviewing Figure 8, it is evident that the differences in the measurement results obtained by the two techniques do not exhibit a systematic distribution around the mean, but instead display a random distribution. According to the Bland–Altman method, the mean of the differences should be dispersed around zero. Therefore, using the Bland–Altman method, it was appropriate to assess the agreement between these two methods. The analysis indicated that there was an agreement between the two methods, and they could be considered as alternatives to each other. The results of the regression analysis of the relevant data sets are depicted in Figure 9. For the values, the fact that 99.7% of the variation in either of the two methods could be explained by the other method (R^2) served as an indicator of the agreement, and the linearity ratio was also utilized to express the agreement.

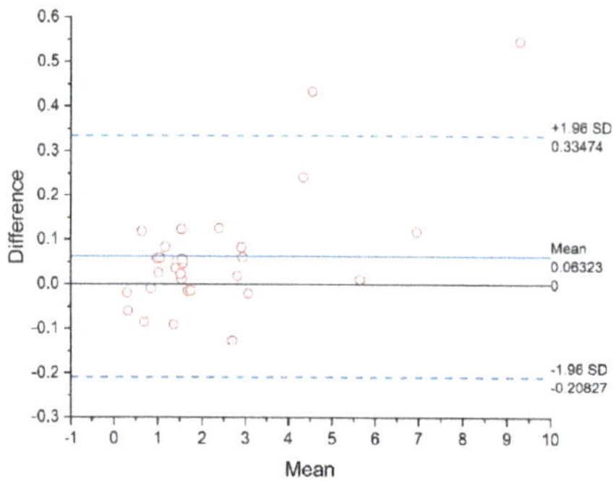

Figure 8. Bland–Altman plot obtained for the phosphate values for the comparison of measurement techniques.

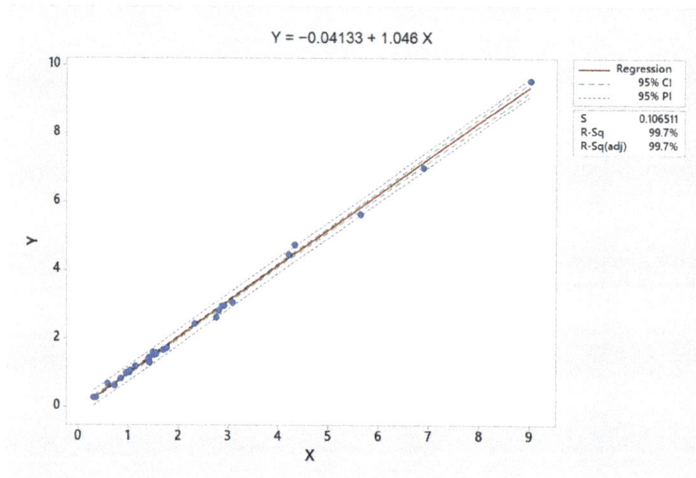

Figure 9. Regression plot obtained for the phosphate values for the comparison of measurement techniques. CI and PI represent 95% confidence intervals and prediction intervals, respectively.

The practical application of the composite phosphate-selective electrode was evaluated by examining soil surface water samples collected from three distinct agricultural regions near Istanbul. Initially, the water samples underwent filtration through 0.45 μm membranes, followed by analysis utilizing the standard addition method (Table 4).

Table 4. Determination of phosphate in soil surface water samples by the composite phosphate-selective electrode.

Sample	Phosphate Conc. (ppm)		Average *n*:7	SD±	CV (%)
	Min.	Max.			
Soil water 1	0.39	0.44	0.41	0.025	3.4
Soil water 2	0.35	0.42	0.39	0.035	5.6
Soil water 3	0.40	0.48	0.44	0.040	6.8

The results obtained from the electrode were compared with those from the conventional spectrophotometric method for determining dissolved phosphate levels (Table 5).

Table 5. Comparison of the results obtained by the composite phosphate-selective electrode and UV-Vis spectrophotometry.

Sample	Phosphate Conc., ppm, *n*:3	
	The Electrode	**UV-Vis**
Soil water 1	0.42	0.45
Soil water 2	0.38	0.40
Soil water 3	0.45	0.42

To validate the correlation between the two methods, statistical tests such as Student's *t*-test and Fisher's F-test were employed, with calculated values provided [36]. Remarkably, the experimental *t*-test and F-test values for phosphate were found to be lower than the theoretical values (t-test = 4.0, n = 7; F-test = 17.0, n = 7), with a confidence level of 95% ($p < 0.05$). This suggests that the proposed composite phosphate-selective electrode yields results that agree with the preferred spectrophotometry method.

The results obtained from the filtered soil surface water samples showed a strong correlation with those obtained from the preferred spectrophotometry method, indicating the high suitability of the proposed composite phosphate-selective electrode for sensing phosphate levels in real agricultural wastewater. This finding highlights its potential for practical applications in monitoring and evaluating phosphate content in soil surface water sources.

Overall, this study presents the development of a novel composite phosphate-selective electrode that employs a matrix of Ag-pGAP and CNT for the on-site and real-time detection of phosphate in various environmental applications [37]. The composite material was synthesized through a chemical reaction involving thiourea dioxide, glutaraldehyde, and silver nitrate ($AgNO_3$), followed by precipitation with disodium monohydrogen phosphate (Na_2HPO_4). The resulting structure integrates silver polyglutaraldehyde phosphate, silver sulfide (Ag_2S), and copper sulfide (Cu_2S) with epoxy as a binding agent, thereby optimizing stability, sensitivity, and selectivity for phosphate detection.

Characterization techniques, including Fourier-transform infrared spectroscopy (FTIR), scanning electron microscopy (SEM), X-ray diffraction (XRD), and thermogravimetric analysis (TGA), confirmed the successful formation of the composite, demonstrating its structural integrity and thermal stability [38]. The electrode exhibited a linear potentiometric response to phosphate ion concentrations ranging from 1.0×10^{-4} M to 1.0×10^{-2} M, showing high selectivity against other anions such as chloride and nitrate, indicating its suitability for application in complex environmental matrices.

The electrode's rapid response time of less than 15 s is essential for real-time monitoring, and its performance was validated against traditional spectrophotometric methods, revealing a strong correlation in phosphate detection from surface water and soil samples [39]. This underscores the electrode's accuracy and its potential to complement or replace conventional laboratory techniques. Its application in agricultural regions near Istanbul highlights its role in mitigating phosphate-related environmental issues, such as eutrophication, while optimizing fertilizer usage.

Moreover, the electrode's portability, cost-effectiveness, minimal sample preparation requirements, and environmental friendliness enhance its practicality for in-field measurements [40]. Statistical analyses, including the Bland–Altman method, regression analysis, Student's *t*-test, and Fisher's F-test, confirmed the reliability of the results, ensuring that observed differences between methods were statistically insignificant [41].

4. Conclusions

The developed CNT/Ag-pGAP composite electrode demonstrated high sensitivity, selectivity, and stability for detecting phosphate ions. It exhibited a Nernstian response across a concentration range of 1.0×10^{-4} M to 1.0×10^{-2} M. This range effectively covers the typical phosphorus concentrations in soil water samples.

The selectivity of the electrode was validated against six common interfering ions: Cl^-, CH_3COO^-, Cu^{2+}, $SO_4{}^{2-}$, $HCO_3{}^-$, and $NO_3{}^-$. The results showed that the electrode responded significantly more to phosphate ions than these interfering ions.

Potentiometric measurements indicated good reproducibility, particularly for Composition 2, which balanced mechanical stability and consistent ion exchange properties, ensuring reliable signal transduction. The optimal performance was observed at pH 7.0, which slightly differs from previous studies; this variation was attributed to differences in electrode composition and surface chemistry.

In real-world applications, the electrode demonstrated robust performance in areas such as water quality management and agricultural monitoring, confirming its suitability for on-site and real-time environmental monitoring. Its potential for scaling and practical deployment positions this electrode as an effective and reliable solution for phosphate detection, thereby contributing to improved environmental management and agricultural productivity.

Author Contributions: O.T.: Conceptualization, data curation, investigation, methodology, visualization, writing—review, editing, and original draft. V.E.: Data curation, methodology, writing—review and editing. F.C.: Investigation, methodology, data curation, writing—review, editing, and original draft. I.I.: Project administration, supervision, writing—review and editing. M.Z.K.: Investigation, methodology, writing—review and editing. All authors have read and agreed to the published version of the manuscript.

Funding: This research received no external funding.

Institutional Review Board Statement: Not applicable.

Data Availability Statement: Data supporting the reported results are available upon reasonable request from the corresponding author.

Acknowledgments: This research was funded by the General Directorate of Agricultural Research and Policies (TA-GEM), grant number TAGEM-19/AR-GE/04.

Conflicts of Interest: The authors declare no conflicts of interest.

References

1. Chen, J.; Li, D.; Ding, X.; Zhang, D. Sensitive and selective electrochemical aptasensing method for the voltammetric determination of dopamine based on AuNPs/PEDOT-ERGO nanocomposites. *Bioelectrochemistry* **2024**, *157*, 108653. [CrossRef] [PubMed]
2. Palo, E.; Zhang, H.; Lastusaari, M.; Salomäki, M. Nanometer-Thick Ion-Selective Polyelectrolyte Multilayer Coatings to Inhibit the Disintegration of Inorganic Upconverting Nanoparticles. *ACS Appl. Nano Mater.* **2020**, *3*, 6892–6898. [CrossRef]
3. Saikrithika, S.; Kumar, S. A selective voltammetric pH sensor using graphitized mesoporous carbon/polyaniline hybrid system. *J. Chem. Sci.* **2021**, *133*, 2. [CrossRef]
4. Daneshgar, S.; Callegari, A.; Capodaglio, A.G.; Vaccari, D. The potential phosphorus crisis: Resource conservation and possible escape technologies: A review. *Resources* **2018**, *7*, 37. [CrossRef]
5. Kates, D.M.; Sherrard, D.J.; Andress, D.L. Evidence that serum phosphate is independently associated with serum PTH in patients with chronic renal failure. *Am. J. Kidney Dis.* **1997**, *30*, 809–813. [CrossRef] [PubMed]
6. Desmeules, S.; Bergeron, M.J. Isenring. Acute Phosphate Nephropathy and Renal Failure. *N. Engl. J. Med.* **2003**, *349*, 1006–1007. [CrossRef]
7. Kim, D.Y.; Kim, D.G.; Jeong, B.; Kim, Y.I.; Heo, J.; Lee, H.K. Reusable and pH-Stable Luminescent Sensors for Highly Selective Detection of Phosphate. *Polymers* **2022**, *14*, 190. [CrossRef]

8. Tang, C.; Fu, D.; Wang, R.; Zhang, X.; Wei, L.; Li, M.; Li, C.; Cao, Q.; Chen, X. An Electrochemical Microfluidic System for on-Site Continuous Monitoring of Soil Phosphate. *IEEE Sens. J.* **2024**, *24*, 6754–6764. [CrossRef]
9. Parra, A.; Ramon, M.; Alonso, J.; Lemos, S.G.; Vieira, E.C.; Nogueira, A.R.A. Flow injection potentiometric system for the simultaneous determination of inositol phosphates and phosphate: Phosphorus nutritional evaluation on seeds and grains. *J. Agric. Food Chem.* **2005**, *53*, 7644–7648. [CrossRef]
10. Ganjali, M.R.; Norouzi, P.; Ghomi, M.; Salavati-Niasari, M. Highly selective and sensitive monohydrogen phosphate membrane sensor based on molybdenum acetylacetonate. *Anal. Chim. Acta* **2006**, *567*, 196–201. [CrossRef]
11. Akyilmaz, E.; Yorganci, E. Construction of an amperometric pyruvate oxidase enzyme electrode for determination of pyruvate and phosphate. *Electrochim. Acta* **2007**, *52*, 7972–7977. [CrossRef]
12. Kwan, R.C.H.; Leung, H.F.; Hon, P.Y.T.; Barford, J.P.; Renneberg, R. A screen-printed biosensor using pyruvate oxidase for rapid determination of phosphate in synthetic wastewater. *Appl. Microbiol. Biotechnol.* **2005**, *66*, 377–383. [CrossRef] [PubMed]
13. Zhao, H.X.; Liu, L.Q.; de Liu, Z.; Wang, Y.; Zhao, X.J.; Huang, C.Z. Highly selective detection of phosphate in very complicated matrixes with an off–on fluorescent probe of europium-adjusted carbon dots. *Chem. Commun.* **2011**, *47*, 2604–2606. [CrossRef] [PubMed]
14. Preechaworapun, A.; Dai, Z.; Xiang, Y.; Chailapakul, O.; Wang, J. Investigation of the enzyme hydrolysis products of the substrates of alkaline phosphatase in electrochemical immunosensing. *Talanta* **2008**, *76*, 424–431. [CrossRef]
15. Mousty, C.; Cosnier, S.; Shan, D.; Mu, S. Trienzymatic biosensor for the determination of inorganic phosphate. *Anal. Chim. Acta* **2001**, *443*, 1–8. [CrossRef]
16. Mahmud, M.A.P.; Ejeian, F.; Azadi, S.; Myers, M.; Pejcic, B.; Abbassi, R.; Razmjou, A.; Asadnia, M. Recent progress in sensing nitrate, nitrite, phosphate, and ammonium in aquatic environment. *Chemosphere* **2020**, 259. [CrossRef]
17. Berchmans, S.; Issa, T.B.; Singh, P. Determination of inorganic phosphate by electroanalytical methods: A review. *Anal. Chim. Acta* **2012**, *729*, 7–20. [CrossRef]
18. Ebadi, M.; Asareh, A.; Yengejeh, R.J.; Hedayat, N. Investigation of electro-coagulation process for phosphate and nitrate removal from sugarcane wastewaters. *Iran. J. Toxicol.* **2021**, *15*, 19–26. [CrossRef]
19. Lawrence, N.S.; Beckett, E.L.; Davis, J.; Compton, R.G. Advances in the voltammetric analysis of small biologically relevant compounds. *Anal. Biochem.* **2002**, *303*, 1–16. [CrossRef]
20. Hart, J.P.; Crew, A.; Crouch, E.; Honeychurch, K.C.; Pemberton, R.M. Some Recent Designs and Developments of Screen-Printed Carbon Electrochemical Sensors/Biosensors for Biomedical, Environmental, and Industrial Analyses. *Anal. Lett.* **2004**, *37*, 789–830. [CrossRef]
21. Pan, D.; Wang, Y.; Chen, Z.; Lou, T.; Qin, W. Nanomaterial/ionophore-based electrode for anodic stripping voltammetric determination of lead: An electrochemical sensing platform toward heavy metals. *Anal. Chem.* **2009**, *81*, 5088–5094. [CrossRef] [PubMed]
22. Manjunatha, J.G.; Deraman, M.; Basri, N.H.; Talib, I.A. Fabrication of poly (Solid Red A) modified carbon nano tube paste electrode and its application for simultaneous determination of epinephrine, uric acid and ascorbic acid. *Arab. J. Chem.* **2018**, *11*, 149–158. [CrossRef]
23. Wardak, C.; Pietrzak, K.; Morawska, K.; Grabarczyk, M. Ion-Selective Electrodes with Solid Contact Based on Composite Materials: A Review. *Sensors* **2023**, *23*, 5839. [CrossRef] [PubMed]
24. Khairuddin, R.; Andarini, N.; Oktriviani, W.; Toda'A, M.; Kaope, V.H. Study of Phosphate Adsorption Using Ferrihydrite with Diffusive Gradient in Thin Films Method. *IOP Conf. Ser. Earth Environ. Sci.* **2022**, *1075*, 1. [CrossRef]
25. Vardar, G.; Altikatoılu, M.; Ortaç, D.; Cemek, M.; Işildak, I. Measuring calcium, potassium, and nitrate in plant nutrient solutions using ion-selective electrodes in hydroponic greenhouse of some vegetables. *Biotechnol. Appl. Biochem.* **2015**, *62*, 663–668. [CrossRef]
26. Dai, J.; Zhu, Y.; Tahini, H.A.; Lin, Q.; Chen, Y.; Guan, D.; Zhou, C.; Hu, Z.; Lin, H.-J.; Chan, T.-S.; et al. Single-phase perovskite oxide with super-exchange induced atomic-scale synergistic active centers enables ultrafast hydrogen evolution. *Nat. Commun.* **2020**, *11*, 1. [CrossRef]
27. Savale, P.A. Comparative Study of POA-PVS-DBS-GOD Electrode in Acetate and Phosphate Buffers for Determination of Glucose. *Adv. Sci. Lett.* **2018**, *24*, 5759–5763. [CrossRef]
28. Hamidah, L.N.; Afkar, K.; Rahmayanti, A.; Fitrianah, L. Performance of slow sand filter reactor using geotextile for reducing total n and phosphate. *in IOP Conf. Ser. Earth Environ. Sci.* **2022**, *1211*, 1. [CrossRef]
29. Guan, D.; Xu, H.; Huang, Y.; Jing, C.; Tsujimoto, Y.; Xu, X.; Lin, Z.; Tang, J.; Wang, Z.; Sun, X.; et al. Operando Studies Redirect Spatiotemporal Restructuration of Model Coordinated Oxides in Electrochemical Oxidation. *Adv. Mater.* **2024**, 2413073. [CrossRef]
30. Staaks, D.; Olynick, D.L.; Rangelow, I.W.; Altoe, M.V.P. Polymer-metal coating for high contrast SEM cross sections at the deep nanoscale. *Nanoscale* **2018**, *10*, 22884–22895. [CrossRef]

31. Kozlovskaya, V.; Zavgorodnya, O.; Chen, Y.; Ellis, K.; Tse, H.M.; Cui, W.; Thompson, J.A.; Kharlampieva, E. Ultrathin polymeric coatings based on hydrogen-bonded polyphenol for protection of pancreatic islet cells. *Adv. Funct. Mater.* **2012**, *22*, 3389–3398. [CrossRef]
32. Prakasham, R.S.; Kumar, B.S.; Kumar, Y.S.; Shankar, G.G. Characterization of silver nanoparticles synthesized by using marine isolate Streptomyces albidoflavus. *J. Microbiol. Biotechnol.* **2012**, *22*, 614–621. [CrossRef] [PubMed]
33. Ebuele, V.O.; Congrave, D.G.; Gwenin, C.D. Fitzsimmons-Thoss. Development of a Cobalt Electrode for the Determination of Phosphate in Soil Extracts and Comparison with Standard Methods. *Anal. Lett.* **2018**, *51*, 834–848. [CrossRef]
34. Forano, C.; Farhat, H. Mousty. Recent trends in electrochemical detection of phosphate in actual waters. *Curr. Opin. Electrochem.* **2018**, *11*, 55–61. [CrossRef]
35. Hall, L.B. A text-book of quantitative chemical analysis. *J. Am. Chem. Soc.* **1903**, *25*, 318–319. [CrossRef]
36. Ziegel, E.R. Statistics and Chemometrics for Analytical Chemistry. *Technometrics* **2004**, *46*, 498–499. [CrossRef]
37. Meruva, R.K.; Meyerhoff, M.E. Mixed potential response mechanism of cobalt electrodes toward inorganic phosphate. *Anal. Chem.* **1996**, *68*, 2022–2026. [CrossRef]
38. Topcu, C.; Caglar, B.; Onder, A.; Coldur, F.; Caglar, S.; Guner, E.K.; Cubuk, O.; Tabak, A. Structural characterization of chitosan-smectite nanocomposite and its application in the development of a novel potentiometric monohydrogen phosphate-selective sensor. *Mater. Res. Bull.* **2018**, *98*, 288–299. [CrossRef]
39. Özkütük, E.B.; Yıldız, B.; Gündüz, M.; Hür, E. Phosphate-imprinted polymer as an efficient modifier for the design of ion-selective electrode. *J. Chem. Technol. Biotechnol.* **2021**, *96*, 2604–2609. [CrossRef]
40. Bhat, K.S.; Nakate, U.T.; Yoo, J.Y.; Wang, Y.; Mahmoudi, T.; Hahn, Y.B. Nozzle-Jet-Printed Silver/Graphene Composite-Based Field-Effect Transistor Sensor for Phosphate Ion Detection. *ACS Omega* **2019**, *4*, 8373–8380. [CrossRef]
41. Ratkovski, G.P.; Nascimento, K.T.O.D.; Pedro, G.C.; Ratkovski, D.R.; Gorza, F.D.S.; da Silva, R.J.; Maciel, B.G.; Mojica-Sánchez, L.C.; de Melo, C.P. Spinel Cobalt Ferrite Nanoparticles for Sensing Phosphate Ions in Aqueous Media and Biological Samples. *Langmuir* **2020**, *36*, 2920–2929. [CrossRef] [PubMed]

C Journal of Carbon Research

Article

Development and Characterization of Biomass-Derived Carbons for the Removal of Cu^{2+} and Pb^{2+} from Aqueous Solutions

Vahid Rahimi [1,2], Catarina Helena Pimentel [1,2], Diego Gómez-Díaz [1], María Sonia Freire [1], Massimo Lazzari [2] and Julia González-Álvarez [1,*]

[1] Department of Chemical Engineering, School of Engineering, Universidade de Santiago de Compostela, Rúa Lope Gómez de Marzoa S/N, 15782 Santiago de Compostela, Spain; vah.rahimi@usc.es (V.R.); catarinahelenada.silveiramirandaguedes@usc.es (C.H.P.); diego.gomez@usc.es (D.G.-D.); mariasonia.freire@usc.es (M.S.F.)
[2] Department of Physical Chemistry, Universidade de Santiago de Compostela, Av. das Ciencias s/n, 15782 Santiago de Compostela, Spain; massimo.lazzari@usc.es
* Correspondence: julia.gonzalez@usc.es

Abstract: This research explores the synthesis and application of carbon-based adsorbents derived from olive stones and almond shells as low-cost biomass precursors through carbonization at 600 °C combined with chemical activation using KOH, H_3PO_4, and $ZnCl_2$ with carbon/activating agent (C/A) ratios of 1:2 and 1:4 (*w*/*w*) at 850 °C for the removal of Cu^{2+} and Pb^{2+} ions from aqueous solutions. The carbons produced were characterized using different techniques including SEM-EDX, FTIR, XRD, BET analysis, CHNS elemental analysis, and point of zero charge determination. Batch-mode adsorption experiments were carried out at adsorbent doses of 2 and 5 g L^{-1}, initial metal concentrations of 100 and 500 mg L^{-1}, and natural pH (around 5) with agitation at 350 rpm and 25 °C for 24 h. KOH-activated carbons, especially at a 1:4 (*w*/*w*) ratio, exhibited superior adsorption performance mainly due to their favorable surface characteristics and functionalities. Pb^{2+} was entirely removed (100%) at the highest initial concentration of 500 mg L^{-1} and an adsorbent dosage of 5 g L^{-1}, while for Cu^{2+}, the maximum adsorption efficiency was 86.29% at an initial concentration of 100 mg L^{-1} and a dosage of 2 g L^{-1}. The results of this study will help advance knowledge in the design and optimization of adsorption processes for heavy metal removal, benefiting industries seeking green technologies to mitigate environmental pollution.

Keywords: biomass; carbon adsorbents; chemical activation; copper; lead; adsorption

1. Introduction

Heavy metals' increasing contamination of water bodies poses a serious environmental threat, particularly due to the toxic nature of metals such as copper (Cu^{2+}) and lead (Pb^{2+}). These metals, commonly found in wastewater and leach liquor streams from mining, metallurgy, battery manufacturing, electronics production, and industrial discharge, are non-biodegradable, persistent, and harmful to both the environment and human health [1,2]. Elevated levels of Cu^{2+} and Pb^{2+} in water sources can lead to bioaccumulation in living organisms, resulting in severe health issues such as neurological disorders, kidney damage, and other chronic conditions [2–4].

Moreover, the global rise in electronic waste (e-waste) has become a critical environmental challenge, with waste printed circuit boards (PCBs) representing a significant portion due to their complex composition and the hazardous substances they contain [5].

PCBs are integral components in electronic devices such as computers, smartphones, and televisions, and they are often rich in valuable and toxic metals including Cu^{2+} and Pb^{2+} [6–8]. The improper disposal or recycling of PCBs can result in the release of these toxic metals into the environment, especially into soil and water systems, through leachates from landfills or unsuitable industrial processing, creating significant environmental and public health risks [9,10]. Therefore, removing these metals from wastewater and PCB leachates is crucial for protecting the environment and recovering valuable resources.

Among the various treatment methods for heavy metal removal, adsorption has gained widespread recognition due to its high efficiency, simplicity, and ability to remove a wide range of pollutants [11,12]. Activated carbon is widely recognized as a highly effective adsorbent for removing heavy metals from aqueous solutions, owing to its well-developed porosity, extensive surface area, diverse surface chemistry properties, strong surface reactivity, and significant adsorptive capacity [2,13]. However, traditional activated carbon is often derived from non-renewable sources such as coal, making its production costly and environmentally unsustainable.

In recent years, the focus has shifted towards the use of biomass-derived activated carbons as an eco-friendly and cost-effective alternative, including orange peels [3], date palm leaflets and seeds [14,15], pinewood sawdust [16], and pistachio green hulls [17], among others, for the removal of Cu^{2+} and Pb^{2+} from aqueous solutions. Carbonization and activation processes can convert biomass materials into highly porous carbons with tailored surface properties suitable for adsorption applications. Moreover, using biomass waste as a precursor for activated carbon production aligns with the principles of waste valorization and a circular economy, making it an environmentally responsible approach [18].

The adsorption capacity of activated carbons has been greatly influenced by the type of raw materials, the methods used in their production, and the specific conditions during fabrication [19]. The carbonization process is typically performed prior to precursor activation, where the material is exposed to pyrolysis, a thermal treatment that enhances its carbon content [20]. During pyrolysis, moisture, hydrogen, volatiles, and aromatics with low molecular weight are released, forming a stable carbon-rich structure [21]. The carbonaceous material obtained from pyrolysis normally has limited porosity, as most of the pores are occupied with tar-like substances [21]. Therefore, an activation process is necessary to enhance and develop the specific and desired carbon properties. Activation treatment is applied physically or chemically to improve porosity and develop oxygen-containing functional groups on the surface of activated carbons [22]. KOH, $ZnCl_2$, and H_3PO_4 as chemical activating agents, and CO_2 and steam as physical activating agents can be employed to activate the carbonaceous materials. Comparing physical and chemical activation techniques for preparing activated carbon and assessing their adsorption performance, chemical activation was found to yield superior results, particularly in terms of the total pore volume, specific surface area, and porosity enhancement [23].

This work is focused on the preparation of carbon-based adsorbents from olive stones (OS) and almond shells (AS) as low-cost, eco-friendly, and available biomass precursors to analyze and evaluate their future potential for the removal of Cu^{2+} and Pb^{2+} from wastewater or leach liquor streams. Utilizing biomass materials as precursors for carbon production promotes sustainability while addressing the dual environmental challenges of e-waste management and wastewater treatment. The influences of the type of activating agent (KOH, H_3PO_4, or $ZnCl_2$) and the carbon/activating agent (C/A) ratio were examined to achieve better adsorption characteristics for each individual metal. A detailed characterization of the biomass-derived carbons was carried out to provide insights into the physical and chemical properties of the adsorbents, such as the surface area, pore structure, functional groups, and crystallinity, which are critical for understanding their adsorption behavior.

2. Materials and Methods

2.1. Materials

Olive stones and almond shells, provided by Fertínez Company (Jaén, Spain) and Biogramasa Company (Granada, Spain), respectively, were utilized as biomass precursors for fabricating carbons. Potassium hydroxide (KOH 85%, Panreac, Darmstadt, Germany), zinc chloride ($ZnCl_2$ 98%, Scharlau, Barcelona, Spain), and phosphoric acid (H_3PO_4 85%, Panreac, Darmstadt, Germany) were used as activating agents. Hydrochloric acid (HCl 37%, Panreac, Darmstadt, Germany) and sodium hydroxide (NaOH 98%, Sigma Aldrich, Saint Louis, MO, USA) were employed to eliminate the excess of activating agents after activation and to determine the point of zero charge of carbons. Cu^{2+} and Pb^{2+} aqueous solutions were prepared by dissolving copper nitrate trihydrate ($Cu(NO_3)_2{\cdot}3H_2O$ 99%, Fluka, Seelze, Germany) and lead nitrate ($Pb(NO_3)_2$ 99.5%, Scharlau, Barcelona, Spain), respectively, in double-distilled water. All chemical reagents were of analytical grade.

2.2. Fabrication of Biomass-Derived Carbons

A total of 14 carbon materials were produced using olive stones and almond shells under various conditions outlined in Table 1. A specific mass of the precursor (200 g of olive stones and 140 g of almond shells) was first placed in an alumina recipient and subjected to pyrolysis inside an oven (Nabertherm GmbH, Lilienthal, Germany) under a flow of N_2 (34 L h^{-1}) with a temperature ramp of 5 °C min^{-1} until reaching a temperature of 600 °C, which was kept for 1 h.

Table 1. Carbons produced from biomass precursors for the removal of Cu^{2+} and Pb^{2+}.

Precursor	Carbon	Activating Agent	C/A Ratio (*w*/*w*)
Olive Stones	OSC	Non-activated	-
	OSACK1	KOH	1:2
	OSACK2		1:4
	OSACP1	H_3PO_4	1:2
	OSACP2		1:4
	OSACZ1	$ZnCl_2$	1:2
	OSACZ2		1:4
Almond Shells	ASC	Non-activated	-
	ASACK1	KOH	1:2
	ASACK2		1:4
	ASACP1	H_3PO_4	1:2
	ASACP2		1:4
	ASACZ1	$ZnCl_2$	1:2
	ASACZ2		1:4

OSC: olive stones carbon; OSACK: olive stones KOH-activated carbon; OSACP: olive stones H_3PO_4-activated carbon; OSACZ: olive stones $ZnCl_2$-activated carbon; ASC: almond shells carbon; ASACK: almond shells KOH-activated carbon; ASACP: almond shells H_3PO_4-activated carbon; ASACZ: almond shells $ZnCl_2$-activated carbon.

KOH, $ZnCl_2$, and H_3PO_4 were then used as activating agents to activate the materials obtained from the carbonization process. The KOH- and $ZnCl_2$-activated carbons were produced by mixing the biochar with the corresponding quantities of KOH and $ZnCl_2$ through milling, achieving C/A ratios of 1:2 and 1:4 (*w*/*w*). Subsequently, the mixtures obtained were introduced into the oven and pyrolyzed at 850 °C for 2 h with the same temperature ramp and N_2 flow rate as on the carbonization process (5 °C min^{-1} and 34 L h^{-1}, respectively). The materials, once cooled, were rinsed with 0.5 M HCl aqueous solutions, followed by double deionized water until a constant pH was achieved. To fabricate H_3PO_4-activated carbons, a certain volume of 85% H_3PO_4 solution was mixed with distilled water to obtain the corresponding C/A ratios (1:2 and 1:4 (*w*/*w*)). Then, a specific mass of pre-milled carbonized materials was put in contact with the diluted H_3PO_4 solution with a solid/liquid ratio of 100 g L^{-1} and agitated at room temperature for 6 h. The materials obtained were then pyrolyzed in the oven using the procedure previously described for KOH and $ZnCl_2$ activation. After cooling, the H_3PO_4-activated carbons were washed using double-distilled water until constant pH was reached. After the washing stage, all of the biomass-based carbons prepared were dried at 105 °C for 24 h.

2.3. *Characterization of Biomass-Derived Carbons*

The carbons prepared in this study were characterized through various approaches that enable the analysis of metal adsorption data. BET surface area and porosity determination was carried out using a volumetric sorption analyzer (Micromeritics ASAP 2020, Norcross, GA, USA), analyzing N_2 adsorption–desorption measurements at −196.15 °C, together with CO_2 adsorption data at 0 °C. Previously, the carbons were degassed under vacuum at 300 °C for 8 h, employing a temperature ramp of 10 °C min^{-1}. The specific surface area was determined from N_2 and CO_2 adsorption isotherms using the BET equation. The total pore volume was obtained based on the quantity of N_2 adsorbed at a relative pressure close to saturation ($P/P_0 \approx 0.99$), ensuring a representative measure of the entire porosity network. To thoroughly characterize the porosity, the pore size distribution and total pore volume were further refined using the two-dimensional non-local density functional theory (2D-NLDFT) model, analyzing both CO_2 and N_2 adsorption data simultaneously. This model was selected for its precision in differentiating micropores (≤2 nm) from mesopores (2–50 nm), leveraging a two-dimensional adsorption framework to account for surface heterogeneity and pore connectivity effects. The micropore volume was then calculated by subtracting the combined mesopore and macropore (≥50 nm) volumes from the total pore volume.

Scanning electron microscopy combined with energy-dispersive X-ray spectroscopy (SEM-EDX) was employed to investigate the carbons' morphological characteristics and surface chemical composition. SEM imaging was carried out using a high-resolution field emission scanning electron microscope (FESEM, Ultra-Plus Zeiss, Oberkochen, Germany), enabling detailed surface analysis. Elemental mapping and point analysis were performed at multiple locations across the carbon surface to ensure homogeneity in the composition profiles.

The samples were exposed to a high-precision CHNS elemental analyzer (Thermo Flashsmart, Waltham, MA, USA) to determine the carbon (C), hydrogen (H), nitrogen (N), sulfur (S), and oxygen (O) (calculated by difference) percentages. The analysis was conducted in a dynamic flash combustion mode, where the samples were introduced into a high-temperature reactor in the presence of pure oxygen.

The X-ray diffraction (XRD) technique was employed to characterize the crystallinity of the samples, and the diffraction measurements were performed using an Empyrean diffractometer (Malvern, Worcestershire, United Kingdom) from the Panalytical brand. X-rays were generated from a sealed tube with a Cu anode (λ ($K_{\alpha 1}$) = 1.5406 Å), and the radiation emitted by the sample was obtained using a solid-state "PIXcel3D" detector (Malvern, Worcestershire, United Kingdom). The electron diffraction patterns were collected in the angular range of $5° < 2\theta < 80°$ with a step size of 0.04° and a counting time of 1.75 s per step. The HighScore Plus Software Version 3.0d was used for data processing.

Fourier-transform infrared (FTIR) spectroscopy (VARIAN FTIR 670 spectrometer, Agilent, Santa Clara, CA, USA) was used to characterize the chemical bonds and surface functional groups generated on the surface of carbons. The sample pellets were prepared by mixing carbon with potassium bromide (KBr) using a carbon/KBr ratio of 0.4:600 (mg/mg), and the FTIR spectra were obtained within a wide spectral range of 400–4000 cm^{-1}, allowing for the analysis of a variety of chemical bonds.

To determine the point of zero charge (pH_{pzc}) of the carbons, several flasks were filled with aqueous solutions (10 mL) of 0.01 M NaCl, adjusted to various pH levels (ranging from 2 to 12) by adding varying concentrations of NaOH and HCl (0.1–2 M). After setting the desired pH, 10 mg of the sample was introduced into each flask and shaken at 350 rpm and 25 °C using an orbital air bath shaker (VWR Cienytech, Santiago de Compostela, Spain) for 48 h to achieve adsorption equilibrium. The final pH was then recorded, and the pH_{pzc} was identified as the intersection of the curve plotting ΔpH (initial pH–final pH) versus initial pH, with the x-axis representing the initial pH.

2.4. Adsorption Experiments

Metal adsorption experiments were carried out in batch mode to assess the capability of the prepared carbon-based adsorbents to remove Cu^{2+} and Pb^{2+} ions from aqueous solutions by analyzing their metal adsorption capacity and efficiency. Stock solutions of Cu^{2+} and Pb^{2+} (1000 mg L^{-1}) were first prepared by dissolving precise amounts of $Cu(NO_3)_2.3H_2O$ and $Pb(NO_3)_2$ in double-distilled water. Then, they were subjected to dilution to obtain the working solutions at specified concentrations. The sorption experiments were conducted in an orbital air bath shaker using 25 mL of the metal solution at initial concentrations of 100 or 500 mg L^{-1}, which was put in contact with the corresponding amount of the adsorbent at adsorption doses of 2 or 5 g L^{-1} and maintained at natural pH (4.94 and 5.33 for Cu^{2+}, and 5.24 and 5.42 for Pb^{2+} solutions corresponding to initial concentrations of 500 and 100 mg L^{-1}, respectively), an agitation rate of 350 rpm and 25 °C for 24 h. This time was selected based on preliminary experiments and prior studies found in the literature [14,24] to ensure sufficient time for adsorption equilibrium to be reached. After reaching adsorption equilibrium, the carbon materials were separated from solutions by filtration, and the metal concentration of the filtrates was determined through flame atomic absorption spectrometry (FAAS, GBC Scientific, Victoria, Australia). A schematic representation of the adsorption process for the removal Cu^{2+} and Pb^{2+} ions from aqueous solutions is illustrated in Figure 1.

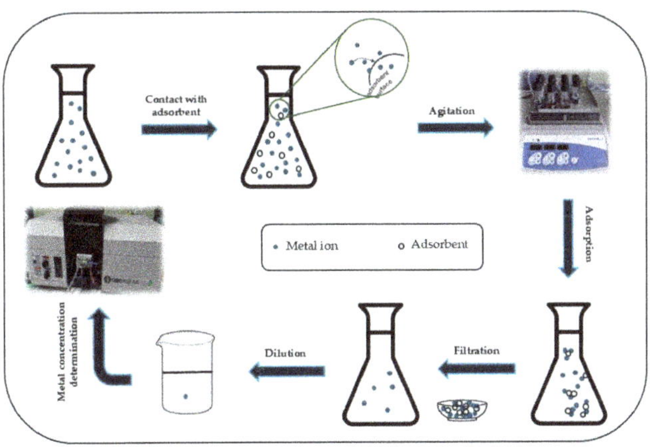

Figure 1. Schematic representation of the adsorption process for the removal of Cu^{2+} and Pb^{2+} ions.

All of the experiments were carried out in triplicate to guarantee the consistency of the adsorption data, and the mean value was calculated for data analysis. The adsorption capacity and efficiency at equilibrium of the carbons prepared were calculated using Equations (1) and (2), respectively.

$$q_e = \left(\frac{C_i - C_e}{m}\right) \times V \quad (1)$$

$$AE(\%) = \left(\frac{C_i - C_e}{C_i}\right) \times 100 \quad (2)$$

where q_e represents the amount of metal adsorbed at equilibrium (mg g^{-1}); C_i is the initial metal concentration at t (time) = 0 (mg L^{-1}); C_e is the final concentration of metals in solution at equilibrium (mg L^{-1}); V is the volume of metal solution (L); m represents the mass of adsorbent used (g); and AE denotes the adsorption efficiency of the adsorbents at equilibrium (%).

2.5. Statistical Analysis

As mentioned earlier, all adsorption experiments were performed in triplicate, and the results were averaged. The presence of significant differences in the values of the adsorption capacity and efficiency of carbons for Cu^{2+} and Pb^{2+} ions were evaluated using one-way analysis of variance (ANOVA), depending on the existence of variance equality and applying the Tukey or T3 Dunnett tests. All statistical analyses were carried out at the 95% confidence level using IBM SPSS Statistics 25 software.

3. Results and Discussion

3.1. Carbon Characterization

The carbons' physical, chemical, and structural characteristics were analyzed through different techniques. The surface morphological features of the carbons were assessed using SEM, and the corresponding images are shown in Figures 2 and 3. These figures illustrate the surface texture and porosity variations of non-activated and activated carbons produced using various activating agents at C/A ratios of 1:2 (*w*/*w*) and 1:4 (*w*/*w*). The results demonstrate the contrasting morphological characteristics and structural changes after activating carbons, especially for the KOH-activated ones, indicating specific modifications in the external surface. Hence, a higher degree of porosity and cavities was observed for activated carbons than for non-activated ones. Moreover, the carbons activated with KOH exhibited a more porous texture than the H_3PO_4- and $ZnCl_2$-activated carbons with scattered cracks generated on their surface, which could result in an enhancement in the number of active surface sites and efficient adsorption of adsorbate molecules [25]. The KOH-activated carbons displayed a significantly porous structure and altered morphology, featuring a monolithic framework and expanded carbon matrix lattices, which could be driven by the mobility and reactivity of metallic potassium (K), as evidenced by the SEM images shown in Figures 2 and 3 (images B and C). The carbons activated with H_3PO_4 exhibited a uniform structure with certain surface modifications, characterized by the formation of small porous regions and randomly distributed pores and craters, which could be due to the diffusion and reaction of H_3PO_4 during the thermochemical activation process at 850 °C, as depicted in Figures 2 and 3 (images D and E) [26]. Some primary pores and an altered porous framework were also identified for $ZnCl_2$-activated carbons (Figures 2 and 3 (images F and G)). All of the activated carbons were characterized by the presence of mesopores and a significant proportion of micropores embedded within their surface structure, demonstrating a clear alignment with prior research findings [27–29].

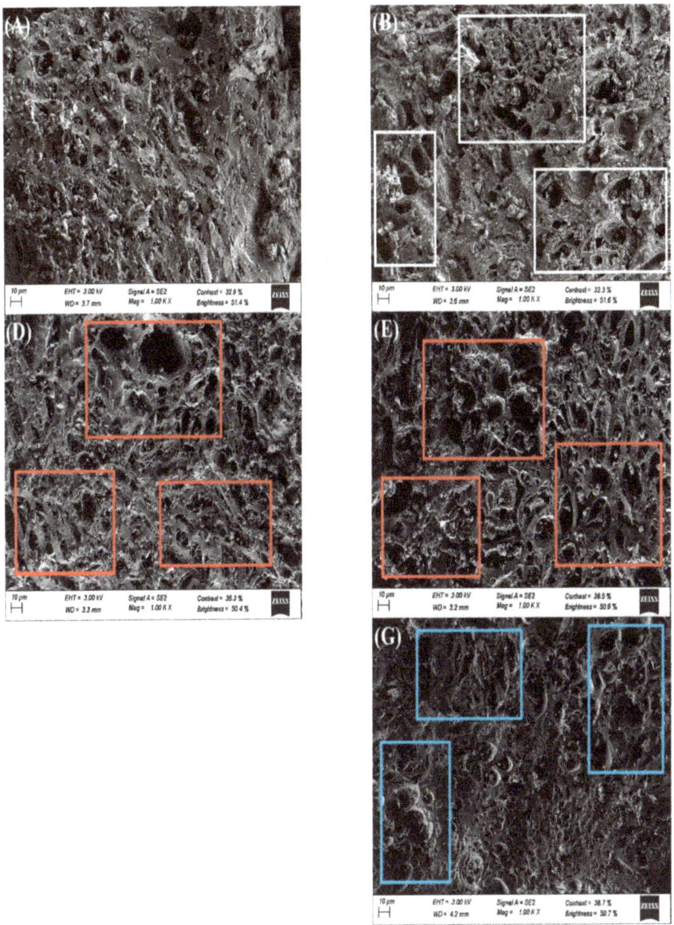

Figure 2. SEM images of carbons produced from olive stones. (**A**) OSC; (**B**) OSACK1; (**C**) OSACK2; (**D**) OSACP1; (**E**) OSACP2; (**F**) OSACZ1; (**G**) OSACZ2. The white, red, and blue marks are indicative of distinct characteristics, namely a porous texture, uniform structural formations, and the presence of primary pores, respectively.

Moreover, the EDX analysis offers the corresponding surface chemical composition, as presented in Table 2. The findings reveal minor quantities of K, P, Zn, and Cl on the surfaces of the KOH-, H_3PO_4-, and $ZnCl_2$-activated carbons, indicating the effective incorporation of these compounds during the fabrication process. KOH-activated carbons were found to have the lowest content of C and the highest content of O, especially for the carbon fabricated from olive stones with a C/A ratio of 1:4 (w/w) (OSACK2, 53.79% and 35.11% of C and O, respectively). This spatial distribution implies the establishment of chemical bonds, particularly carbon- and oxygen-containing functional groups, with the aromatic structure of the carbon, highlighting the effective integration of these elements at the molecular level [30].

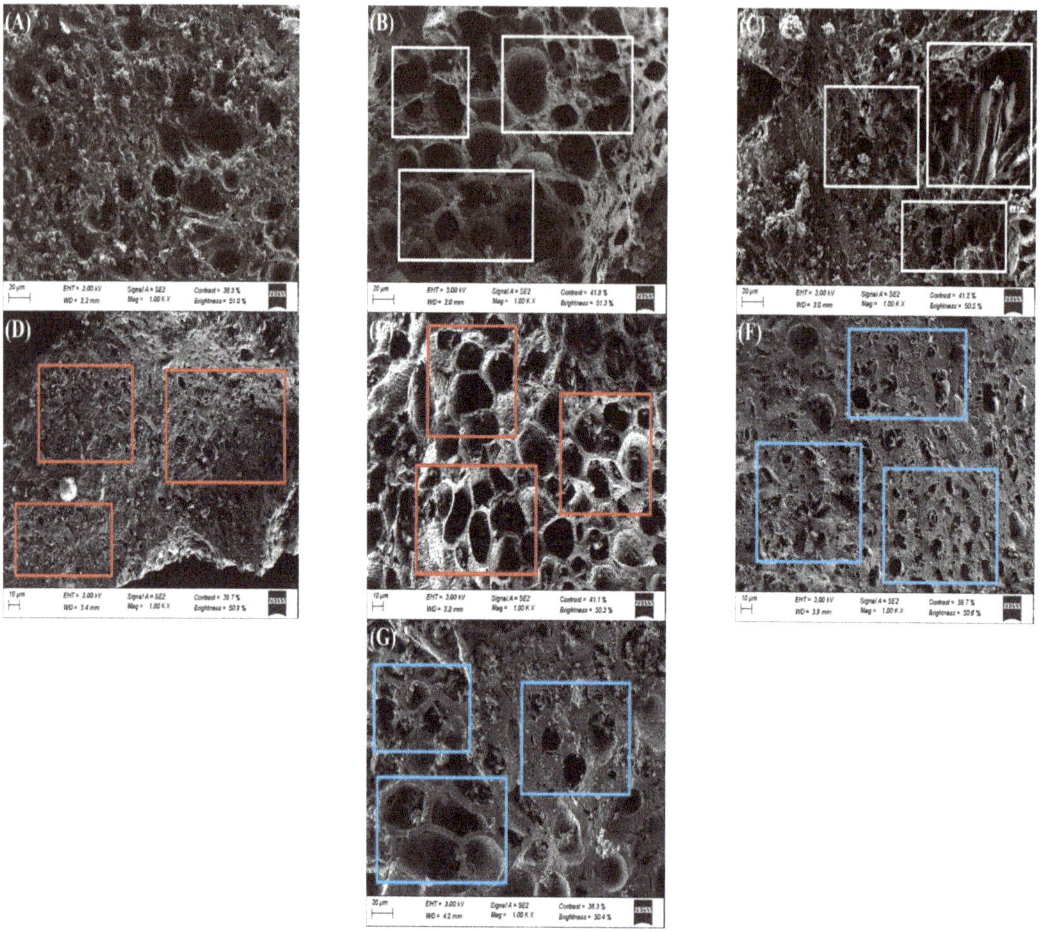

Figure 3. SEM images of carbons produced from almond shells. (**A**) ASC; (**B**) ASACK1; (**C**) ASACK2; (**D**) ASACP1; (**E**) ASACP2; (**F**) ASACZ1; (**G**) ASACZ2. The white, red, and blue marks are indicative of distinct characteristics, namely a porous texture, uniform structural formations, and the presence of primary pores, respectively.

Table 2. EDX surface chemical composition of non-activated and activated carbons produced from biomass precursors.

Carbon	% C	% O	% K	% Ca	% Al	% P	% Cl	Zn
OSC	90.80	8.74	0.16	0.31	-	-	-	-
OSACK1	66.67	26.44	2.96	0.47	3.47	-	-	-
OSACK2	53.79	35.11	3.86	0.24	7.01	-	-	-
OSACP1	84.29	14.51	-	0.08	-	1.12	-	-
OSACP2	83.45	14.99	-	0.15	-	1.42	-	-
OSACZ1	89.94	9.71	-	-	0.05	-	0.15	0.15
OSACZ2	88.90	10.69	-	-	0.09	-	0.11	0.20
ASC	87.80	10.65	0.87	0.67	-	-	-	-
ASACK1	72.03	23.52	1.60	0.40	2.36	-	-	-
ASACK2	75.65	19.37	3.44	0.39	1.16	-	-	-
ASACP1	82.80	16.73	-	0.15	-	0.32	-	-
ASACP2	83.87	15.39	-	0.05	-	0.70	-	-
ASACZ1	82.83	14.92	-	0.20	0.18	-	1.36	0.46
ASACZ2	86.48	11.89	-	0.07	0.13	-	1.09	0.27

Based on the N_2 adsorption–desorption isotherms at −196.15 °C, shown in Figure 4, certain differences in the shape and magnitude of the isotherms can be observed depending on the activating agent used to produce the carbon adsorbents. Regarding the N_2 adsorption capacity, the activated carbons treated with KOH demonstrated a high amount of N_2 adsorbed on the surface, indicating that these materials possessed the largest surface area. In all cases, the N_2 adsorption–desorption isotherms fell under type I, which suggests a significant presence of microporosity in the solids. The isotherms of the KOH-activated carbons displayed a slightly different shape compared to the non-activated ones and those activated with other agents, particularly showing a broader knee. This suggests that the microporosity was generated by micropores in the highest range of the pore diameter (supramicroporosity), while the carbons activated with H_3PO_4 and $ZnCl_2$ appeared to have smaller pores. Moreover, the use of olive stone as a precursor tended to generate a narrower knee, especially when using a 1:2 (w/w) C/A ratio.

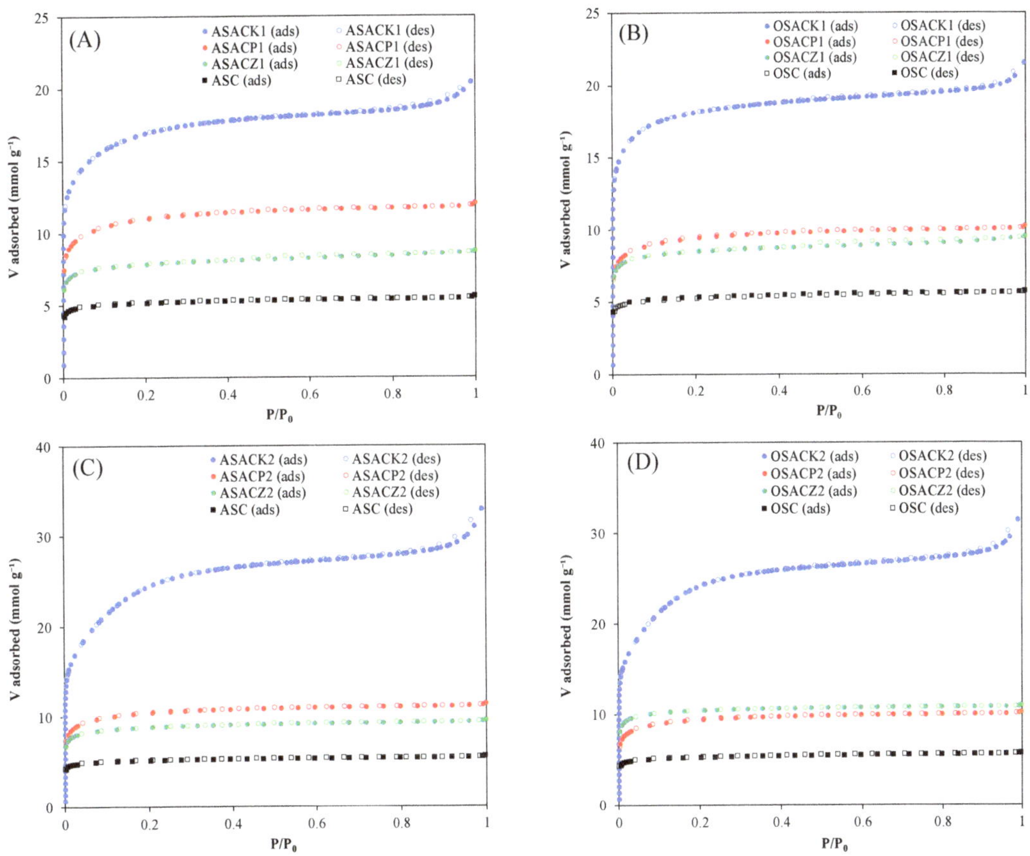

Figure 4. N_2 adsorption–desorption isotherms at −196.15 °C for carbons produced from biomass precursors. (**A**,**C**) almond shell-based carbons; (**B**,**D**) olive stone-based carbons.

On the other hand, the plateau in the isotherms for KOH-activated carbons did not reach a constant value, and a slight increase in the amount of N_2 adsorbed could indicate the presence of a certain degree of mesoporosity.

Regarding the amount of activating agent used during the fabrication process, it is evident that the isotherms for the KOH-activated carbons changed noticeably as the quantity of KOH was increased, with the knee extending over a larger relative pressure range and an increase in the magnitude of adsorbed nitrogen. The activated carbons treated with $ZnCl_2$ exhibited a similar trend to that observed for the KOH-activated carbons, with an increase in the amount of N_2 adsorbed, whereas no influence was observed for the H_3PO_4-activated carbons.

CO_2 adsorption isotherms at 0 °C were obtained to supplement the conclusions drawn from the N_2 adsorption studies at −196.15 °C. The experimental data (Figure 5) show that the KOH-activated carbons achieved higher CO_2 adsorption values, although this difference is mainly observed in the high-pressure range. The presence of ultramicropores is associated with the amount of CO_2 adsorbed [31], but preferably at relative pressures below 0.02. Carbons activated with H_3PO_4 exhibited similar isotherms, suggesting a similar microporosity. For carbons activated with $ZnCl_2$, a substantial increase in the amount of CO_2 adsorbed was observed as the C/A ratio increased, matching the behavior seen in the N_2 adsorption studies.

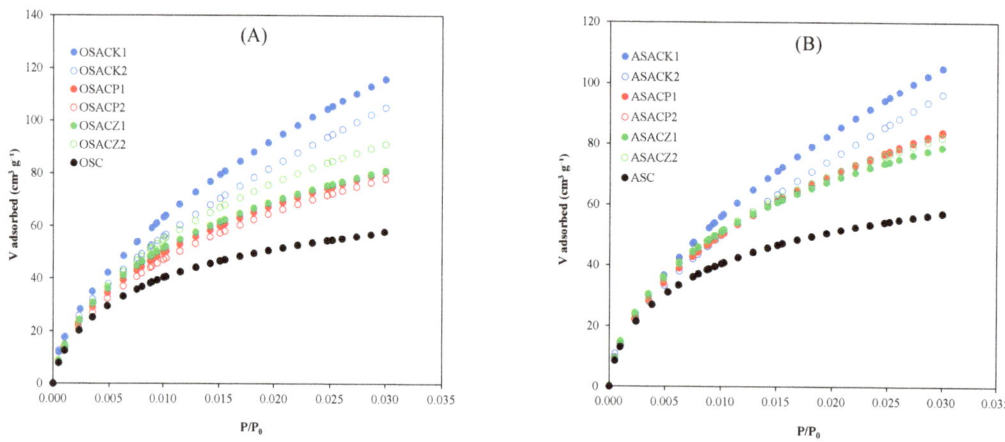

Figure 5. CO_2 adsorption isotherms at 0 °C. (**A**) olive stone-based carbons; (**B**) almond shell-based carbons.

As previously discussed, the experimental data obtained for N_2 and CO_2 adsorption were used to generate the information presented in Table 3. As mentioned earlier, the use of KOH tended to produce carbons with significantly greater surface areas (determined using N_2), which increased as more KOH was used in the activation process for both biomass precursors. This is consistent with the well-known ability of KOH activation to develop a highly porous network, resulting in substantial increases in both microporosity and mesoporosity. Considering the effect of the activating agent on the surface area magnitude, the trend observed was KOH > H_3PO_4 > $ZnCl_2$ for the almond shell-based activated carbons and KOH > $ZnCl_2$ > H_3PO_4 for the olive stone-based activated carbons. While KOH and $ZnCl_2$ allowed the modification of the surface area value when the C/A ratio was varied, H_3PO_4 did not show that influence. The H_3PO_4- and $ZnCl_2$-activated carbons yielded moderate increases in surface area, promoting micropore formation without significantly enlarging mesopores.

Moreover, the total pore volume and pore size provide further insight into the structural transformations induced by activation. The KOH-activated carbons had the highest total pore volumes compared to the non-activated and H_3PO_4- and $ZnCl_2$-activated carbons (Table 3). This trend highlights that KOH activation increases the surface area and substantially amplifies the overall pore volume, enhancing the material's capacity to retain adsorbates. The greatest average pore diameter was obtained for KOH-treated samples in the range of 4.59 to 5.34 nm, indicating the introduction of wider mesopores alongside micropores. This bimodal pore size distribution is advantageous for applications requiring both a high surface area for adsorption and mesoporous channels for rapid molecular transport. The difference in pore size between the almond shell- and olive stone-derived carbons can probably be attributed to the intrinsic characteristics of the original materials and their behavior during the carbonization and activation processes [32].

Table 3. Textural characteristics of carbons produced from biomass precursors.

			P/P_0 = 0.99	DFT	2DNLDFT			
Carbon	S_{BET} (N_2) (m^2 g^{-1})	S_{BET} (CO_2) (m^2 g^{-1})	V_{total} (cm^3 g^{-1})	D_p (nm)	V_{total} (cm^3 g^{-1})	V_{micro} (cm^3 g^{-1})	V_{meso} (cm^3 g^{-1})	Micro Φ (%)
OSC	427.5	294.8	0.197	1.84	0.183	0.171	0.012	93.4
OSACK1	1443.5	760.8	0.747	5.34	0.676	0.575	0.090	85.1
OSACK2	1955.3	672.8	1.091	4.72	1.024	0.695	0.308	67.9
OSACP1	714.9	464.2	0.352	2.53	0.323	0.302	0.021	93.5
OSACP2	711.8	458.0	0.351	2.53	0.322	0.299	0.023	92.9
OSACZ1	712.0	459.2	0.328	2.87	0.304	0.275	0.029	90.5
OSACZ2	849.5	546.0	0.377	2.31	0.348	0.340	0.008	97.7
ASC	405.3	298.2	0.190	2.73	0.177	0.170	0.007	96.0
ASACK1	1350.1	713.6	0.733	5.04	0.729	0.578	0.151	79.3
ASACK2	1979.4	662.4	1.142	4.59	1.033	0.686	0.320	66.4
ASACP1	833.1	500.3	0.412	2.36	0.377	0.350	0.027	92.8
ASACP2	791.2	496.5	0.388	2.48	0.355	0.332	0.023	93.5
ASACZ1	635.5	441.1	0.300	2.80	0.278	0.252	0.026	90.6
ASACZ2	719.6	473.0	0.327	2.25	0.302	0.285	0.017	94.4

S_{BET}, V_{total}, D_p, V_{micro}, V_{meso}, and Φ denote the BET surface area, total pore volume, pore diameter, micropore volume, mesopore volume, and porosity, respectively.

XRD analysis was employed to provide a more detailed understanding of the changes in the crystalline structure of the biomass-derived carbon adsorbents, and the corresponding patterns are presented in Figure 6. From the figure, two prominent diffraction peaks can be observed at around 23–25° and 43.6° for the non-activated carbons and the H_3PO_4- and $ZnCl_2$-activated carbons from both precursors, representing the existence of the (002) and (100) crystal planes of the graphite layer, respectively, caused by weak crystalline phases [33]. The peak observed at around 23–25° confirms the hexagonal and developed aromatic crystallite characteristics, whereas the broad and weak band at about 43.6° represents the microcrystalline and amorphous nature of the carbons with a very low degree of graphitization [27,34]. However, different patterns were detected for the KOH-activated carbons, especially for the carbons activated with a C/A ratio of 1:4 (*w*/*w*). The (002) diffraction plane characteristic of graphite is missing in all KOH-activated carbons, which may be due to the structural disintegration of the carbon matrix, potentially leading to the disruption of the graphite-like ordering [35]. On the other hand, compared to the other carbons, the intensity of the (100) plane peak (43.8°) was significantly reduced for the KOH-activated carbons, particularly for those with a 1:4 (*w*/*w*) ratio, which almost disappeared, signifying a diminishing degree of graphitization order within the samples [36]. In addition, the diffraction peak observed at roughly 21.5° is likely associated with cellulose. In comparison, the peak at 26.3° is due to the presence of crystalline inorganic phases, which can be associated with the trace elements found within the KOH-activated carbons [36]. The sharp peaks observed in the KOH-activated carbons with a ratio of 1:4 (*w*/*w*) may be ascribed to the formation of additional chemical species within the interlayers of the collapsed hexagonal graphite (carbon) framework [35].

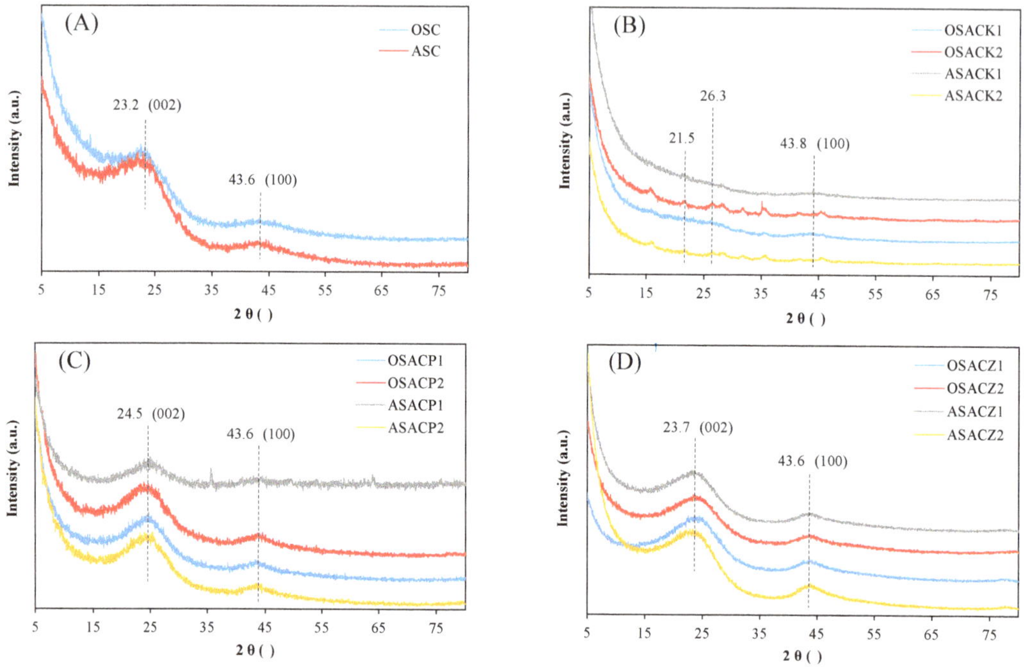

Figure 6. XRD patterns of carbons produced from biomass precursors. (**A**) Non-activated carbons; (**B**) KOH-activated carbons; (**C**) H_3PO_4-activated carbons; (**D**) $ZnCl_2$-activated carbons.

FTIR analysis was carried out to provide insight into the surface chemistry and functional groups present on the surface of the carbons prepared in this study. The FTIR spectra of the samples exhibited several characteristic peaks, indicating the presence of a variety of functional groups, as depicted in Figure 7. The peak observed at around 3433 cm^{-1} is typically attributed to the O-H stretching vibrations of hydroxyl groups, including those associated with alcohols, phenols, and carboxylic acids [2,37]. The non-activated and KOH-activated carbons exhibited a broader intensity at this peak compared to the H_3PO_4- and $ZnCl_2$-activated carbons, suggesting a significant concentration of O–H groups due to their alkaline nature. This observation aligns with the pH_{pzc} data of the carbons, as presented in Table 4, corroborating the consistency between the spectral findings and the surface charge characteristics of the materials. The band located at about 3136–3149 cm^{-1} can be assigned to the N–H stretching mode observed in compounds such as amines or to the O–H bond [35,38]. Moreover, the dual peaks detected at 2920 cm^{-1} and 2852–2862 cm^{-1} correspond to the stretching vibrations associated with aliphatic C–H bonds [39], whereas the peaks obtained within the spectral region of 1711–1733 cm^{-1} are indicative of the C=O stretching vibrations, which are characteristic of carbonyl functionalities found in compounds such as esters and carboxylic acids [35]. The transmittance bands at roughly 1595–1637 cm^{-1} can be ascribed to the in-plane deformation vibrations of the N=H bond or the C=C stretching vibrations, particularly those inherent to aromatic rings [36]. A sharp peak was detected at 1404 cm^{-1}, which corresponds to bending vibrations of alkyl C–H bonds, reflecting the presence of methyl ($-CH_3$) or methylene ($-CH_2-$) groups [40]. The prominent and well-defined band at 1099 cm^{-1} observed in the non-activated and KOH- and H_3PO_4-activated carbons contrasted with the relatively weak and broad peak detected at 1124 cm^{-1} in $ZnCl_2$-activated carbons that can be attributed to C–O stretching vibrations, which are typically associated with alcohols, phenols, ethers, or esters [41,42].

Furthermore, the peaks identified at 796 cm^{-1} and 470 cm^{-1} in the non-activated and KOH- and H_3PO_4-activated carbons are likely attributed to out-of-plane bending vibrations of aromatic C–H bonds [35]. The peaks appearing at 1099 cm^{-1} and 470 cm^{-1} in the carbons activated with H_3PO_4 can also be assigned to the stretching vibrations of the O–P–O and P=O bands, respectively [43]. By contrast, the characteristics peak observed at 592 cm^{-1} for the $ZnCl_2$-activated carbons can be ascribed to Zn–O stretching vibrations [44].

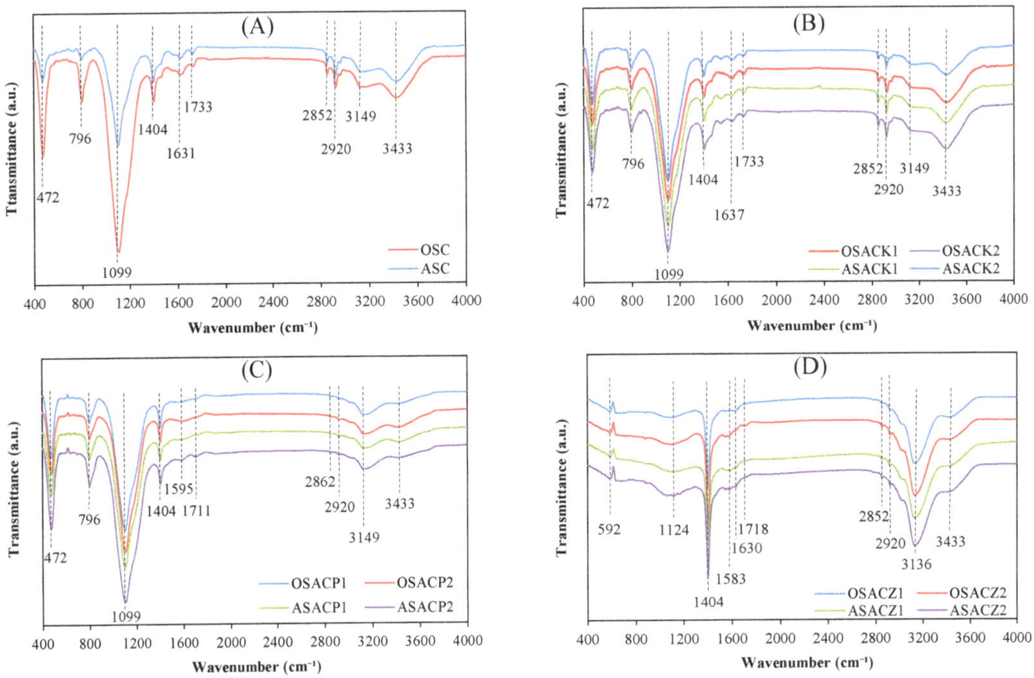

Figure 7. FTIR spectra of carbons produced from biomass precursors. (**A**) Non-activated carbons; (**B**) KOH-activated carbons; (**C**) H_3PO_4-activated carbons; (**D**) $ZnCl_2$-activated carbons.

Table 4. Points of zero charge of carbons produced from biomass precursors.

Carbon	pH_{pzc}
OSC	7.3
OSACK1	8.1
OSACK2	8.4
OSACP1	3.5
OSACP2	2.9
OSACZ1	6.1
OSACZ2	5.5
ASC	7.7
ASACK1	8.0
ASACK2	8.4
ASACP1	3.4
ASACP2	3.4
ASACZ1	6.2
ASACZ2	5.9

Point of zero charge (pH_{pzc}) is a key property in understanding how carbon-based materials interact with different ions in aqueous solutions, which reflects the acidity or basicity of the material's surface. The pH_{pzc} is the pH at which the material's surface has a neutral net charge. If the pH of the solution is below the pH_{pzc}, the surface of the carbon material will be positively charged. On the contrary, if the pH is above the pH_{pzc}, the surface will be negatively charged [24]. The points of zero charge of the carbons prepared are given in Table 4. As observed, the non-activated and KOH-activated carbons possessed pH_{pzc} values within a range of 7.3–8.4, indicating that their surface was in slightly basic conditions. This could be due to the introduction of electron-donating groups such as amine or hydroxyl groups on the carbon surface [45]. By contrast, the carbons activated with H_3PO_4 exhibited the lowest pH_{pzc} values (2.9–3.5), suggesting that phosphorus modification greatly increased their surface acidity. The $ZnCl_2$-activated carbons showed intermediate pH_{pzc} values (5.5–6.2), confirming that zinc modification made their surface slightly acidic but not as much as phosphorus, as zinc oxide is amphoteric and can behave as both an acid and a base [46].

The CHNS elemental analysis provides valuable insights into the changes in elemental composition following the chemical activation of olive stone- and almond shell-derived carbons, and the corresponding results are given in Table 5. These changes are critical, as they influence the surface chemistry and the potential applications of the activated carbons. As observed in Table 5, the non-activated carbons (OSC and ASC) exhibited the highest carbon contents, namely 83.36% and 83.14%, respectively, with comparable hydrogen, nitrogen, and oxygen levels. These high carbon percentages reflect the inherent carbon-rich nature of the biochar produced from biomass prior to activation. The KOH activation induced a pronounced carbon content reduction alongside a significant oxygen level increase, a trend consistently observed across both the OSACK- and ASACK-series samples. For instance, within the OSACK series, the carbon content declined markedly from 83.36% in OSC to 66.44% in OSACK1 and further to 55.02% in OSACK2, while oxygen levels increased from 14.40% in OSC to 32.02% and 43.64% in OSACK1 and OSACK2, respectively. Similar patterns were observed in the ASACK series, where carbon content in ASACK1 and ASACK2 was reduced to 66.48% and 56.72%, with corresponding oxygen amounts rising to 32.23% and 41.75%, respectively. The substantial decrease in carbon content and concurrent oxygen increase, which is consistent with the findings obtained from the EDX analysis (Table 2), suggests the incorporation of oxygen-containing functional groups, likely because of KOH's strong oxidizing effect, which enhances pore development and functional group introduction [47]. Activation with H_3PO_4 also reduced the carbon content (70.60–74.18%), albeit not as drastically as KOH activation, and introduced a moderate increase in the oxygen amount (23.54–27.20%). The less-pronounced carbon depletion and oxygenation in H_3PO_4-activated carbons suggest that H_3PO_4 is milder compared to KOH, promoting the formation of phosphorus-containing functional groups that enhance the surface polarity without a significant carbon loss [48]. The $ZnCl_2$-activated carbons demonstrated the highest carbon contents among the activated samples, containing 80.65–82.12%. This indicates that $ZnCl_2$ is less aggressive in depleting carbon compared to KOH and H_3PO_4, likely because $ZnCl_2$ primarily acts as a dehydrating agent, promoting carbonization and cross-linking without extensive oxygen incorporation [49]. Correspondingly, the relatively low oxygen contents (15.57–17.41%) were obtained for $ZnCl_2$-activated carbons. In addition, the significantly low hydrogen and nitrogen contents could be associated with promoting carbon- and oxygen-dominant structures, possibly through gasification reactions that remove other elements [50].

Table 5. CHNS elemental analysis of non-activated and activated carbons produced from biomass precursors.

Carbon	% C	% H	% N	% O *
OSC	83.36	2.00	0.24	14.40
OSAC1	66.44	1.30	0.24	32.02
OSAC2	55.02	1.16	0.18	43.64
OSACP1	74.18	1.90	0.38	23.54

Table 5. *Cont.*

Carbon	% C	% H	% N	% O *
OSACP2	70.66	1.79	0.35	27.20
OSACZ1	80.70	1.20	0.69	17.41
OSACZ2	81.70	1.67	0.22	16.41
ASC	83.14	2.02	0.30	14.54
ASAC1	66.48	1.09	0.20	32.23
ASAC2	56.72	1.42	0.11	41.75
ASACP1	70.60	2.02	0.53	26.85
ASACP2	73.38	2.08	0.43	24.11
ASACZ1	82.12	1.40	0.91	15.57
ASACZ2	80.65	1.72	0.60	17.03

* Calculated by difference.

3.2. Cu^{2+} Adsorption Performance of Biomass-Derived Carbons

The Cu^{2+} adsorption efficiency (AE) and capacity (q_e) of the non-activated and activated carbons were assessed under varying conditions, specifically at two Cu^{2+} concentrations (100 and 500 mg L^{-1}) and two adsorbent dosages (2 and 5 g LL^{-1}), to identify the most effective adsorbent for the removal of Cu^{2+} ions from aqueous solutions (Table 6). The variations in adsorption performance across carbons activated with KOH, H_3PO_4, and $ZnCl_2$ reveal the influence of the activation method and the acquired surface characteristics on the Cu^{2+} removal efficiency. As observed, the non-activated carbons (OSC and ASC) displayed relatively lower AE and q_e values than the KOH- and H_3PO_4-activated carbons under all conditions essayed. At a low Cu^{2+} concentration (100 mg L^{-1}) with an adsorbent dosage of 2 g L^{-1}, OSC and ASC achieved moderate AE values of 29.92% and 43.25%, with q_e values of 14.75 mg g^{-1} and 21.32 mg g^{-1}, respectively. However, increasing the adsorbent dosage to 5 g L^{-1} at 100 mg L^{-1} resulted in a slight increase in AE values (33.44% for OSC and 51.40% for ASC), but significantly lower q_e values (7.21 mg g^{-1} and 11.10 mg g^{-1}, respectively), following a trend similar to that observed in previous studies [51]. At 500 mg L^{-1}, both OSC and ASC showed a decrease in AE values (12.81% and 13.06% at 2 g L^{-1}; 8.80% and 15.33% at 5 g L^{-1}, respectively) with correspondingly lower q_e values. This indicates that non-activated carbons face challenges in efficiently removing Cu^{2+} at high initial concentrations, likely due to limited active sites [52]. ASC's superior Cu^{2+} adsorption efficiency compared to OSC, particularly at the lower concentration of 100 mg LL^{-1}, may be ascribed to its relatively larger average pore size (Table 3).

Table 6. Adsorption performance of biomass-derived carbons for the removal of Cu^{2+} from aqueous solutions.

			C_i = 100 mg L^{-1}–AD = 2 g L^{-1}		C_i = 100 mg L^{-1}–AD = 5 g L^{-1}	
Carbon	Activating Agent	C/A Ratio (*w/w*)	AE, %	q_e, mg g^{-1}	AE, %	q_e, mg g^{-1}
OSC	-	-	29.92 ± 1.15 [bc]	14.75 ± 0.68 [bc]	33.44 ± 2.16 [b]	7.21 ± 0.33 [b]
OSACK1	KOH	1:2	71.66 ± 3.59 [e]	35.33 ± 2.16 [e]	98.74 ± 0.78 [f]	21.33 ± 0.41 [f]
OSACK2		1:4	86.29 ± 2.73 [f]	42.52 ± 1.42 [f]	99.86 ± 0.13 [f]	21.58 ± 0.60 [f]
OSACP1	H_3PO_4	1:2	36.29 ± 2.15 [cd]	17.87 ± 1.33 [cd]	44.27 ± 0.73 [c]	9.14 ± 0.26 [c]
OSACP2		1:4	38.97 ± 1.07 [cd]	19.17 ± 0.44 [cd]	50.05 ± 1.76 [cd]	10.33 ± 0.15 [cd]
OSACZ1	$ZnCl_2$	1:2	23.70 ± 2.07 [ab]	12.33 ± 0.94 [ab]	29.29 ± 2.87 [ab]	5.82 ± 0.65 [ab]
OSACZ2		1:4	22.56 ± 2.56 [ab]	11.80 ± 1.58 [ab]	32.35 ± 2.02 [b]	6.41 ± 0.36 [b]
ASC	-	-	43.25 ± 2.65 [d]	21.32 ± 1.22 [d]	51.40 ± 0.90 [d]	11.10 ± 0.35 [de]
ASACK1	KOH	1:2	66.22 ± 2.47 [e]	32.64 ± 1.47 [e]	96.55 ± 1.58 [f]	20.85 ± 0.44 [f]
ASACK2		1:4	84.63 ± 1.35 [f]	41.69 ± 0.55 [f]	98.18 ± 2.23 [f]	21.20 ± 0.10 [f]
ASACP1	H_3PO_4	1:2	44.12 ± 2.25 [d]	21.73 ± 1.43 [d]	58.77 ± 1.10 [e]	12.13 ± 0.25 [e]
ASACP2		1:4	42.15 ± 3.08 [d]	20.70 ± 0.86 [d]	54.95 ± 0.82 [de]	11.34 ± 0.31 [de]
ASACZ1	$ZcCl_2$	1:2	24.29 ± 2.57 [ab]	12.49 ± 0.86 [ab]	33.14 ± 1.25 [b]	6.58 ± 0.33 [b]
ASACZ2		1:4	15.45 ± 4.22 [a]	8.07 ± 2.30 [a]	25.11 ± 1.83 [a]	4.98 ± 0.43 [a]

Table 6. *Cont.*

Carbon	Activating Agent	C/A Ratio (w/w)	C_i = 500 mg L^{-1}—AD = 2 g L^{-1}		C_i = 500 mg L^{-1}—AD = 5 g L^{-1}	
			AE, %	q_e, mg g^{-1}	AE, %	q_e, mg g^{-1}
OSC	-	-	12.81 ± 0.44 c	32.01 ± 1.78 bcd	8.80 ± 1.98 ab	10.02 ± 2.28 ab
OSACK1	KOH	1:2	21.68 ± 1.73 d	53.75 ± 5.97 e	38.75 ± 3.44 fg	39.34 ± 3.20 ef
OSACK2		1:4	26.88 ± 1.3 d	66.62 ± 5.40 e	48.65 ± 3.56 h	49.40 ± 2.82 g
OSACP1	H_3PO_4	1:2	10.01 ± 0.68 abc	24.40 ± 1.91 abc	19.43 ± 3.73 cd	19.67 ± 3.87 bc
OSACP2		1:4	9.84 ± 1.41 abc	23.94 ± 3.33 abc	22.14 ± 1.99 cd	22.39 ± 2.03 c
OSACZ1	$ZnCl_2$	1:2	7.73 ± 1.39 abc	20.27 ± 3.37 abc	6.57 ± 1.34 ab	6.61 ± 1.21 a
OSACZ2		1:4	4.32 ± 2.15 a	11.40 ± 5.77 a	6.32 ± 2.58 ab	6.34 ± 2.48 a
ASC	-	-	13.06 ± 2.82 c	32.65 ± 7.29 cd	15.33 ± 2.00 bc	16.64 ± 3.07 bc
ASACK1	KOH	1:2	20.42 ± 2.55 d	51.13 ± 8.09 de	34.40 ± 3.19 ef	34.95 ± 3.27 de
ASACK2		1:4	24.92 ± 3.33 d	62.33 ± 10.14 e	45.15 ± 4.11 h	45.83 ± 3.43 fg
ASACP1	H_3PO_4	1:2	12.37 ± 2.90 bc	29.82 ± 5.82 abc	25.27 ± 2.10 de	25.58 ± 2.26 cd
ASACP2		1:4	12.82 ± 2.00 c	31.29 ± 5.22 abcd	25.07 ± 1.48 cde	25.36 ± 1.52 cd
ASACZ1	$ZnCl_2$	1:2	5.21 ± 1.78 ab	13.63 ± 4.41 abc	6.82 ± 3.09 ab	6.83 ± 2.92 a
ASACZ2		1:4	4.42 ± 1.07 a	11.62 ± 2.77 ab	1.34 ± 0.95 a	1.35 ± 0.96 a

AE: adsorption efficiency at equilibrium; q_e: adsorption capacity at equilibrium; C_i: cation initial concentration; AD: adsorbent dose. In each column and at each concentration, different letters imply significant differences among means ($p < 0.05$).

The KOH-activated carbons were found to possess superior adsorption characteristics among all samples, especially for the carbons fabricated with a C/A ratio of 1:4 (*w*/*w*) (OSACK2 and ASACK2) with the highest AE and q_e, underscoring the effectiveness of KOH activation in enhancing the Cu^{2+} adsorption potential. At a 100 mg L^{-1} Cu^{2+} concentration and 2 g L^{-1} dosage, OSACK2 and ASACK2 achieved notable AEs of 86.29% and 84.63% with elevated q_e values of 42.52 mg g^{-1} and 41.69 mg g^{-1}, respectively. Increasing the dosage to 5 g L^{-1} further raised the AE values to near-complete levels (99.86% and 98.18%), although the q_e values decreased to 21.58 mg g^{-1} and 21.20 mg g^{-1}, respectively. At a higher Cu^{2+} concentration (500 mg L^{-1}), the KOH-activated carbons maintained effective performance, particularly OSACK2 and ASACK2, which achieved AEs of 26.88% and 24.92% at 2 g L^{-1}, with q_e values of 66.62 mg g^{-1} and 62.33 mg g^{-1}, respectively. The adsorption efficiencies nearly doubled with an increased dosage of 5 g L^{-1} (48.65% and 45.15% for OSACK2 and ASACK2, respectively), accompanied by high q_e values of 49.40 mg g^{-1} and 45.83 mg g^{-1}, respectively. This implies that KOH activation effectively generates high-energy adsorption sites capable of accommodating greater Cu^{2+} loadings. The findings also demonstrate that the C/A ratio significantly affects the structural and chemical characteristics of the carbon activated by KOH, thereby influencing its Cu^{2+} adsorption performance. When increasing the C/A ratio from 1:2 (*w*/*w*) to 1:4 (*w*/*w*), the amount of copper adsorbed increased, especially under two specific conditions: 100 mg L^{-1} with a 2 g L^{-1} adsorbent dose and 500 mg L^{-1} with a 5 g L^{-1} adsorbent dose, as confirmed by the analysis of significant differences. As shown in Table 3, the surface area and total pore volume of the KOH-activated carbon increased substantially with a higher C/A ratio for both precursors, as the activating agent helped create a more porous framework by etching and expanding existing pores. This high porosity facilitates Cu^{2+} ion diffusion into the carbon matrix, thereby improving the adsorption capacity [53]. Furthermore, the C/A ratio directly impacted the diversity of functional groups on the carbon surface, which plays a critical role in Cu^{2+} binding through chemisorption mechanisms [25]. EDX and CHNS elemental analyses (Tables 2 and 5) revealed a considerably high oxygen content in the KOH-activated carbons, particularly in the olive stone-derived activated carbon produced at a 1:4 (*w*/*w*) ratio. This oxygen concentration could be associated with a significant presence of oxygen-containing functional groups, which exhibit a strong affinity for metal ions, enhancing adsorption through ion exchange and surface complexation mechanisms (Figure 8) [18,54,55]. In other words, the oxygen atoms within these functional groups possess lone electron pairs that can coordinate with metal ions like Cu^{2+} and Pb^{2+} through a mechanism known as surface complexation or chelation (Figure 8). This process involves the donation of electrons from the oxygen atoms to the vacant orbitals of the metal ions, forming stable complexes and binding the ions to the surface [56]. Additionally, these functional groups may also facilitate ion exchange, where hydrogen ions (H+) or other cations on the surface are exchanged with metal ions in solution [57]. Electrostatic interaction and physical adsorption represent additional key mechanisms of carbon-based materials for the removal of Cu^{2+} and Pb^{2+} ions, as illustrated in Figure 8.

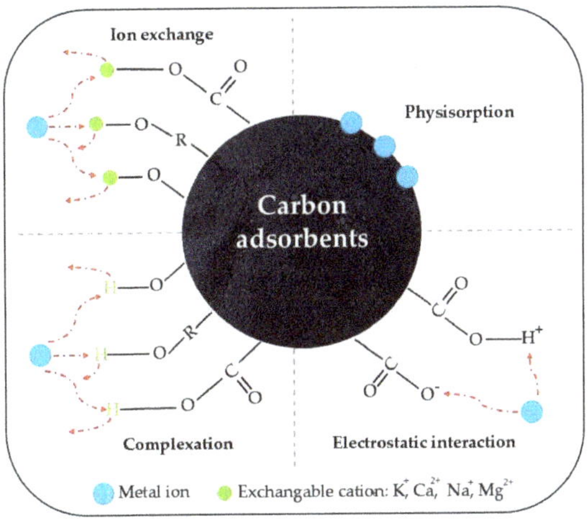

Figure 8. Main adsorption mechanisms of carbon adsorbents for the removal of Cu^{2+} and Pb^{2+} ions.

The carbons activated with H_3PO_4 showed moderate adsorption performance, with the AE and q_e values falling between those of non-activated and KOH-activated carbons. For instance, at a 100 mg L^{-1} Cu^{2+} concentration and 2 g L^{-1} dosage, OSACP2 and ASACP1 achieved AE values of 38.97% and 44.12%, with q_e values of 19.17 mg g^{-1} and 21.73 mg g^{-1}, respectively. When the dosage was increased to 5 g L^{-1}, the AE improved to 50.05% and 58.77%, although the q_e decreased to 10.33 mg g^{-1} and 12.13 mg g^{-1}, respectively. A similar trend was observed at a higher Cu^{2+} concentration, with both the AE and q_e values decreasing. This suggests that while H_3PO_4 activation moderately enhances adsorption sites, it is less effective than KOH at producing the high-affinity sites required for higher Cu^{2+} loadings. Among the H_3PO_4-activated carbons, the almond shell-derived activated carbon fabricated at a 1:2 (w/w) ratio (ASACP1) demonstrated slightly more favorable adsorption performance, especially at a Cu^{2+} concentration of 100 mg L^{-1} and an adsorbent dose of 5 g L^{-1}, which may be attributed to its relatively higher surface area and pore volume (Table 3). Moreover, increasing the C/A ratio from 1:2 (w/w) to 1:4 (w/w) did not result in a substantial enhancement in adsorption capability for either precursor, suggesting no significant differences in the adsorption efficacy between these two ratios, as outlined in Table 6. This outcome may be attributed to minimal changes in the surface chemistry and pore structure despite the higher C/A ratio.

The $ZnCl_2$-activated carbons had the lowest AE and q_e values among the non-activated and activated carbons. For example, at a 100 mg L^{-1} Cu^{2+} concentration and 2 g L^{-1} dosage, OSACZ1 and OSACZ2 achieved AEs of only 23.70% and 22.56%, with corresponding q_e values of 12.33 mg g^{-1} and 11.80 mg g^{-1}, respectively. A similar behavior was observed when the dosage was increased to 5 g L^{-1}, where the AE slightly improved but the q_e remained low, indicating limited adsorption efficiency. At 500 mg L^{-1}, the $ZnCl_2$-activated samples displayed poor adsorption performance (e.g., 7.73% and 4.32% AE for OSACZ1 and OSACZ2, respectively, using an adsorbent dosage of 2 g L^{-1}), highlighting the $ZnCl_2$ activation's limitations in creating high-capacity adsorption sites for copper. This result suggests that $ZnCl_2$ may primarily enhance microporosity rather than the specific types of active sites that favor Cu^{2+} ion adsorption.

In general, the KOH-activated carbons (especially OSACK2 and ASACK2) demonstrated the highest Cu^{2+} adsorption efficiency and capacity, making them suitable for high-performance Cu^{2+} adsorption applications. This suggests that KOH activation produces a more suitable porous structure and surface chemistry that enhances interactions with Cu^{2+} ions. H_3PO_4 activation resulted in moderate improvements in adsorption, which makes it suitable for applications with lower concentration requirements, while $ZnCl_2$ activation was less effective for Cu^{2+} adsorption under the tested conditions. These results highlight the importance of activation method selection for optimizing carbon materials for specific heavy metal adsorption applications.

Based on a previous analysis about the influence of the average pore size upon Cu^{2+} adsorption in the different carbon-based adsorbents and considering possible size-exclusion phenomena, an evaluation of the role of each pore type (total, mesopore, and micropore) was carried out, observing a clear impact of the mesopore volume upon the adsorption efficiency and capacity for all the adsorbents obtained. Figure 9 shows examples of this type of influence, and it is possible to observe that an increase in the mesopore volume in the adsorbent enhances Cu^{2+} adsorption. This behavior is not observed for the micropore volume, indicating that the pore size plays an important role and concluding that part of the specific surface area of the adsorbent is not available for metal adsorption.

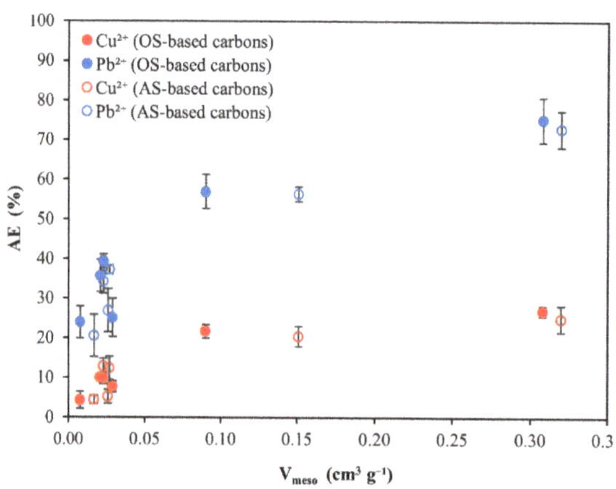

Figure 9. Effect of mesopore volume upon adsorption efficiency (initial metal concentration of 500 mg L^{-1} and adsorbent dose of 2 g L^{-1}).

3.3. *Pb^{2+} Adsorption Performance of Biomass-Derived Carbons*

As in the previous section, the adsorption efficiency (AE, %) and capacity (q_e, mg g^{-1}) of biomass-derived carbons were evaluated for removing Pb^{2+} ions from aqueous solutions under different initial concentrations (100 mg L^{-1} and 500 mg L^{-1}) and adsorbent dosages (2 g L^{-1} and 5 g L^{-1}), and the corresponding results are presented in Table 7. At a 100 mg L^{-1} Pb^{2+} initial concentration and 2 g L^{-1} adsorbent dosage, the KOH-activated carbons displayed the highest adsorption efficiency (AE = 100%) and adsorption capacity (q_e = 52.25–52.28 mg g^{-1}), showing their potential for high removal rates at low dosages. This performance indicates that these activated carbons possess a high surface area and effective pore distribution for Pb^{2+} adsorption. Other carbons, like the non-activated (OSC and ASC) and H_3PO_4-activated carbons (OSACP and ASACP series), also exhibited high AE values of 62.58–75.56% and 73.04–79.28%, respectively, but with slightly lower q_e values (34.55–41.45 mg g^{-1} and 37.83–41.18 mg g^{-1}, respectively). This indicates that while these carbons are effective for Pb^{2+} removal, their adsorption capacity is lower than those of KOH-activated carbons, possibly due to a less-favorable pore structure (in agreement with data in Figure 9) or fewer available functional groups.

Table 7. Adsorption performance of biomass-derived carbons for the removal of Pb^{2+} from aqueous solutions.

			C_i = 100 mg L^{-1}—AD = 2 g L^{-1}		C_i = 100 mg L^{-1}—AD = 5 g L^{-1}	
Carbon	**Activating Agent**	**C/A Ratio (*w*/*w*)**	**AE, %**	**q_e, mg g^{-1}**	**AE, %**	**q_e, mg g^{-1}**
OSC	-	-	62.58 ± 5.70 ^{ab}	34.55 ± 1.98 ^{abcd}	73.28 ± 1.65 ^{ab}	16.34 ± 0.37 ^{abc}
OSACK1	KOH	1:2	100.00 ^{d}	52.25 ± 2.15 ^{e}	–	–
OSACK2		1:4	100.00 ^{d}	52.28 ± 1.98 ^{e}	–	–
OSACP1	H_3PO_4	1:2	75.97 ± 2.83 ^{c}	39.59 ± 0.72 ^{cd}	83.82 ± 1.50 ^{cd}	17.99 ± 1.33 ^{bcd}
OSACP2		1:4	74.44 ± 5.00 ^{c}	38.86 ± 2.60 ^{cd}	86.94 ± 2.07 ^{d}	18.64 ± 1.33 ^{cd}
OSACZ1	$ZnCl_2$	1:2	58.97 ± 2.85 ^{a}	30.82 ± 1.33 ^{ab}	68.30 ± 0.23 ^{a}	14.81 ± 1.15 ^{ab}
OSACZ2		1:4	63.47 ± 2.59 ^{ab}	33.29 ± 2.47 ^{abc}	76.38 ± 2.97 ^{bc}	16.39 ± 1.25 ^{abc}
ASC	-	-	75.56 ± 4.04 ^{c}	41.45 ± 1.28 ^{d}	100.00 ^{e}	22.30 ± 0.01 ^{e}
ASACK1	KOH	1:2	100.00 ^{d}	52.28 ± 1.98 ^{e}	–	–
ASACK2		1:4	100.00 ^{d}	52.25 ± 2.04 ^{e}	–	–
ASACP1	H_3PO_4	1:2	79.28 ± 3.67 ^{c}	41.18 ± 2.62 ^{d}	89.65 ± 2.91 ^{d}	19.25 ± 1.51 ^{cde}
ASACP2		1:4	73.04 ± 1.86 ^{bc}	37.83 ± 0.98 ^{bcd}	87.83 ± 2.96 ^{d}	18.85 ± 1.41 ^{cd}
ASACZ1	$ZcCl_2$	1:2	70.31 ± 1.84 ^{bc}	36.80 ± 2.06 ^{bcd}	78.90 ± 2.66 ^{bc}	16.92 ± 0.91 ^{abcd}
ASACZ2		1:4	53.21 ± 2.76 ^{a}	27.80 ± 1.55 ^{a}	68.05 ± 4.35 ^{a}	14.55 ± 0.57 ^{a}
			C_i = 500 mg L^{-1}—AD = 2 g L^{-1}		**C_i = 500 mg L^{-1}—AD = 5 g L^{-1}**	
Carbon	**Activating Agent**	**C/A Ratio (*w*/*w*)**	**AE, %**	**q_e, mg g^{-1}**	**AE, %**	**q_e, mg g^{-1}**
OSC	-	-	27.02 ± 3.05 ^{abc}	71.21 ± 8.58 ^{a}	33.84 ± 2.07 ^{a}	37.00 ± 2.32 ^{a}
OSACK1	KOH	1:2	56.94 ± 4.29 ^{d}	141.77 ± 14.37 ^{cd}	97.61 ± 0.69 ^{e}	99.22 ± 5.72 ^{d}
OSACK2		1:4	75.14 ± 5.69 ^{e}	186.16 ± 18.96 ^{d}	100.00 ^{e}	101.41 ± 5.24 ^{d}
OSACP1	H_3PO_4	1:2	35.70 ± 4.06 ^{bc}	89.07 ± 14.03 ^{ab}	58.63 ± 2.72 ^{cd}	61.32 ± 2.40 ^{bc}
OSACP2		1:4	39.40 ± 1.73 ^{c}	97.94 ± 8.45 ^{abc}	62.84 ± 1.08 ^{d}	65.81 ± 2.07 ^{c}
OSACZ1	$ZnCl_2$	1:2	25.06 ± 4.79 ^{abc}	62.60 ± 14.36 ^{a}	34.56 ± 3.52 ^{a}	36.81 ± 3.23 ^{a}
OSACZ2		1:4	23.93 ± 4.05 ^{ab}	59.81 ± 12.25 ^{a}	32.58 ± 2.18 ^{a}	34.13 ± 2.60 ^{a}
ASC	-	-	33.35 ± 3.02 ^{abc}	88.06 ± 8.39 ^{a}	47.23 ± 1.42 ^{b}	51.67 ± 1.60 ^{b}
ASACK1	KOH	1:2	56.34 ± 1.87 ^{d}	140.27 ± 13.88 ^{bcd}	93.58 ± 0.93 ^{e}	95.08 ± 5.85 ^{d}
ASACK2		1:4	72.83 ± 4.67 ^{e}	182.69 ± 21.56 ^{d}	100.00 ^{e}	101.68 ^{d}
ASACP1	H_3PO_4	1:2	37.26 ± 1.16 ^{bc}	92.88 ± 8.49 ^{abc}	52.38 ± 3.09 ^{bc}	55.21 ± 3.03 ^{bc}
ASACP2		1:4	34.20 ± 2.96 ^{abc}	85.45 ± 12.30 ^{a}	55.96 ± 3.92 ^{cd}	59.47 ± 2.73 ^{bc}
ASACZ1	$ZnCl_2$	1:2	26.88 ± 5.41 ^{abc}	67.59 ± 18.12 ^{a}	35.64 ± 1.93 ^{a}	37.54 ± 1.32 ^{a}
ASACZ2		1:4	20.54 ± 5.33 ^{a}	51.77 ± 16.72 ^{a}	29.95 ± 2.02 ^{a}	31.57 ± 2.61 ^{a}

AE: adsorption efficiency at equilibrium; q_e: adsorption capacity at equilibrium; C_i: cation initial concentration; AD: adsorbent dose. In each column and at each concentration different letters imply significant differences among means ($p < 0.05$).

Increasing the adsorbent dosage to 5 g L^{-1} at the same Pb^{2+} concentration led to a decreased q_e across all samples. Apart from the KOH-activated carbons (OSACK and ASACK series), ASC was the only carbon with a complete adsorption level (AE = 100%). For the other carbons, the adsorption efficiency improved slightly, reaching 73.28%, 83.82–89.65%, and 68.05–78.90% for OSC, the H_3PO_4-activated carbons, and the $ZnCl_2$-activated carbons, respectively.

At the highest concentration (500 mg L^{-1}) and lower adsorbent dosage (2 g L^{-1}), the carbons activated with KOH at a C/A ratio of 1:4 (*w*/*w*) (OSACK2 and ASACK2) showed a notable adsorption performance with AE values of 75.14% and 72.83%, respectively, and the highest q_e values (186.16 and 182.69 mg g^{-1}). This highlights their strong capacity for Pb^{2+} adsorption at higher initial metal concentrations, likely due to their high surface area and well-developed porosity. OSACK1 and ASACK1 also demonstrated good performance, achieving AE values around 56% and q_e values near 140 mg g^{-1}, suggesting that they effectively accommodate higher Pb^{2+} loads even at moderate dosages. In comparison, the non-activated carbons and those activated with H_3PO_4 and $ZnCl_2$ exhibited lower AE values (27.02–33.35%, 34.20–39.40%, and 20.54–26.88%, respectively) and q_e values ranging from 71.21 to 88.06 mg g^{-1}, 85.45 to 97.94 mg g^{-1}, and 51.77 to 67.59 mg g^{-1}, respectively, indicating reduced effectiveness at high Pb^{2+} concentrations. This likely reflects limitations in the available surface area or pore structure for these specific carbons.

At an increased dosage of 5 g L^{-1} and 500 mg L^{-1} Pb^{2+} concentration, OSACK2 and ASACK2 again reached maximum AE values (100%) with high q_e values of approximately 101 mg g^{-1}. These results highlight the strong metal uptake potential of these carbons, regardless of adsorbent dosage, making them highly efficient for Pb^{2+} removal under varying operational conditions. The improvement in the AE across all carbons such as OSACK1 (97.61%), ASACK1 (93.58%), and OSACP2 (62.84%), reflects the advantage of a higher adsorbent dosage in enhancing the adsorption efficiency for systems with high Pb^{2+} concentrations. The superior AE and q_e values of the KOH-activated carbons compared to other carbons can be attributed to their higher surface area, pore volume, and pore size (as shown in Table 3 and supported by Figure 9), along with the presence of more surface functionalities, particularly oxygen-containing functional groups, which facilitate Pb^{2+} adsorption. The findings further reveal that the carbon materials synthesized in this study exhibited a significantly enhanced capacity for Pb^{2+} ion recovery compared to Cu^{2+}, implying a pronounced selectivity for Pb^{2+} over Cu^{2+}. This behavior may be attributed to the larger ionic radius of Pb^{2+} compared to Cu^{2+}, which improves the attraction and binding affinity of Pb^{+2} ions to the functional groups present on the surface of carbons, especially oxygen-containing groups such as hydroxyl (–OH) and carboxyl (–COOH) groups, through intensified electrostatic interactions [58,59].

The data also indicate that base and chemical activation are effective strategies for enhancing adsorption performance. However, the optimal conditions for maximum Pb^{2+} removal vary among carbon types, likely due to specific interactions between Pb^{2+} ions and carbon surface properties. The high efficiency of the OSACK and ASACK series carbons, especially those activated with a 1:4 (*w*/*w*) ratio, highlights their potential as viable options for Pb^{2+} removal applications, particularly at high pollutant concentrations and moderate dosages.

4. Conclusions

In this study, biomass-based carbon adsorbents were produced from olive stones and almond shells using various activating agents including KOH, H_3PO_4, and $ZnCl_2$. The activation method influenced the physical and chemical properties of the biomass-derived carbons, leading to differences in the surface area, pore structure, and functional group distributions that impacted their adsorption capabilities. Compared to H_3PO_4- and $ZnCl_2$-activated carbons with a high degree of microporosity, activation with KOH yielded carbons with highly developed micro- and mesoporous structures, greater surface area and pore size, and more introduction of oxygen-containing functional groups, exhibiting significant potential for the removal of heavy metal ions, specifically Cu^{2+} and Pb^{2+}, from aqueous solutions. The KOH-activated carbons prepared using a C/A ratio of 1:4 (*w*/*w*) from both precursors (OSACK2 and ASACK2) were found to have the highest adsorption performance for the removal of Cu^{2+} and Pb^{2+}, achieving adsorption efficiencies of 84.63–86.29% (41.69–42.52 mg g^{-1}) at 100 mg L^{-1} and 2 g L^{-1} dosage and 100% (101.41–101.68 mg g^{-1}) at 500 mg L^{-1} and 5 g L^{-1} dosage. Moreover, the carbons activated with H_3PO_4 and $ZnCl_2$ did not significantly influence adsorption, likely due to their specific texture, surface properties, and adsorption potential. The results also indicated that the carbons fabricated in this work have a significantly higher capacity for Pb^{2+} adsorption than for Cu^{2+} adsorption.

This research aims to contribute to sustainable e-waste and wastewater management by developing biomass-derived activated carbons as cost-effective, renewable adsorbents for heavy metal removal. The study offers an environmentally friendly solution to the growing problem of heavy metal pollution, specifically targeting wastewater and leach liquors from waste PCB processing. By valorizing biomass waste into high-performance adsorbents, the research aligns with circular economy principles, reduces the environmental footprint of both e-waste and wastewater treatment, and provides a scalable method for mitigating the adverse effects of toxic metal contamination.

Author Contributions: Conceptualization, V.R., D.G.-D., M.S.F. and J.G.-Á.; methodology, V.R., D.G.-D. and M.S.F.; Validation, V.R.; formal analysis, V.R.; investigation, V.R. and C.H.P.; data curation, V.R. and C.H.P.; writing—original draft preparation, V.R.; writing—review and editing, V.R., D.G.-D., M.S.F., M.L. and J.G.-Á.; Visualization, V.R.; supervision, D.G.-D., M.S.F. and J.G.-Á.; Project administration, M.L. and J.G.-Á.; funding acquisition, M.L. and J.G.-Á. All authors have read and agreed to the published version of the manuscript.

Funding: This work is part of the I+D+i project, Reference PID2021-122923NB-I00, financed by MCIN/AEI/10.13039/501100011033/FEDER, UE.

Data Availability Statement: The data presented in this study are available from the authors.

Conflicts of Interest: The authors declare no conflicts of interest.

References

1. Gutwiński, P.; Cema, G.; Ziembińska-Buczyńska, A.; Wyszyńska, K.; Surmacz-Górska, J. Long-term effect of heavy metals Cr(III), Zn(II), Cd(II), Cu(II), Ni(II), Pb(II) on the anammox process performance. *J. Water Process Eng.* **2021**, *39*, 101668. [CrossRef]
2. Demiral, H.; Güngör, C. Adsorption of copper(II) from aqueous solutions on activated carbon prepared from grape bagasse. *J. Clean. Prod.* **2016**, *124*, 103–113. [CrossRef]
3. Neisan, R.S.; Saady, N.M.C.; Bazan, C.; Zendehboudi, S.; Albayati, T.M. Adsorption of copper from water using TiO2-modified activated carbon derived from orange peels and date seeds: Response surface methodology optimization. *Heliyon* **2023**, *9*, e21420. [CrossRef] [PubMed]
4. Jaishankar, M.; Tseten, T.; Anbalagan, N.; Mathew, B.B.; Beeregowda, K.N. Toxicity, mechanism and health effects of some heavy metals. *Interdiscip. Toxicol.* **2014**, *7*, 60–72. [CrossRef] [PubMed]
5. Hadi, P.; Gao, P.; Barford, J.P.; Mckay, G. Novel application of the nonmetallic fraction of the recycled printed circuit boards as a toxic heavy metal adsorbent. *J. Hazard. Mater.* **2013**, *252–253*, 166–170. [CrossRef]
6. Wang, H.; Zhang, S.; Li, B.; Pan, D.; Wu, Y.; Zuo, T. Recovery of waste printed circuit boards through pyrometallurgical processing: A review. *Resour. Conserv. Recycl.* **2017**, *126*, 209–218. [CrossRef]
7. Ning, C.; Lin, C.S.K.; Hui, D.C.W.; McKay, G. Waste printed circuit board (PCB) recycling techniques. *Top. Curr. Chem.* **2017**, *375*, 43. [CrossRef]
8. Rahimi, V.; Inzulza-Moraga, E.A.; Gómez-Díaz, D.; Freire, M.S.; González-Álvarez, J. Screening of variables affecting the selective leaching of valuable metals from waste motherboards' PCBs. *Environ. Sci. Pollut. Res.* **2024**, *31*, 1–16. [CrossRef]
9. Hossain, R.; Nekouei, R.K.; Mansuri, I.; Sahajwalla, V. Sustainable recovery of Cu and Sn from problematic global waste: Exploring value from waste printed circuit boards. *ACS Sustain. Chem. Eng.* **2019**, *7*, 1006–1017. [CrossRef]
10. Rao, M.D.; Singh, K.K.; Morrison, C.A.; Love, J.B. Optimization of process parameters for the selective leaching of copper, nickel and isolation of gold from obsolete mobile phone PCBs. *Clean. Eng. Technol.* **2021**, *4*, 100180. [CrossRef]
11. Rajendran, S.; Priya, A.; Kumar, P.S.; Hoang, T.K.; Sekar, K.; Chong, K.Y.; Khoo, K.S.; Ng, H.S.; Show, P.L. A critical and recent developments on adsorption technique for removal of heavy metals from wastewater—A review. *Chemosphere* **2022**, *303*, 135146. [CrossRef] [PubMed]
12. Zhang, Y.; Yue, X.; Xu, W.; Zhang, H.; Li, F. Amino modification of rice straw-derived biochar for enhancing its cadmium (II) ions adsorption from water. *J. Hazard. Mater.* **2019**, *379*, 120783. [CrossRef]
13. Zhou, Y.; Li, W.; Qi, W.; Chen, S.; Tan, Q.; Wei, Z.; Gong, L.; Chen, J.; Zhou, W. The comprehensive evaluation model and optimization selection of activated carbon in the O_3-BAC treatment process. *J. Water Process Eng.* **2021**, *40*, 101931. [CrossRef]
14. Alhawtali, S.; El-Harbawi, M.; El Blidi, L.; Alrashed, M.M.; Alzobidi, A.; Yin, C.-Y. Date Palm Leaflet-Derived Carbon Microspheres Activated Using Phosphoric Acid for Efficient Lead (II) Adsorption. *C-J. Carbon Res.* **2024**, *10*, 26. [CrossRef]

15. Azmi, S.N.H.; Al-Balushi, M.; Al-Siyabi, F.; Al-Hinai, N.; Khurshid, S. Adsorptive removal of Pb(II) ions from groundwater samples in Oman using carbonized *Phoenix dactylifera* seed (Date stone). *J. King. Saud. Univ. Sci.* **2020**, *32*, 2931–2938. [CrossRef]
16. Gao, X.; Wu, L.; Xu, Q.; Tian, W.; Li, Z.; Kobayashi, N. Adsorption kinetics and mechanisms of copper ions on activated carbons derived from pinewood sawdust by fast H_3PO_4 activation. *Environ. Sci. Pollut. Res.* **2018**, *25*, 7907–7915. [CrossRef]
17. Teğin, İ.; Öc, S.; Saka, C. Adsorption of copper (II) from aqueous solutions using adsorbent obtained with sodium hydroxide activation of biochar prepared by microwave pyrolysis. *Biomass Conv. Bioref.* **2024**, *14*, 1–12. [CrossRef]
18. Hoang, A.T.; Kumar, S.; Lichtfouse, E.; Cheng, C.K.; Varma, R.S.; Senthilkumar, N.; Nguyen, P.Q.P.; Nguyen, X.P. Remediation of heavy metal polluted waters using activated carbon from lignocellulosic biomass: An update of recent trends. *Chemosphere* **2022**, *302*, 134825. [CrossRef]
19. Ali, I.; Asim, M.; Khan, T.A. Low cost adsorbents for the removal of organic pollutants from wastewater. *J. Environ. Manag.* **2012**, *113*, 170–183. [CrossRef]
20. González-García, P.; Centeno, T.A.; Urones-Garrote, E.; Ávila-Brande, D.; Otero-Díaz, L.C. Microstructure and surface properties of lignocellulosic-based activated carbons. *Appl. Surf. Sci.* **2013**, *265*, 731–737. [CrossRef]
21. Mariana, M.; Khalil, A.; Mistar, E.M.; Yahya, E.B.; Alfatah, T.; Danish, M.; Amayreh, M. Recent advances in activated carbon modification techniques for enhanced heavy metal adsorption. *J. Water Process Eng.* **2021**, *43*, 102221. [CrossRef]
22. Pérez-Mayoral, E.; Matos, I.; Bernardo, M.; Fonseca, I.M. New and advanced porous carbon materials in fine chemical synthesis. Emerging precursors of porous carbons. *Catalysts* **2019**, *9*, 133. [CrossRef]
23. Song, M.; Jin, B.; Xiao, R.; Yang, L.; Wu, Y.; Zhong, Z.; Huang, Y. The comparison of two activation techniques to prepare activated carbon from corn cob. *Catalysts* **2013**, *48*, 250–256. [CrossRef]
24. Chen, W.-S.; Chen, Y.-C.; Lee, C.H. Modified activated carbon for copper ion removal from aqueous solution. *Processes* **2022**, *10*, 150. [CrossRef]
25. Razak, M.R.; Yusof, N.A.; Haron, M.J.; Ibrahim, N.; Mohammad, F.; Kamaruzaman, S.; Al-Lohedan, H.A. Iminodiacetic acid modified kenaf fiber for waste water treatment. *Int. J. Biol. Macromol.* **2018**, *112*, 754–760. [CrossRef]
26. Chen, J.; Zhang, L.; Yang, G.; Wang, Q.; Li, R.; Lucia, L.A. Preparation and characterization of activated carbon from hydrochar by phosphoric acid activation and its adsorption performance in prehydrolysis liquor. *Bioresources* **2017**, *12*, 5928–5941. [CrossRef]
27. Sarwar, A.; Ali, M.; Khoja, A.H.; Nawar, A.; Waqas, A.; Liaquat, R.; Naqvi, S.R.; Asjid, M. Synthesis and characterization of biomass-derived surface-modified activated carbon for enhanced CO_2 adsorption. *J. CO2 Util.* **2021**, *46*, 101476. [CrossRef]
28. Yakout, S.M.; El-Deen, G.S. Characterization of activated carbon prepared by phosphoric acid activation of olive stones. *Arab. J. Chem.* **2016**, *9*, S1155–S1162. [CrossRef]
29. Chomiak, K.; Gryglewicz, S.; Kierzek, K.; Machnikowski, J. Optimizing the properties of granular walnut-shell based KOH activated carbons for carbon dioxide adsorption. *J. CO2 Util.* **2017**, *21*, 436–443. [CrossRef]
30. Zhou, J.; Yang, P.; Kots, P.A.; Cohen, M.; Chen, Y.; Quinn, C.M.; de Mello, M.D.; Boscoboinik, J.A.; Shaw, W.J.; Caratzoulas, S.; et al. Tuning the reactivity of carbon surfaces with oxygen-containing functional groups. *Nat. Commun.* **2023**, *14*, 2293. [CrossRef]
31. Serafin, J.; Narkiewicz, U.; Morawski, A.W.; Wróbel, R.J.; Michalkiewicz, B. Highly microporous activated carbons from biomass for CO_2 capture and effective micropores at different conditions. *J. CO2. Util.* **2017**, *18*, 73–79. [CrossRef]
32. Danish, M.; Ahmad, T. A review on utilization of wood biomass as a sustainable precursor for activated carbon production and application. *Renew. Sustain. Energy Rev.* **2018**, *87*, 1–21. [CrossRef]
33. Li, Q.; Liu, S.; Peng, W.; Zhu, W.; Wang, L.; Chen, F.; Shao, J.; Hu, X. Preparation of biomass-derived porous carbons by a facile method and application to CO_2 adsorption. *J. Taiwan Inst. Chem. Eng.* **2020**, *116*, 128–136. [CrossRef]
34. Liu, X.; Sun, Y.; Liu, J.; Sun, C.; Liu, H.; Xue, Q.; Smith, E.; Snape, C. Potassium and zeolitic structure modified ultra-microporous adsorbent materials from a renewable feedstock with favorable surface chemistry for CO_2 capture. *ACS Appl. Mater. Interfaces* **2017**, *32*, 26826–26839. [CrossRef]
35. Sher, F.; Iqbal, S.Z.; Albazzaz, S.; Ali, U.; Mortari, D.A.; Rashid, T. Development of biomass derived highly porous fast adsorbents for post-combustion CO_2 capture. *Fuel* **2020**, *282*, 118506. [CrossRef]
36. Wang, K.; Zhao, N.; Lei, S.; Yan, R.; Tian, X.; Wang, J.; Song, Y.; Xu, D.; Guo, Q.; Liu, L. Promising biomass-based activated carbons derived from willow catkins for high performance supercapacitors. *Electrochim. Acta* **2015**, *166*, 1–11. [CrossRef]
37. Piccirillo, C.; Moreira, I.S.; Novais, R.M.; Fernandes, A.J.S.; Pullar, R.C.; Castro, P.M.L. Biphasic apatite-carbon materials derived from pyrolysed fish bones for effective adsorption of persistent pollutants and heavy metals. *J. Environ. Chem. Eng.* **2017**, *5*, 4884–4894. [CrossRef]
38. Vohl, S.; Ban, I.; Drofenik, M.; Buksek, H.; Gyergyek, S.; Petrinic, I.; Hélix-Nielsen, C.; Stergar, J. Microwave synthesis of poly(acrylic) acid-coated magnetic nanoparticles as draw solutes in forward osmosis. *Materials* **2023**, *16*, 4138. [CrossRef]
39. Hao, G.P.; Li, W.C.; Qian, D.; Lu, A.H. Rapid synthesis of nitrogen-doped porous carbon monolith for CO_2 capture. *Adv. Mater.* **2010**, *22*, 853–857. [CrossRef]
40. Daffalla, S.B.; Mukhtar, H.; Shaharun, M.S. Properties of activated carbon prepared from rice husk with chemical activation. *Int. J. Glob. Environ. Issues* **2012**, *12*, 107–129. [CrossRef]

41. Liou, T.H.; Wu, S.J. Characteristics of microporous/mesoporous carbons prepared from rice husk under base- and acid-treated conditions. *J. Hazard. Mater.* **2009**, *171*, 693–703. [CrossRef] [PubMed]
42. Zhao, H.; Zhong, H.; Jiang, Y.; Li, H.; Tang, P.; Li, D.; Feng, Y. Porous $ZnCl_2$-activated carbon from shaddock peel: Methylene blue adsorption behavior. *Materials* **2022**, *15*, 895. [CrossRef] [PubMed]
43. Zhang, J.; Xia, X.; Li, K.; Shen, Y.; Xue, Y. New insights into temperature-induced mechanisms of copper adsorption enhancement on hydroxyapatite-in situ self-doped fluffy bread-like biochar. *Chem. Eng. J.* **2024**, *479*, 147657. [CrossRef]
44. Alanazi, A.G.; Habila, M.A.; ALOthman, Z.A.; Badjah-Hadj-Ahmed, A.Y. Synthesis and characterization of zinc oxide nanoparticle anchored carbon as hybrid adsorbent materials for effective heavy metals uptake from wastewater. *Crystals* **2024**, *14*, 447. [CrossRef]
45. Al-Degs, Y.S.; El-Barghouthi, M.I.; El-Sheikh, A.H.; Walker, G.M. Effect of solution pH, ionic strength, and temperature on adsorption behavior of reactive dyes on activated carbon. *Dyes Pigm.* **2008**, *77*, 16–23. [CrossRef]
46. Kikuchi, Y.; Qian, Q.; Machida, M.; Tatsumoto, H. Effect of ZnO loading to activated carbon on Pb (II) adsorption from aqueous solution. *Carbon* **2006**, *44*, 195–202. [CrossRef]
47. Rahimi, V.; Ferreiro-Salgado, A.; Gómez-Díaz, D.; Freire, M.S.; González-Álvarez, J. Evaluating the performance of carbon-based adsorbents fabricated from renewable biomass precursors for post-combustion CO_2 capture. *Sep. Purif. Technol.* **2024**, *344*, 127110. [CrossRef]
48. Siipola, V.; Tamminen, T.; Källi, A.; Lahti, R.; Romar, H.; Rasa, K.; Keskinen, R.; Hyväluoma, J.; Hannula, M.; Wikberg, H. Effects of biomass type, carbonization process, and activation method on the properties of bio-based activated carbons. *BioResources* **2018**, *13*, 5976–6002. [CrossRef]
49. Liu, Z.; Huang, Y.; Zhao, G. Preparation and characterization of activated carbon fibers from liquefied wood by $ZnCl_2$ activation. *BioResources* **2016**, *11*, 3178–3190. [CrossRef]
50. Singh, G.; Kim, I.Y.; Lakhi, K.S.; Srivastava, P.; Naidu, R.; Vinu, A. Single step synthesis of activated bio-carbons with a high surface area and their excellent CO_2 adsorption capacity. *Carbon* **2017**, *116*, 448–455. [CrossRef]
51. Rukayat, O.O.; Usman, M.F.; Elizabeth, O.M.; Abosede, O.O.; Faith, I.U. Kinetic adsorption of heavy metal (copper) on rubber (*Hevea Brasiliensis*) leaf powder. *S. Afr. J. Chem. Eng.* **2021**, *37*, 74–80. [CrossRef]
52. Darweesh, M.A.; Elgendy, M.Y.; Ayad, M.I.; Ahmed, A.M.; Elsayed, N.M.K.; Hammad, W.A. Adsorption isotherm, kinetic, and optimization studies for copper (II) removal from aqueous solutions by banana leaves and derived activated carbon. *S. Afr. J. Chem. Eng.* **2022**, *40*, 10–20. [CrossRef]
53. Girgis, B.S.; Alaya, M.N.; Sherif, I.E.; Fathy, N.A. Development of porosity and copper (II) ion adsorption capacity by activated nano-carbon xerogels in relation to treatment schemes. *Adsorpt. Sci. Technol.* **2011**, *29*, 943–961. [CrossRef]
54. Chen, H.; Pan, Y.; Zhang, W.; Long, A.; Chen, M.; Xiao, X.; Wang, Z.; Tang, M.; Peng, Y.; Sun, S.; et al. Adsorption-desorption of copper (II) by temperature-sensitive nano-biochar@PNIPAM/alginate double-network composite hydrogel: Enhanced mechanisms and application potentials. *Chem. Eng. J.* **2024**, *495*, 153356. [CrossRef]
55. Fei, Y.; Hu, Y.H. Design, synthesis, and performance of adsorbents for heavy metal removal from wastewater: A review. *J. Mater. Chem. A* **2022**, *10*, 1047–1085. [CrossRef]
56. Hotová, G.; Slovák, V.; Zelenka, T.; Maršálek, R.; Parchaňská, A. The role of the oxygen functional groups in adsorption of copper (II) on carbon surface. *Sci. Total Environ.* **2020**, *711*, 135435. [CrossRef]
57. Shahrokhi-Shahraki, R.; Benally, C.; El-Din, M.G.; Park, J. High efficiency removal of heavy metals using tire-derived activated carbon vs commercial activated carbon: Insights into the adsorption mechanisms. *Chemosphere* **2021**, *264*, 128455. [CrossRef]
58. Dowlatshahi, S.; Torbati, A.R.H.; Loloei, M. Adsorption of copper, lead and cadmium from aqueous solutions by activated carbon prepared from saffron leaves. *Environ. Health Eng. Manag. J.* **2014**, *1*, 37–44.
59. Zhou, X.-Y.; Wu, Y.-W.; Cai, Q.; Mi, T.-G.; Zhang, B.; Zhao, L.; Lu, Q. Interaction mechanism between lead species and activated carbon in MSW incineration flue gas: Role of different functional groups. *Chem. Eng. J.* **2022**, *436*, 135252. [CrossRef]

C Journal of Carbon Research

MDPI

Article

Photocatalytic Purification of Orange-II-Dye-Polluted Wastewater Using TiO_2-Modified g-C_3N_4 Composite

Bin Guo [1,*], Monir Uzzaman [1], Mai Furukawa [1], Ikki Tateishi [2], Hideyuki Katsumata [1] and Satoshi Kaneco [1,*]

[1] Department of Applied Chemistry, Graduate School of Engineering, Mie University, Tsu 514-8507, Japan; monircu92@gmail.com (M.U.); maif@chem.mie-u.ac.jp (M.F.); hidek@chem.mie-u.ac.jp (H.K.)

[2] Mie Global Environment Center for Education & Research, Mie University, Tsu 514-8507, Japan; tateishi@gecer.mie-u.ac.jp

* Correspondence: 423db03@m.mie-u.ac.jp (B.G.); kaneco@chem.mie-u.ac.jp (S.K.); Tel.: +81-59-231-9427 (S.K.)

Abstract: Textile dyes are considered to be one of the major sources of water pollution due to their complex chemical structures. Photocatalytic dye degradation is an eco-friendly method that uses light-activated catalysts to break down complex dye molecules in wastewater into harmless byproducts. In the present study, pure and N-doped g-C_3N_4 were synthesized using one-pot calcination. The fabrication of g-C_3N_4/TiO_2 and TC-g-C_3N_4/TiO_2 was accomplished by physical mixing. The prepared photocatalysts were used to treat the Orange-II-polluted wastewater. The structural properties, surface morphology, light-harvesting capability, and electrochemical properties were assessed using XRD, XPS, SEM, TEM, BET, DRS, PL, EIS, Mott–Schottky, and transient photocurrent response (TPCR) analyses. The improved charge separation and migration properties of g-C_3N_4/TiO_2 were confirmed by the PL, EIS, and TPCR results. TEM and XPS data confirmed the formation of a g-C_3N_4/TiO_2 composite. Enhanced photocatalytic decolorization was achieved for g-C_3N_4/TiO_2 compared with pure g-C_3N_4 or TiO_2. The 1 wt% g-C_3N_4/TiO_2 composite exhibited the highest (92.1%) decolorization efficiency during 60 min of UV irradiation. Furthermore, the pseudo-first-order reaction rate constant (k) of 1 wt% g-C_3N_4/TiO_2 was calculated to be 0.0213 min^{-1}, which was 2.5 times better than that obtained with bare TiO_2. Finally, for a better understanding of the g-C_3N_4/TiO_2 decolorization process, a possible reaction mechanism was proposed.

Keywords: g-C_3N_4/TiO_2; photocatalytic purification; Orange II dye; wastewater

1. Introduction

Water pollution presents a pressing global concern, with its adverse effects on human life, the animal kingdom, and the environment escalating alongside the rapid expansion of industrialization. Water reserves are contaminated by a variety of industrial and domestic effluents such as pesticides, herbicides, fungicides, organic solvents, paints, toxic heavy metals, household chemicals, and pharmaceuticals waste [1]. These pollutants pose a wide range of public health hazards, contributing to respiratory problems, cancer, reproductive and endocrine disfunction, hepatic and renal injury, diarrhea, neurological disorders, and cardiovascular disease [2]. Numerous chemical, electrochemical, biological, and physical methods have been utilized for wastewater treatment, including filtration, coagulation, reverse osmosis, sedimentation, flotation, oxidation, precipitation, evaporation, and adsorption [1–3]. Each method possesses unique advantages, and their suitability determines their application in targeting specific classes of toxins.

The dyeing industry has gained substantial importance in modern times, yet it generates a large volume of dye wastewater, which significantly contributes to environmental pollution. Of particular concern is the widespread use of C.I. Acid Orange 7 (Orange II), a representative of azo (–N=N–) and an acid dye. Its incomplete degradation can also produce harmful degradation products, particularly aromatic amine (aniline), which could be carcinogenic and poses significant health risks [4]. Therefore, the development of an effective treatment method for this dye is urgently needed [5]. Although numerous approaches exist for treating dye wastewater, implementing semiconductor-based photocatalytic decolorization and eco-friendly processes are promising due to their cost-effectiveness and simplicity [1–3].

In recent years, there has been a surge in research activity dedicated to advancing the synthesis and functionalization of semiconductor materials and metal nanoparticles. The primary aim of these studies is to improve the performance, stability, and versatility of nanoparticles across a wide range of applications, including sensing devices, photonic systems, molecular electronics, and advanced oxidation techniques [1–3]. The optical and electronic properties of these nanoparticles depend on their size and shape. They present an intriguing opportunity for leveraging them in light-induced chemical reactions, particularly within advanced oxidation processes. Photocatalysis and radiocatalysis are common applications of these nanoparticles [6–8]. However, maximizing the efficiency of photoinduced charge separation in semiconductor systems remains a significant challenge for the scientific community. Gaining an insight into charge transfer processes is crucial for enhancing the photoconversion efficiencies in semiconductor-based nanoassemblies. Extending the photo-response of the particles is a shared objective, and researchers frequently explore efficiency enhancements by experimenting with various combinations of particle materials [9–11]. Quantization and its effects on charge transfer processes, especially those induced by charge separation at the particle's surface, represent a central focus of many current advancements in the field. Moreover, there is growing interest in the dual use of these materials and the integration of advanced oxidation techniques to further exploit their potential.

The principle of photocatalysts relies on their redox capacity when exposed to specific conditions, enabling the purification of pollutants and the synthesis and conversion of substances. Photocatalytic oxidation reactions commonly employ semiconductors as catalysts and utilize light energy to degrade organic matter into carbon dioxide (CO_2) and water. Titanium dioxide (TiO_2) is a well-known photocatalyst material that absorbs light beyond its band gap (3.2 eV), generating electron–hole pairs in the valence band (VB) and conduction band (CB) and facilitating oxidation and reduction reactions. Due to its wide band gap, it requires a larger amount of energy to excite and transport the electron from the VB to the CB. As a result, it responds only under UV light. Further modifications such as metal doping or composite formation may increase its light absorption capability and photocatalytic efficiency [1,2,11].

When a semiconductor absorbs photons with energy greater than its band gap, it undergoes excitation, creating electron–hole pairs in the CB. These photogenerated holes possess the ability to oxidize various chemical species. For instance, n-type titanium dioxide serves as a potent oxidizer when the photoenergy exceeds 3.2 eV (~400 nm) [12–14]. In an aqueous solution, photoexcited titanium dioxide generates electrons and holes, which can interact with the targeted species on its surface. Specifically, the holes can oxidize water molecules to form hydroxyl radicals ($^{\bullet}OH$), while the electrons can reduce oxygen to produce superoxide anions ($^{\bullet}O_2^{-}$) or hydrogen peroxide. These reactive oxygen species play a pivotal role in driving various chemical reactions. The photoelectrochemical reactions of titanium dioxide particles have undergone extensive study for several photo-technologies, including solar energy conversion to chemical energy, photosynthesis, and photocatalysis [2,3].

Recently, g-C_3N_4 has garnered significant attention as a metal-free photocatalyst for the degradation and decolorization of environmental pollutants owing to its narrow bandgap, low cost, excellent chemical stability, easy preparation using affordable materials, environmental friendliness, and unique electron band structure [14,15]. This graphite-like structure consists of weak van der Waals forces between C–N layers, with each layer containing aromatic tri-s-triazine units connected by tertiary amines, and it can respond under visible light [2,14,15].

The preparation of photocatalysts involves skillfully combining two or more materials to create new composite materials that effectively regulate the properties of individual components and generate novel photochemical and photophysical characteristics [16–19]. Constructing heterojunctions is a highly effective technique for enhancing photocatalytic efficiency. Due to variations in band structures among different materials, heterojunction photocatalysts can significantly broaden the wavelength range of light absorption and facilitate the separation of photogenerated electron–hole pairs, resulting in increased photocatalytic activity and efficiency. Among the various types of heterojunction photocatalysts, the emerging "step-scheme" (S-scheme) heterojunction photocatalysts have been attracting increasing attention [3,14,20]. S-scheme heterojunctions can address the defects associated with type-II heterojunctions and exhibit a strong redox ability. Limited information exists on the photocatalytic degradation of Orange II in aqueous solutions using g-C_3N_4/TiO_2 composites [12,21,22].

In this study, a g-C_3N_4/TiO_2 composite was prepared using a very simple calcination and physical mixing method. The photocatalytic purification of Orange-II-polluted wastewater was investigated under UV light with a simply prepared g-C_3N_4/TiO_2 composite. The photocatalytic degradation kinetics of the composite were examined and compared.

2. Materials and Methods

2.1. Materials

All the chemicals were analytical grade and used as received. ST-01 TiO_2 (average diameter 7 nm; specific surface area 300 $m^2\ g^{-1}$) was purchased from ISHIHARA SANGYO KAISHA, Ltd., Japan. Urea ($CO(NH_2)_2$), trisodium citrate ($C_6H_5O_7Na_3$), and C.I. Acid Orange 7 ($C_{16}H_{11}N_2O_4SNa$) were obtained from FUJIFILM Wako Pure Chemical Corporation, Japan.

2.2. Synthesis

2.2.1. Preparation of g-C_3N_4

The g-C_3N_4 powders were synthesized using a poly-condensation method. Typically, 10 g urea was transferred into a 30 mL ceramic crucible with a lid. Then, the crucible was covered with aluminum foil, placed in a muffle furnace, heated to 550 °C at a heating increment rate of 2 °C min^{-1}, and held for 2 h at 550 °C. After natural cooling to room temperature, the sample was manually ground in an agate mortar and the product was collected as a light-yellow powder.

2.2.2. Synthesis of TC-g-C_3N_4

The N-doped g-C_3N_4 (TC-g-C_3N_4) was prepared using a previously reported method [23]. Briefly, 10 g urea and 100 mg trisodium citrate (TC) were dissolved into 5 mL distilled water. The mixture was transferred into a ceramic crucible with a cover, heated to 550 °C at a heating increment rate of 2 °C min^{-1}, and maintained at this temperature for 2 h. After cooling to room temperature, the obtained TC-g-C_3N_4 was repeatedly washed with pure water and ethanol and dried at 60 °C for 12 h. Using TC as an auxiliary material in the synthesis of N-doped g-C_3N_4 can optimize the final material's structure and performance by providing a templating effect, improving nitrogen doping uniformity, enhancing the surface area, and promoting photocatalytic performance [23].

2.2.3. Composite Synthesis

Various amounts of pure g-C_3N_4 or TC-g-C_3N_4 were physically mixed with TiO_2 to fabricate g-C_3N_4/TiO_2 and TC-g-C_3N_4/TiO_2 composites. The mass ratios of g-C_3N_4 or TC-g-C_3N_4 to TiO_2 were set as 0.1, 0.5, 1, and 5 wt%, respectively, and the composites were named xg-C_3N_4/TiO_2 and xTC-g-C_3N_4/TiO_2, respectively (x represents the wt% of g-C_3N_4 or TC-g-C_3N_4).

2.2.4. Characterization Techniques

The crystal structures of the synthesized photocatalysts were assessed using powder X-ray diffraction (XRD) analyses by utilizing a Rigaku RINT Ultima-IV diffractometer (Tokyo, Japan) with Cu Kα radiation in a scan range of 10–80° at a scan rate of 0.04° s^{-1}. X-ray photoelectron spectroscopy (XPS) was employed to identify the chemical state of the studied photocatalysts using a PHI Quantera SXM instrument (Ulvac-Phi) with Al-Kα radiation. The surface morphology was analyzed using scanning electron microscopy (SEM; S-4000, Hitachi) and transmitted electron microscopy (TEM; JEM-1011, JEOL), respectively. The Brunauer-Emmett-Teller (BET) surface area was calculated utilizing BELPREP-Vac II (BEL, Tokyo, Japan). Optical properties were measured using a UV-Vis DRS spectroscopic instrument (JASCO V-750) equipped with an integrating sphere adaptor and a photoluminescence (PL) spectroscopic measurement instrument (Shimadzu RF-5300PC, Kyoto, Japan). The electrochemical impedance spectra (EIS) measurement of the photocatalysts was performed using an electrochemical Versa STAT 3 workstation (Princeton Applied Research, USA) equipped with a conventional three-electrode system. A uniform photocatalyst slurry combined with a Nafion solution (Aldrich, USA) was coated on a fluorine-doped tin oxide glass plate to prepare the working electrode and a 0.5 mol L^{-1} Na_2SO_4 aqueous solution was used as the electrolyte.

2.2.5. Photocatalytic Performance Test

The photocatalytic decolorization reaction was carried out in a Pyrex glass vessel (inner volume of 123 mL). Typically, 100 mL of a 5 ppm Orange II solution and 30 mg of the g-C_3N_4/TiO_2 composite were transferred to the glass vessel and magnetically stirred before and during irradiation. Before irradiation, the photocatalyst suspension containing Orange II was kept in the dark for 30 min to reach an adsorption–desorption equilibrium. Then, the reaction mixture was irradiated for 90 min with an ultraviolet black light (352 nm). The light intensity was measured using a UV radiometer with a 320–410 nm wavelength sensor (UVR-400; Luchi Co., Osaka, Japan), and the value was 1.0 mW cm^{-2}. During the decolorization reaction, 2 mL of the sample solution was withdrawn at different intervals and then centrifuged for 5 min at 10,000 rpm to isolate the photocatalyst. Then, the absorbance of the supernatant was analyzed using a UV–Vis spectrophotometer (UV-2450; Shimadzu, Kyoto, Japan) at 485 nm to assess the concentration of the Orange II solution.

3. Results and Discussion

3.1. Structural Analyses (XRD and XPS)

The prepared samples were subjected to an XRD analysis to examine their crystal structure and phase composition (Figure 1a). The XRD pattern of g-C_3N_4 exhibited two distinctive diffraction peaks at 27.4° and 13.1°, corresponding with the (002) and (100) crystalline planes of g-C_3N_4. The (002) crystal plane of g-C_3N_4 represents inter-planar stacking peaks of the aromatic systems (heptazine unit). The peak at about 13.1° (100) indexed to the in-plane packing of the tri-s-triazine structure was less pronounced, implying that the order of the structure in the plane became worse because of a loss of lattice carbon or nitrogen. The XRD pattern of g-C_3N_4/TiO_2 revealed numerous peaks at 25.4°, 37.7°, 48.2°, 54.0°, 55.1°, 62.8°, 68.9°, 70.2°, and 75.8° corresponding with the diffraction planes of (101), (004), (200), (105), (211), (204), (116), and (220), which corresponded with the anatase phase [24]. TiO_2 displayed a characteristic diffraction peak at 25.4°, indicating the presence of (101) crystal planes, thus confirming its cubic crystal structure. However, due to the low content of g-C_3N_4, diffraction peaks around 13.1° and 27.4° were not detected in the g-C_3N_4/TiO_2 composite [24–26].

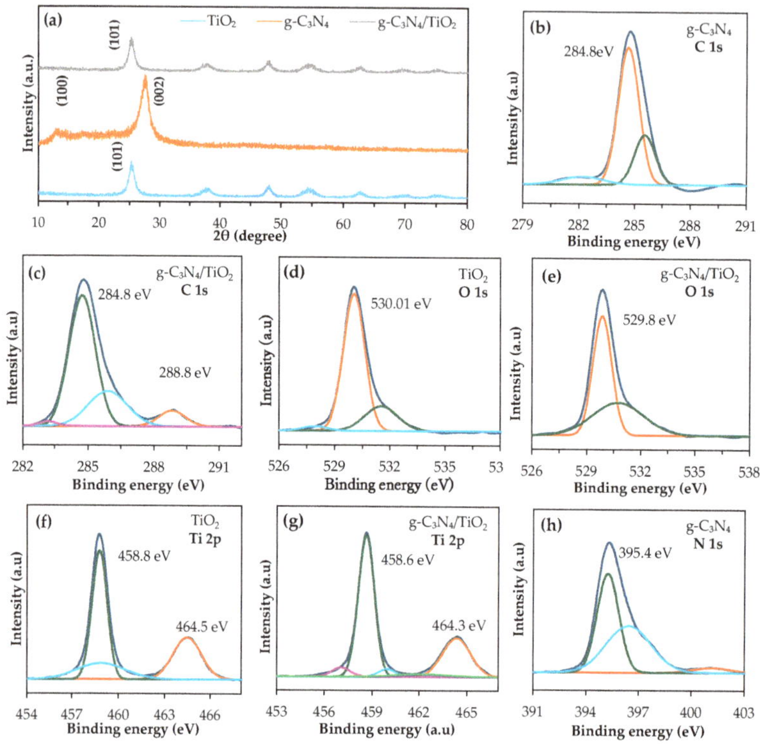

Figure 1. (**a**) XRD patterns and XPS high-resolution spectra. (**b**,**c**) C 1s, (**d**,**e**) O 1s, (**f**,**g**) Ti 2p and (**h**) N 1s of the studied photocatalysts.

The XPS analysis confirmed the presence of carbon (C), oxygen (O), and titanium (Ti) in the g-C_3N_4/TiO_2 composite, and nitrogen (N) was detected only in pristine g-C_3N_4 (Figure 1b–h). This could have been due to a low concentration of g-C_3N_4 that was insufficient to produce a detectable nitrogen peak in the XPS spectrum [27–29]. The detected peak at 284.8 eV in pure g-C_3N_4 and g-C_3N_4/TiO_2 corresponded with C–C or C–H bonds. However, a new peak at 288.8 eV was found in the composite, which corresponded with the C=O bond. This peak suggested the oxidation or functionalization of C due to interactions with TiO_2 during composite formation [30]. There was no distinct Ti-C(N) coordination peak observed, indicating that nitrogen and carbon were not incorporated into the TiO_2 lattice. This suggests that g-C_3N_4 was merely adsorbed onto the surface of TiO_2, with no chemical interaction between the two materials [31]. Both TiO_2 and the g-C_3N_4/TiO_2 composite showed firm peaks for Ti^{4+} at nearly 464 eV (Ti $2p_3/_2$) and 458 eV (Ti $2p_1/_2$), indicating that the overall oxidation state of titanium in both materials was Ti^{4+}. The peak positions and intensities were very similar, suggesting that the addition of g-C_3N_4 did not significantly affect the Ti^{4+} state in TiO_2. A small peak related to Ti^{3+} was present in both spectra, but it was more noticeable in the composite. In the composite, a peak at 459.92 eV was likely associated with Ti^{3+}, indicating that the presence of g-C_3N_4 induced some reduction in TiO_2 at the surface. In contrast, TiO_2 had a smaller contribution from Ti^{3+}, but the composite showed a shift and a higher proportion of reduced Ti species (which was species 3/4). The position of the Ti–O peak was almost identical in both materials (530.01 eV for TiO_2 and 529.8 eV for the composite), indicating that the titanium–oxygen bonding in TiO_2 remained relatively unchanged in the composite. However, there was a slight shift, possibly due to the influence of g-C_3N_4 on the surface structure of TiO_2. There was no apparent shift in the composite's oxygen vacancy peak, which suggests that g-C_3N_4 remained stable on the surface of TiO_2, reducing the presence of oxygen vacancies [16,17,31,32].

3.2. Morphological Analyses (SEM, TEM, and BET)

The surface morphology and microstructural properties of the synthesized photocatalysts were investigated using SEM and TEM techniques. Figure 2a shows that the shapes of the TiO_2 particles were sphere-like structures with an average diameter of about 7 nm. Figure 2b depicts the irregularly spaced layered structure of g-C_3N_4 and loosened thinner nanosheets. TiO_2 is primarily shown in the form of aggregated particles. Figure 2c depicts the changed morphology of g-C_3N_4/TiO_2, suggesting that TiO_2 nanoparticles were distributed on the surface of g-C_3N_4 to construct a heterostructure, which provided a transport channel for photogenerated electrons. This was further confirmed by the TEM analysis [25,26].

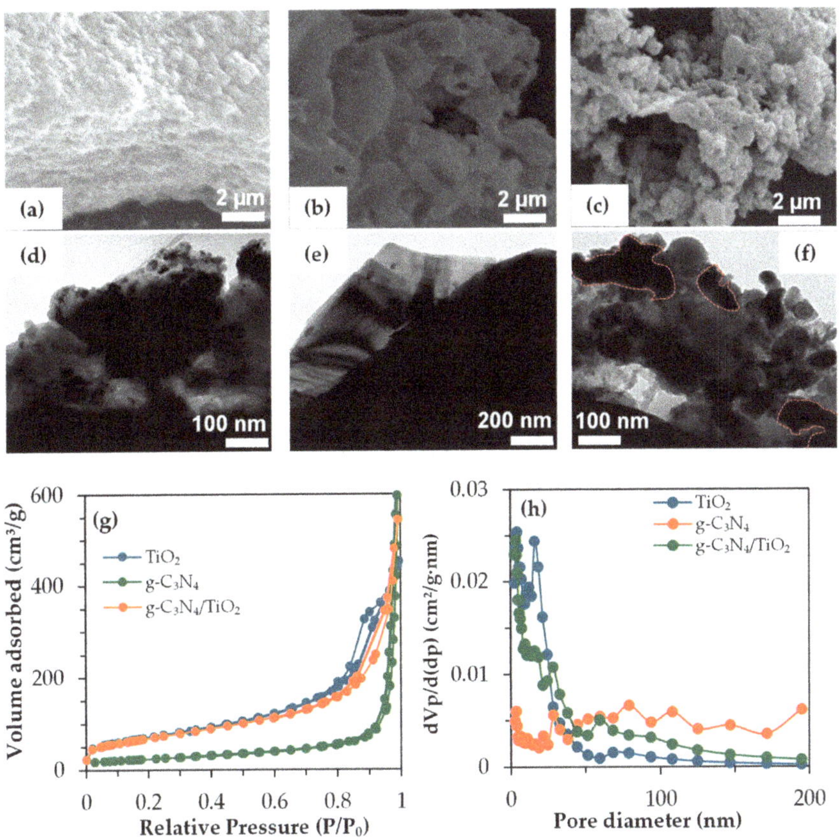

Figure 2. SEM and TEM images of (**a**,**d**) TiO_2, (**b**,**e**) g-C_3N_4, and (**c**,**f**) g-C_3N_4/TiO_2. (**g**) N_2 adsorption–desorption isotherm and (**h**) corresponding pore-size distribution of TiO_2, g-C_3N_4, and g-C_3N_4/TiO_2.

From the composite TEM image (Figure 2f), the distribution of TiO_2 on the g-C_3N_4 surface was observed. The TEM image indicates that TiO_2 comprised agglomerated nanoparticles in the g-C_3N_4/TiO_2 composite. The smaller-sized TiO_2 particles are shown in a dark shade (dot-marked), while g-C_3N_4 is presented as a light shade with irregular edges. The characterization verified the adequate combination between each of the compounds of TiO_2 and g-C_3N_4, which could promote the carrier's transfer between interfaces.

The BET surface area of $g\text{-}C_3N_4/TiO_2$ was 244 m^2/g, while that of $g\text{-}C_3N_4$ was 824 m^2/g, and TiO_2 was 253 m^2/g. The pore diameter of $g\text{-}C_3N_4/TiO_2$ was the smallest among $g\text{-}C_3N_4$, TiO_2, and $g\text{-}C_3N_4/TiO_2$ (Table 1; Figure 2h). The reduction in surface area for the composite was likely due to the interaction between $g\text{-}C_3N_4$ and TiO_2, which may have caused partial aggregation or the covering of active surface sites [26,28,33]. This suggests that the composite material had a less available surface area for reactions than pure $g\text{-}C_3N_4$ but still retained a reasonable amount of surface area for photocatalytic activity. As shown in Figure 2g, all the studied photocatalysts exhibited IV-type adsorption–desorption isotherms, with H3-type hysteresis loops appearing between the P/P_0 values of 0.4 and 1.0, suggesting their mesoporous structure [26,27,34].

Table 1. BET surface areas and pore volumes of TiO_2, $g\text{-}C_3N_4$, and $g\text{-}C_3N_4/TiO_2$.

Photocatalyst	Surface Area (m^2/g)	Pore Volume (cm^3/g)
TiO_2	253	0.6841
$g\text{-}C_3N_4$	824	0.9362
$g\text{-}C_3N_4/TiO_2$	244	0.8294

3.3. DRS Analysis

Figure 3a illustrates the UV–Vis diffuse reflectance spectra of the synthesized photocatalysts, providing insights into their light absorption properties. Furthermore, the enhanced Kubelka–Munk function was utilized to determine the energy band gap (E_g) of $g\text{-}C_3N_4$, TiO_2, and $g\text{-}C_3N_4/TiO_2$, employing a Tauc diagram (Figure 3b). Figure 3a demonstrates that $g\text{-}C_3N_4$ exhibited robust absorption within the ultraviolet–visible region. In contrast, the $g\text{-}C_3N_4/TiO_2$ photocatalyst displayed a notable blueshift and exhibited substantial absorption primarily in the ultraviolet region, resembling the behavior of TiO_2. These outcomes indicate the significant impact of the $g\text{-}C_3N_4$ compound on the optical properties of $g\text{-}C_3N_4/TiO_2$, further enhancing its photocatalytic activity under UV light.

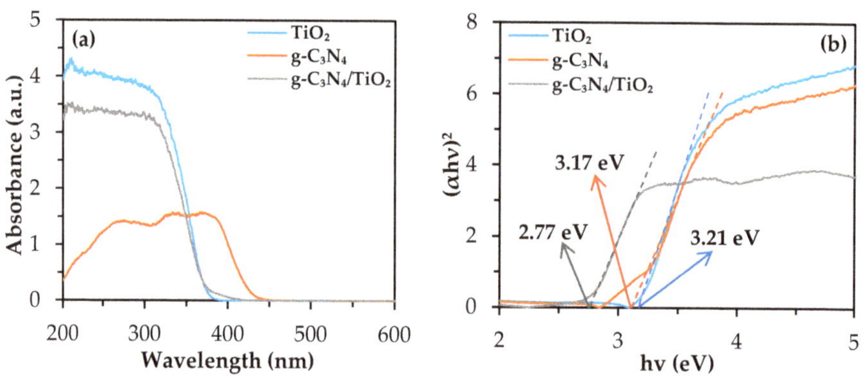

Figure 3. (**a**) DRS spectra and (**b**) Tauc plot of TiO_2, $g\text{-}C_3N_4$, and $g\text{-}C_3N_4/TiO_2$ photocatalysts.

The band gap energies (E_g) of the single and composite materials were calculated using the Kubelka–Munk formula based on the UV–Vis DRS spectra as follows [6,7]:

$$(\alpha h\upsilon) = A(h\upsilon - E_g)^{n/2} \quad (1)$$

where α, h, υ, A, and E_g are the absorption coefficient, Planck constant, light frequency, constant, and band gap, respectively. The value of n depends on the type of semiconductor (1 for semiconductors with a direct band gap and 4 for semiconductors with an indirect band gap). In the present study, the value of n was 4 because $g\text{-}C_3N_4$ and TiO_2 are indirect semiconductors.

The calculated band gap energies for g-C_3N_4, TiO_2, and g-C_3N_4/TiO_2 were 2.77 eV, 3.21 eV, and 3.17 eV, respectively, as depicted in Figure 3b, aligning with Equation (3). The broadening of the UV light range and the reduction in the energy band gap of the g-C_3N_4/TiO_2 photocatalyst contributed to enhancing its photocatalytic efficiency.

3.4. Mott–Schottky and Band Structure Analyses

The Mott–Schottky analysis is commonly used to determine the flat-band potential of semiconductors. This analysis gives an insight into the electrostatic potential at the surface of a semiconductor when no external voltage is applied. This potential is associated with the energy levels of the semiconductor's CB and VB [35]. The positive slope proved that the TiO_2, g-C_3N_4, and g-C_3N_4/TiO_2 photocatalysts belonged to the n-type semiconductor category.

$$E_{NHE} = E_{Ag/AgCl} + 0.197 \quad (2)$$

The flat-band potentials of TiO_2, g-C_3N_4, and g-C_3N_4/TiO_2 were −0.71 V, −0.73 V, and −0.87 V, respectively (V vs. Ag/AgCl; sat. KCl at pH 7). The value was further converted to −0.53V, −0.53 V, and −0.67 V, respectively, using Equation (2), which was V vs. the standard hydrogen electrode (NHE) (Figure 4a–c). Based on a previous report, n-type semiconductor materials have a CB potential of approximately 0.1 eV lower than the E_{fb} [36]. Therefore, the conduction band positions of TiO_2, g-C_3N_4, and g-C_3N_4/TiO_2 were −0.41 eV, −0.43 eV, and −0.57 eV, respectively.

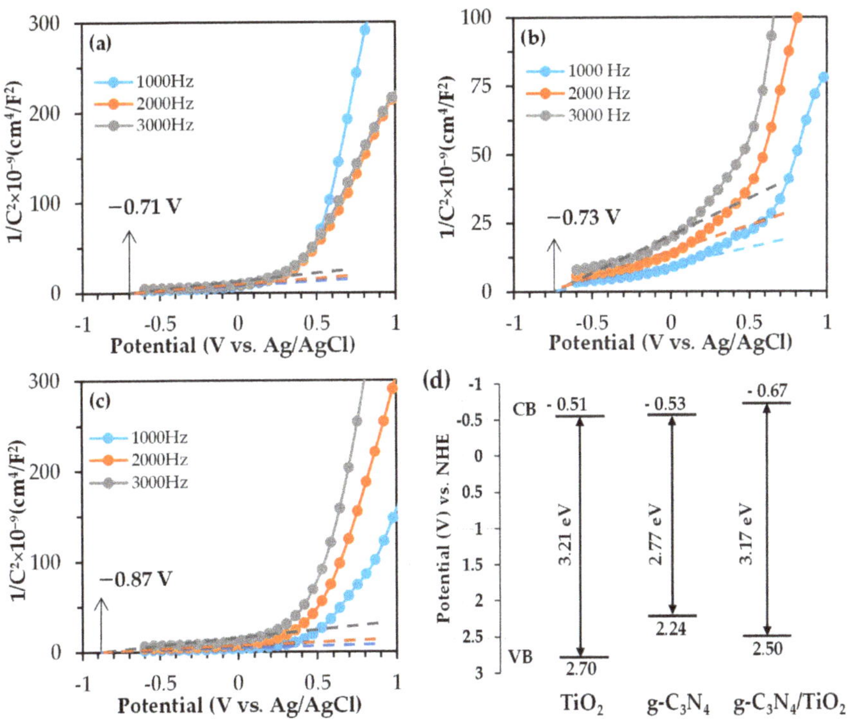

Figure 4. Mott–Schottky plots: (**a**) TiO_2, (**b**) g-C_3N_4, and (**c**) g-C_3N_4/TiO_2. (**d**) Band structure.

The CB and VB positions are critical for understanding how electrons and holes behave in a material under light excitation. For instance, a more negative CB implies that the material can more readily accept electrons, and the VB position determines the energy level of the holes generated during photoexcitation. The g-C_3N_4/TiO_2 composite had a band structure that could facilitate a more efficient electron transfer between the two materials, which could be beneficial for applications such as photocatalysis, where electron–hole separation and charge transfer are critical for improving performance (Figure 4d) [35,36]. The shift in the flat-band potential and changes in the band positions in the g-C_3N_4/TiO_2 material suggested enhanced electronic interactions, which could improve photocatalytic performance due to better electron transfer between the two components.

3.5. PL and Electrochemical Analyses

The performance of photocatalysts is significantly influenced by the rate of electron–hole recombination, which can be examined using photoluminescence (PL) spectra. A weaker peak intensity in the PL spectra indicates a stronger suppression of electron–hole recombination, leading to enhanced photocatalytic activity. In Figure 5a, the PL spectra of the two photocatalysts, g-C_3N_4 and g-C_3N_4/TiO_2, are presented with an excitation wavelength of 350 nm. In the case of pure g-C_3N_4, the strong peak was located at about 450 nm, which corresponded with the band gap of g-C_3N_4 (2.68 eV) [6,36], indicating a large number of detected electron–hole recombinations in the PL spectra. Moreover, the PL spectrum was similar to that of pure g-C_3N_4, but the emission intensity of g-C_3N_4/TiO_2 sharply decreased due to the charge transfer between g-C_3N_4 and TiO_2 microspheres, implying an increase in the separation efficiency of the photogenerated electron–hole pairs and suggesting efficient electron transfer within the formed g-C_3N_4/TiO_2 heterojunction structures. Therefore, it is easy to explain that g-C_3N_4/TiO_2 exhibited enhanced photocatalytic performance for the photodegradation (decolorization) of Orange II.

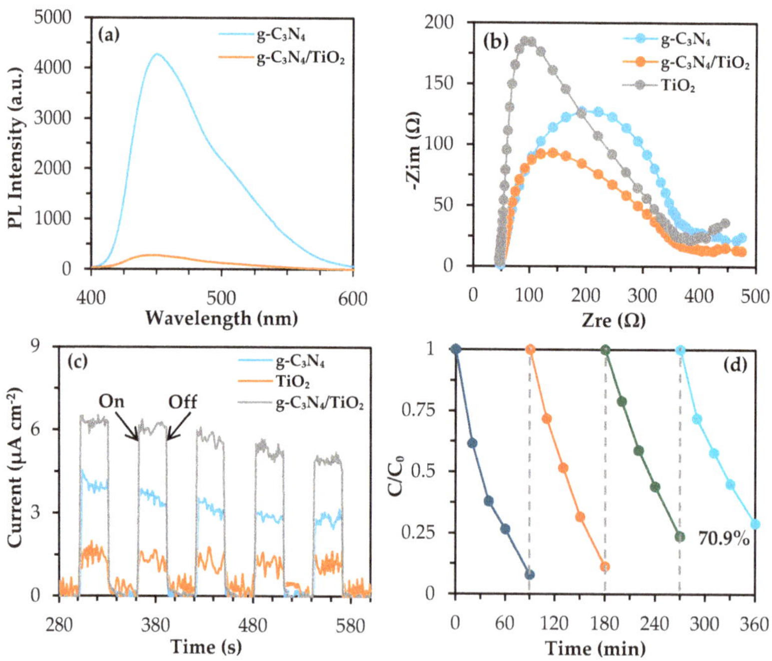

Figure 5. (**a**) Photoluminescence spectra, (**b**) EIS Nyquist plot, (**c**) TPCR plot, and (**d**) stability test result of the reported photocatalysts.

In electrochemical impedance spectroscopy (EIS), a Nyquist plot is commonly used to analyze the impedance characteristics of electrochemical systems. In these plots, the x-axis typically represents the real part of the impedance (resistance) and the y-axis represents the imaginary part (reactance). The shape of the plot and the size of the semicircle are related to the electrochemical properties, including charge transfer resistance and the diffusion processes in the system [36,37]. In the EIS results (Figure 5b), the radius of the g-C_3N_4/TiO_2 composite was smaller than that of TiO_2 and g-C_3N_4 alone. The smaller radius observed for the g-C_3N_4/TiO_2 composite suggests that the combination of these two materials enhanced the charge transfer process. This composite likely improved the electron transfer or provided better charge separation compared with the individual components (TiO_2 or g-C_3N_4).

Figure 5c illustrates the photocurrent–time curve. Similar to the transient photocurrent response (TPCR) measurements, the photocurrent intensities of the as-prepared photocatalysts presented in the order of TiO_2 (about 1.5 μA cm^{-2}) < g-C_3N_4 (about 4.0 μA cm^{-2}) < g-C_3N_4/TiO_2 (about 6.2 μA cm^{-2}). The photocurrent density value of g-C_3N_4/TiO_2 was 4.13-fold higher than TiO_2 and 1.55-fold higher than g-C_3N_4. These results proved the important role of the introduction of g-C_3N_4 nanosheets, where the high charge-carrier mobility acted as an efficient photoelectron accepter from the TiO_2 CB and as a low resistive highway for transferring the photoinduced charges through a π–π conjugated network [38,39]. Consequently, the photocurrent response of g-C_3N_4/TiO_2 surpassed that obtained for both TiO_2 or g-C_3N_4 individually.

The findings from the PL and photocurrent analyses demonstrated that the prepared g-C_3N_4/TiO_2 heterojunction enhanced the separation of photogenerated electron–hole pairs and delayed their recombination, which showed superior photocatalytic activity.

3.6. Stability Test

A cyclic dye degradation reaction was performed for up to four consecutive cycles using 0.1 wt% of the g-C_3N_4/TiO_2 composite under the same reaction conditions to assess the photocatalytic stability and reusability. The degradation rate of Orange II was 70.9%, which showed a decrease compared with the initial rate of 92.1% (Figure 5d). However, it still maintained good catalytic efficiency. Despite the drop in efficiency, a 70.9% degradation rate in the fourth cycle is still considered to be high, indicating that the photocatalyst retained good catalytic performance after multiple cycles. This suggests that the material exhibited a reasonable degree of stability and could continue to effectively function over several uses, which is important for practical applications in environmental remediation.

3.7. Photocatalytic Degradation

3.7.1. Effect of g-C_3N_4 Addition

The improvement in the photocatalytic efficiency of TiO_2 due to the addition of g-C_3N_4 is evident in Figure 6. Figure 6a illustrates the results of photocatalytic decolorization when different weight percentages of TC-g-C_3N_4 were added to the composites. The decolorization efficiency reached its peak at a mass percentage of 0.5 wt% TC-g-C_3N_4. Additionally, the highest decolorization rate was achieved when the weight percentage of g-C_3N_4 was 5 wt%. However, Figure 6b indicates that the decolorization rate remained nearly constant at a g-C_3N_4 content of 1 wt%. In Figure 6c, a comparison is made between the highest decolorization rates of TC-g-C_3N_4 and the g-C_3N_4 composites. The decolorization rates of 0.5 wt% TC-g-C_3N_4/TiO_2 and 1 wt% g-C_3N_4/TiO_2 were equal, but higher than that of 1 wt% TC-g-C_3N_4/TiO_2. Therefore, it was found that the addition of trisodium citrate was not effective for the improvement of decolorization efficiency with the composite.

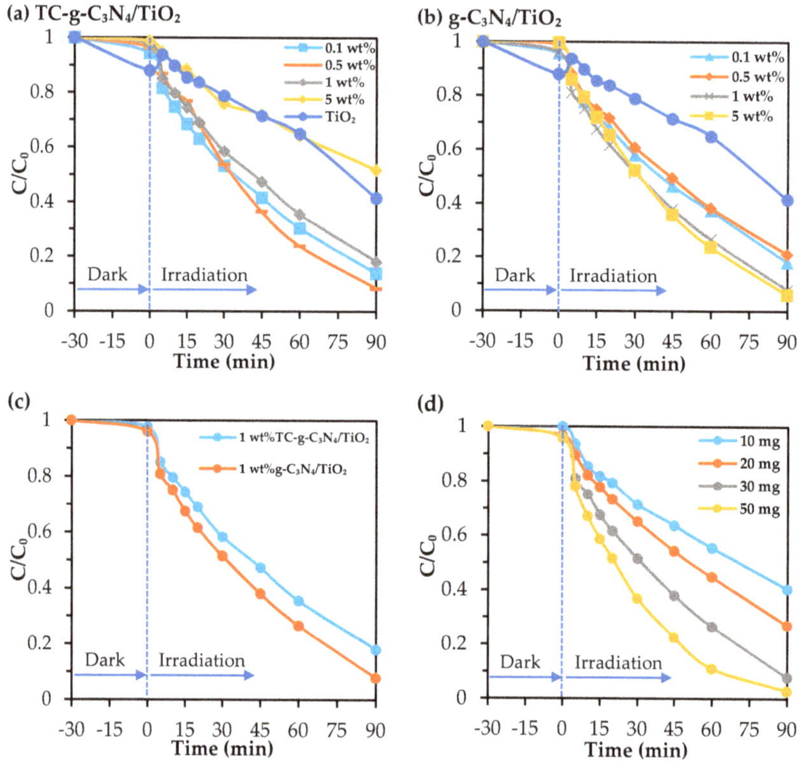

Figure 6. Effect of (**a**) TC−g-C_3N_4 and (**b**) g-C_3N_4 amounts during composite formation on the photocatalytic decolorization of Orange II. Comparison of (**c**) 1 pure g-C_3N_4 and TC−g-C_3N_4 composite. (**d**) Effect of photocatalyst loading on photocatalytic decolorization with 1 wt% g-C_3N_4/TiO_2.

3.7.2. Effect of Photocatalyst Amount

The photocatalytic decolorization efficiency of the composites was evaluated using 1 wt% g-C_3N_4/TiO_2 at various dosages of 10, 20, 30, and 50 mg, as depicted in Figure 6d. It can be observed that the addition amount of the composite material ranged from 10 mg to 50 mg. With an increase in the usage of the composite photocatalyst, the decolorization rate of Orange II dye showed an increase.

3.7.3. Reaction Rate

The Langmuir–Hinshelwood model is often used in the reaction rate theory of photocatalysts [15]. The Langmuir–Hinshelwood formula shown here is a generalization of Langmuir's equation of energy-absorbing isotherms applied to surface reactions.

$$r = -\frac{dC}{dt} = \frac{kKC}{1+KC} \tag{3}$$

where k is the reaction rate constant (mg (L min)$^{-1}$), K is the adsorption coefficient of the reactant (L mg^{-1}), and C is the concentration of the reactant (mg L^{-1}).

When the concentration of the reactant C is minimal, it can be approximated by the following formula [9,15,39]:

$$-\frac{dC}{dt} = kKC \tag{4}$$

$$-Ln\frac{C}{C_0} = k_{app}t \tag{5}$$

where k_{app} is the apparent first-order rate constant.

As Figure 7 presents an almost straight line, this decomposition followed the Langmuir–Hinshelwood model, and decomposition was carried out by a pseudo-first-order reaction.

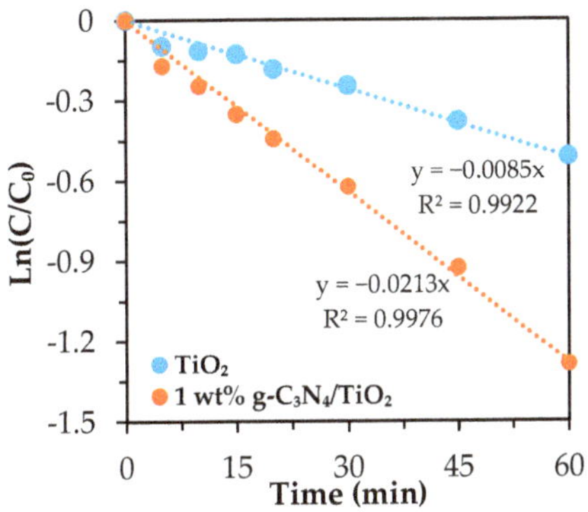

Figure 7. Ln (C/C_0) versus time in photocatalytic decolorization with 1 wt% $g\text{-}C_3N_4/TiO_2$.

As shown in Table 2, using the composite photocatalyst $g\text{-}C_3N_4/TiO_2$, the rate theoretic parameters were as follows: the pseudo-first-order reaction rate constant was 0.0213 min^{-1}, the linear R^2 was 0.99, the time when the concentration changed to a half-life was 32.5 min, and the reaction rate was 2.5 times that of TiO_2.

Table 2. Photocatalytic degradation kinetic parameters (rate constants, correlation, and half-lives).

	k_{obs} (min^{-1})	R^2	$t_{1/2}$ (min)
1 wt% $g\text{-}C_3N_4/TiO_2$	0.0213	0.994	32.5
TiO_2	0.0085	0.978	81.5

3.8. Reaction Mechanism

According to the reference papers [12,17,40–42], the S-scheme charge transfer mechanism of $g\text{-}C_3N_4/TiO_2$ can be considered to propose the photocatalytic decolorization process of Orange II. The Fermi level of TiO_2 microspheres is lower compared with those of $g\text{-}C_3N_4$. When $g\text{-}C_3N_4$ is modified on the surface of TiO_2 microspheres, electrons can be transferred from $g\text{-}C_3N_4$ to titanium dioxide microspheres until both Fermi levels reach the same level.

When exposed to light illumination, the VB electrons of TiO_2 particles and $g\text{-}C_3N_4$ become excited, moving to their respective CB while leaving holes in their VB [15,43,44]. This process led to the formation of photogenerated electron–hole pairs in both the TiO_2 particles and $g\text{-}C_3N_4$ (Figure 8). The photogenerated electrons from the TiO_2 particles were transferred to $g\text{-}C_3N_4$ and combined with the photogenerated holes present in $g\text{-}C_3N_4$. This arrangement left photogenerated holes with strong oxidation ability on the VB of the TiO_2 particles and the photogenerated electrons with strong reduction ability on the CB of $g\text{-}C_3N_4$. Specifically, the photogenerated holes on the VB of the TiO_2 particles reacted with H_2O or OH^- to form hydroxyl radicals ($^\bullet OH$), while the photogenerated electrons on the CB of $g\text{-}C_3N_4$ reacted with O_2 to generate superoxide anion radicals ($^\bullet O_2^-$). Both $^\bullet OH$ and $^\bullet O_2^-$ demonstrate enhanced oxidation and reduction abilities, respectively [45–47].

If conventional type-II heterojunctions are constructed between the two components of $g\text{-}C_3N_4$ and TiO_2, the electrons in the CB of $g\text{-}C_3N_4$ will transfer to the CB of TiO_2, and the h^+ in the VB of TiO_2 will transfer to the VB of $g\text{-}C_3N_4$. However, the h^+ on the VB of $g\text{-}C_3N_4$ cannot react with H_2O or OH^- to produce $^\bullet OH$ according to previous studies [20,34,47], which is contrary to the above conclusion that $^\bullet OH$ was one of the main active species. Therefore, it could be concluded that an S-scheme heterojunction was constructed in the composite, greatly improving the separation efficiency of the photogenerated electron–hole pairs and the photodegradation performance of Orange II.

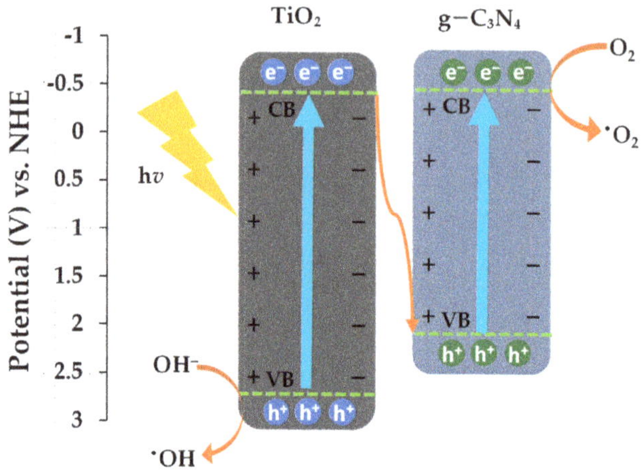

Figure 8. Schematic diagram of the photocatalytic decolorization of Orange II in an aqueous solution with a g-C_3N_4/TiO_2 composite.

When exposed to UV irradiation, the internal electric field promotes the combination of electrons in the TiO_2 CB with holes in the g-C_3N_4 VB. This process facilitates the separation of holes in the TiO_2 VB and electrons in the g-CB, thereby enhancing the decolorization rate [20,25,34]. The holes in the VB of TiO_2 undergo oxidation with H_2O, producing $^{\bullet}OH$ radicals. Simultaneously, the electrons in the CB of g-C_3N_4 react with O_2, forming $^{\bullet}O_2^-$ species. Reactive oxygen species (ROS) are highly reactive and can degrade dye molecules by breaking down their chemical bonds.

The g-C_3N_4/TiO_2 composite system enhanced the photocatalytic activity due to the synergistic effect between the two materials. TiO_2 contributed to the generation of ROS under UV light, while g-C_3N_4 helped to extend the absorption into the visible light range and promoted charge-carrier separation. This combination significantly improved the photocatalytic degradation of organic pollutants.

Compared with other studies, the g-C_3N_4/TiO_2 composite demonstrated a higher degradation rate for orange–yellow dye under UV light, achieving a higher decomposition efficiency within a shorter period (Table 3).

Table 3. Comparison of photocatalytic dye degradation.

Photocatalyst (mg/mL)	Dye	Light Source (nm)	Irrad. Time (min)	Decoloration Rate (%)	Ref.
g-C_3N_4/TCNE-40 (0.5)	OR II	Visible (≥420)	60	90.16	[4]
FeOCl/WS_2 (F1W2) (0.6)	OR II	Visible (≥420)	150	93.4	[5]
Bi-doped g-C_3N_4/Bi_2WO_6 (20 GBW)	RO-16	Visible (≥420)	300	89	[37]
g-C_3N_5/g-C_3N_4 (0.4)	MO	Visible (≥420)	30	99.6	[38]
TiO_2/Al_2O_3 (0.2)	RO-16	UV (365/254)	225	85	[48]
10 wt% g-C_3N_4/ZnO (2)	MO	Solar Light	120	98	[49]
BiOI/TiO_2	OR II	Visible (<400 nm)	120	60	[50]
1 wt% g-C_3N_4/TiO_2 (0.3)	OR II	UV (352)	60	92.1	This work

OR: Orange II; RO-16: Reactive Orange 16; MO: Methyl Orange.

4. Conclusions

In summary, a g-C_3N_4/TiO_2 composite was prepared by the simple physical mixing of TiO_2 and g-C_3N_4 photocatalysts. This composite exhibited enhanced photodecomposition of Orange II dye. The XPS and TEM analyses confirmed the composite formation. PL and electrochemical measurements disclosed the improved charge separation, migration, and delayed recombination of photogenerated electron–hole pairs. The pseudo-first-order reaction rate constant (k) of g-C_3N_4/TiO_2 was 0.0213 min^{-1}, which was 2.5 times higher than pure TiO_2. The best performance was achieved using 1 wt% of the g-C_3N_4/TiO_2 composite. In total, 92.1% of Orange II dye was decomposed within 60 min of irradiation. The g-C_3N_4/TiO_2 composites show great promise in the photodegradation of pollutants, with the advantages of low cost, nontoxicity, a simple preparation process, the efficient use of solar energy, and fast electron–hole pair separation with high redox ability.

Author Contributions: B.G.: conceptualization, data curation, writing—original draft preparation; M.U.: formal analysis, investigation, writing—original draft preparation, writing—review and editing; M.F.: investigation and data curation; I.T.: investigation and data curation; H.K.: writing—review and editing; S.K.: writing—review and editing and supervision. All authors have read and agreed to the published version of the manuscript.

Funding: This research was partially funded by Grant-in-Aid for Scientific Research (B) 21H03642 from the Ministry of Education, Culture, Sports, Science, and Technology of Japan.

Data Availability Statement: Data are contained within the article.

Acknowledgments: The authors acknowledge Erina Takase for experimental support.

Conflicts of Interest: All experiments were conducted at Mie University. Any opinions, findings, conclusions, or recommendations expressed in this paper are those of the authors and do not necessarily reflect the view of the supporting organizations.

References

1. Islam, M.R.; Islam, J.B.; Furukawa, M.; Tateishi, I.; Katsumata, H.; Kaneco, S. Photocatalytic Degradation of a Systemic Herbicide: Picloram from Aqueous Solution Using Titanium Oxide (TiO_2) under Sunlight. *ChemEngineering* **2020**, *4*, 58. [CrossRef]
2. Choi, S.; Do, J.Y.; Lee, J.H.; Ra, C.S.; Kim, S.K.; Kang, M. Optical Properties of Cu-Incorporated ZnO (Cu_xZn_yO) Nanoparticles and Their Photocatalytic Hydrogen Production Performances. *Mater. Chem. Phys.* **2018**, *205*, 206–209. [CrossRef]
3. Chang, C.J.; Lin, Y.G.; Weng, H.T.; Wei, Y.H. Photocatalytic Hydrogen Production from Glycerol Solution at Room Temperature by ZnO-ZnS/Graphene Photocatalysts. *Appl. Surf. Sci.* **2018**, *451*, 198–206. [CrossRef]
4. Wang, Z.; Huo, Y.; Fan, Y.; Wu, R.; Wu, H.; Wang, F.; Xu, X. Facile synthesis of carbon-rich g-C_3N_4 by copolymerization of urea and tetracyanoethylene for photocatalytic degradation of Orange II. *J. Photochem. Photobiol. A Chem.* **2018**, *358*, 61–69. [CrossRef]
5. Xie, D.; Tang, C.; Li, D.; Yuan, J.; Xu, F. FeOCl/WS_2 composite as a heterogeneous Fenton catalyst to efficiently degrade acid orange II. *Inorg. Chem. Commun.* **2023**, *156*, 111085. [CrossRef]
6. Wang, J.; Wang, S. A critical review on graphitic carbon nitride (g-C_3N_4)-based materials: Preparation, modification and environmental application. *Coord. Chem. Rev.* **2022**, *453*, 214338. [CrossRef]
7. Zhang, M.; Yang, Y.; An, X.; Hou, L.-A. A critical review of g-C_3N_4-based photocatalytic membrane for water purification. *Chem. Eng. J.* **2021**, *412*, 128663. [CrossRef]
8. Shen, X.; Song, L.; Luo, L.; Zhang, Y.; Zhu, B.; Liu, J.; Chen, Z.; Zhang, L. Preparation of TiO_2/C_3N_4 heterojunctions on carbon-fiber cloth as efficient filter-membrane-shaped photocatalyst for removing various pollutants from the flowing wastewater. *J. Colloid Interface Sci.* **2018**, *532*, 798–807. [CrossRef]
9. Zhu, B.; Cheng, B.; Fan, J.; Ho, W.; Yu, J. g-C_3N_4-Based 2D/2D Composite Heterojunction Photocatalyst. *Small Struct.* **2021**, *2*, 2100086. [CrossRef]
10. Chouchene, B.; Gries, T.; Balan, L.; Medjahdi, G.; Schneider, R. Graphitic carbon nitride/$SmFeO_3$ composite Z-scheme photocatalyst with high visible light activity. *Nanotechnology* **2020**, *31*, 465704. [CrossRef]
11. Acharya, R.; Parida, K. A review on TiO_2/g-C_3N_4 visible-light-responsive photocatalysts for sustainable energy generation and environmental remediation. *J. Environ. Chem. Eng.* **2020**, *8*, 103896. [CrossRef]
12. Kane, A.; Chafiq, L.; Dalhatou, S.; Bonnet, P.; Nasr, M.; Gaillard, N.; Dangwang Dikdim, J.M.; Monier, G.; Amine Assadi, A.; Zeghioud, H. g-C_3N_4/TiO_2 S-scheme heterojunction photocatalyst with enhanced photocatalytic Carbamazepine degradation and mineralization. *J. Photochem. Photobiol. A Chem.* **2022**, *430*, 113971. [CrossRef]
13. Fujita, S.; Kawamori, H.; Honda, D.; Yoshida, H.; Arai, M. Photocatalytic Hydrogen Production from Aqueous Glycerol Solution Using NiO/TiO_2 Catalysts: Effects of Preparation and Reaction Conditions. *Appl. Catal. B Environ.* **2016**, *181*, 818–824. [CrossRef]
14. Ismael, M. A review on graphitic carbon nitride (g-C_3N_4) based nanocomposites: Synthesis, categories, and their application in photocatalysis. *J. Alloys Compd.* **2020**, *846*, 156446. [CrossRef]

15. Ouedraogo, S.; Chouchene, B.; Desmarets, C.; Gries, T.; Balan, L.; Fournet, R.; Medjahdi, G.; Bayo, K.; Schneider, R. Copper octacarboxyphthalocyanine as sensitizer of graphitic carbon nitride for efficient dye degradation under visible light irradiation. *Appl. Catal. A Gen.* **2018**, *563*, 127–136. [CrossRef]
16. Zhang, Y.; Xu, J.; Mei, J.; Sarina, S.; Wu, Z.; Liao, T.; Yan, C.; Suna, Z. Strongly interfacial-coupled 2D-2D $TiO_2/g\text{-}C_3N_4$ heterostructure for enhanced visible-light induced synthesis and conversion. *J. Hazard. Mater.* **2020**, *394*, 122529. [CrossRef]
17. Jiang, Y.; Sun, Z.; Chen, Q.; Cao, C.; Zhao, Y.; Yang, W.; Zeng, L.; Huang, L. Fabrication of 0D/2D TiO_2 Nanodots/g-C_3N_4 S-scheme heterojunction photocatalyst for efficient photocatalytic overall water splitting. *Appl. Surf. Sci.* **2022**, *571*, 151287. [CrossRef]
18. Yang, J.; Wu, X.; Mei, Z.; Zhou, S.; Su, Y.; Wang, G. CVD Assisted Synthesis of Macro/Mesoporous $TiO_2/g\text{-}C_3N_4$ S-Scheme Heterojunction for Enhanced Photocatalytic Hydrogen Evolution. *Adv. Sustain. Syst.* **2002**, *6*, 220005. [CrossRef]
19. Zhang, X. Layered g-C_3N_4/TiO_2 nanocomposites for efficient photocatalytic water splitting and CO_2 reduction: A review. *Mater. Today Energy* **2022**, *23*, 100904. [CrossRef]
20. Li, Y.; Zhou, M.; Cheng, B.; Shao, Y. Recent advances in g-C_3N_4-based heterojunction photocatalysts. *J. Mater. Sci. Technol.* **2020**, *56*, 1–17. [CrossRef]
21. Ni, S.; Fu, Z.; Li, L.; Ma, M.; Liu, Y. Step-scheme heterojunction g-C_3N_4/TiO_2 for efficient photocatalytic degradation of tetracycline hydrochloride under UV light. *Colloids Surf. A Physicochem. Eng.* **2022**, *649*, 129475. [CrossRef]
22. Hossain, S.M.; Park, H.; Kang, H.-J.; Mun, J.S.; Tijing, L.; Rhee, I.; Kim, J.-H.; Jun, Y.-S.; Shon, H.K. Modified Hydrothermal Route for Synthesis of Photoactive Anatase TiO_2/g-CN Nanotubes from Sludge Generated TiO_2. *Catalysts* **2020**, *10*, 1350. [CrossRef]
23. Muhammad, A.; Hassan, Z.; Mohammad, S.M.; Rajamanickam, S.; Shitu, I.G. Effect of trisodium citrate on morphological, structural, and optical properties of fluorine-doped ZnO structures. *Ceram. Int.* **2022**, *48*, 13431–13439. [CrossRef]
24. Wang, W.; Liu, Y.; Qu, J.; Chen, Y.; Tad, M.O.; Zhou, W. Synthesis of Hierarchical TiO_2–C_3N_4 Hybrid Microspheres with Enhanced Photocatalytic and Photovoltaic Activities by Maximizing the Synergistic Effect. *ChemPhotoChem* **2017**, *1*, 35–45. [CrossRef]
25. Hao, R.; Wang, G.; Tang, H.; Sun, L.; Xu, C.; Han, D. Template-free preparation of macro/mesoporous g-C_3N_4/TiO_2 heterojunction photocatalysts with enhanced visible light photocatalytic activity. *Appl. Catal. B Environ.* **2016**, *187*, 47–58. [CrossRef]
26. Cao, Y.; Yuan, G.; Guo, Y.; Hu, X.; Fang, G.; Wang, S. Facile synthesis of $TiO_2/g\text{-}C_3N_4$ nanosheet heterojunctions for efficient photocatalytic degradation of tartrazine under simulated sunlight. *Appl. Surf. Sci.* **2022**, *600*, 154169. [CrossRef]
27. Rudzani, R.; Potlako, J.M.; Olayemi, J.F.; Ajay, K.M.; Alex, T.K. Ag doped TiO_2 anchored on metal free g-C_3N_4 for enhanced solar light activated photodegradation of a dye. *Opt. Mater.* **2024**, *157*, 116125. [CrossRef]
28. Hu, K.; Li, R.; Ye, C.; Wang, A.; Wei, W.; Hu, D.; Qiu, R.; Yan, K. Facile synthesis of Z-scheme composite of TiO_2 nanorod/g-C_3N_4 nanosheet efficient for photocatalytic degradation of ciprofloxacin. *J. Clean. Prod.* **2020**, *253*, 120055. [CrossRef]
29. Reli, M.; Huo, P.; Sihor, M.; Ambrozova, N.; Troppova, I.; Matejova, L.; Lang, J.; Svoboda, L.; Kustrowski, P.; Ritz, M.; et al. Novel TiO_2/C_3N_4 Photocatalysts for Photocatalytic Reduction of CO_2 and for Photocatalytic Decomposition of N_2O. *J. Phys. Chem. A* **2016**, *120*, 8564–8573. [CrossRef]
30. Liu, X.; Chen, N.; Li, Y.; Deng, D.; Xing, X.; Wang, Y. A general nonaqueous sol-gel route to g-C_3N_4-coupling photocatalysts: The case of Z-scheme g-C_3N_4/TiO_2 with enhanced photodegradation toward RhB under visible-light. *Sci. Rep.* **2016**, *6*, 39531. [CrossRef]
31. Ting, T.L.; Meng, M.X.; Wang, H.-Y.; Wang, J.-S.; Lin, J.-H.; Lou, C.-W. An S-scheme $TiO_2/g\text{-}C_3N_4$ nanocomposite effectively degrades phenanthrene under visible light. *Colloids Surf. A Physicochem. Eng. Asp.* **2025**, *705*, 135554.
32. Dahiya, S.; Shoran, S.; Sharma, D.; Rao, V.; Chaudhary, S.; Nehra, S.; Sharma, A. Bioengineered sustainable phytofabrication of anatase TiO_2-adorned g-C_3N_4 nanocomposites and unveiling their photocatalytic potential towards advanced environmental remediation. *Chemosphere* **2024**, *362*, 142456. [CrossRef] [PubMed]
33. Pal, D.B.; Singh, A.; Bhatnagar, A. A Review on Biomass Based Hydrogen Production Technologies. *Int. J. Hydrogen Energy* **2022**, *47*, 1461–1480. [CrossRef]
34. Zhang, B.; He, X.; Ma, X.; Chen, Q.; Liu, G.; Zhou, Y.; Ma, D.; Cui, C.; Ma, J.; Xin, Y. In situ synthesis of ultrafine TiO_2 nanoparticles modified g-C3N4 heterojunction photocatalyst with enhanced photocatalytic activity. *Sep. Purif. Technol.* **2020**, *247*, 116932. [CrossRef]
35. Takahashi, J.-I. In-situ evaluation of the carrier mobility using Mott-Schottky-equivalent analysis in organic light emitting diodes. *J. Appl. Phys.* **2020**, *59*, 071005. [CrossRef]
36. Uzzaman, M.; Afrin, M.F.; Furukawa, M.; Tateishi, I.; Katsumata, H.; Kaneco, S. Visible Light-Driven Photocatalytic CH_4 Production from an Acetic Acid Solution with Cetyltrimethylammonium Bromide-Assisted $ZnIn_2S_4$. *ChemEngineering* **2024**, *8*, 75. [CrossRef]
37. Priti, R.; Bonamali, P.; Raj, K.D. Bi-doped g-C_3N_4/Bi_2WO_6 ternary composites for superior photocatalytic degradation of reactive orange 16 under visible light irradiation. *J. Ind. Eng. Chem.* **2025**, *141*, 456–467. [CrossRef]
38. Liu, S.; Bu, Y.; Cheng, S.; Tao, Y.; Hong, W. Preparation of g-C_3N_5/g-C_3N_4 heterojunction for methyl orange photocatalytic degradation: Mechanism analysis. *J. Water Process Eng.* **2023**, *54*, 104019. [CrossRef]
39. Liu, S.; Wang, X.; Wang, K.; Lv, R.; Xu, Y. ZnO/ZnS-PdS Core/Shell Nanorods: Synthesis, Characterization and Application for Photocatalytic Hydrogen Production from a Glycerol/Water Solution. *Appl. Surf. Sci.* **2013**, *283*, 732–739. [CrossRef]
40. Deluga, G.A.; Salge, J.R.; Schmidt, L.D.; Verykios, X.E. Renewable Hydrogen from Ethanol by Autothermal Reforming. *Science* **2004**, *303*, 993–997. [CrossRef]

41. Zhou, C.H.; Beltramini, J.N.; Fan, Y.X.; Lu, G.Q. Chemoselective Catalytic Conversion of Glycerol as a Biorenewable Source to Valuable Commodity Chemicals. *Chem. Soc. Rev.* **2008**, *37*, 527–549. [CrossRef] [PubMed]
42. Davda, R.R.; Shabaker, J.W.; Huber, G.W.; Cortright, R.D.; Dumesic, J.A. A Review of Catalytic Issues and Process Conditions for Renewable Hydrogen and Alkanes by Aqueous-Phase Reforming of Oxygenated Hydrocarbons over Supported Metal Catalysts. *Appl. Catal. B Environ.* **2005**, *56*, 171–186. [CrossRef]
43. Byrd, A.J.; Pant, K.K.; Gupta, R.B. Hydrogen Production from Glycerol by Reforming in Supercritical Water over Ru/Al_2O_3 Catalyst. *Fuel* **2008**, *87*, 2956–2960. [CrossRef]
44. Dauenhauer, P.J.; Salge, J.R.; Schmidt, L.D. Renewable Hydrogen by Autothermal Steam Reforming of Volatile Carbohydrates. *J. Catal.* **2006**, *244*, 238–247. [CrossRef]
45. Keskin, T.; Hallenbeck, P.C. Hydrogen Production from Sugar Industry Wastes Using Single-Stage Photofermentation. *Bioresour. Technol.* **2012**, *112*, 131–136. [CrossRef] [PubMed]
46. Lyubina, T.P.; Markovskaya, D.V.; Kozlova, E.A.; Parmon, V.N. Photocatalytic Hydrogen Evolution from Aqueous Solutions of Glycerol under Visible Light Irradiation. *Int. J. Hydrogen Energy* **2013**, *38*, 14172–14179. [CrossRef]
47. Hashaikeh, R.; Butler, I.S.; Kozinski, J.A. Selective Promotion of Catalytic Reactions during Biomass Gasification to Hydrogen. *Energy Fuels* **2006**, *20*, 2743–2747. [CrossRef]
48. Yildiz, T.; Yatmaz, H.C.; Öztürk, K. Anatase TiO_2 powder immobilized on reticulated Al_2O_3 ceramics as a photocatalyst for degradation of RO16 azo dye. *Ceram. Int.* **2020**, *46*, 8651–8657. [CrossRef]
49. Rajachandrasekar, T.; Abina, K.; Subramaniyan, R.; Sakthivel, K.; Mohammad, A.; Tae, H.O.; Srishankar, J.; Sobana, N.; Thirunarayanan, G.; Muthuvel, I. Radiant synergy: Illuminating methyl orange dye removal with g-C_3N_4/ZnO heterojunction photocatalyst. *Diam. Relat. Mater.* **2024**, *147*, 111325. [CrossRef]
50. Yuan, X.; Yi, J.; Wang, H.; Yu, H.; Zhang, S.; Peng, F. New route of fabricating BiOI and Bi_2O_3 supported TiO_2 nanotube arrays via the electrodeposition of bismuth nanoparticles for photocatalytic degradation of acid orange II. *Mater. Chem. Phys.* **2017**, *196*, 237–244. [CrossRef]

C Journal of Carbon Research

MDPI

Article

Synthesis of Ni@SiC/CNFs Composite and Its Microwave-Induced Catalytic Activity

Haibo Ouyang *, Jiaqi Liu , Cuiyan Li, Leer Bao, Tianzhan Shen and Yanlei Li

Key Laboratory for Green Manufacturing & Functional Application of Inorganic Materials, School of Materials Science and Engineering, Shaanxi University of Science and Technology, Xi'an 710021, China; JiaqiLiu068@163.com (J.L.); licuiyan@sust.edu.cn (C.L.); theory2w@163.com (L.B.); tzshen2243@163.com (T.S.); 15829291050@163.com (Y.L.)

* Correspondence: ouyanghb@sust.edu.cn

Abstract: Carbon nanomaterials are promising microwave catalytic materials due to their abundant inhomogeneous interfaces capable of producing ideal interfacial polarization and multiple relaxation, which are favorable for microwave attenuation and dissipation. However, the microwave absorption performance of carbon materials is not ideal in practical applications due to poor impedance matching and single dielectric loss. To solve this problem, a ternary system of "carbon-magnetic" Ni@SiC/CNFs (C/Ni, C/SiC) composites was synthesized by electrostatic spinning, and they efficiently degraded methylene blue under microwave radiation. The results imply that the catalyst Ni@SiC/CNFs with a double-shell structure gave a 99.99% removal rate in 90 s for the degradation of methylene blue under microwave irradiation, outperforming the C/Ni and C/SiC and most other reported catalysts in similar studies. On the one hand, the possible mechanism of the methylene blue degradation should be ascribed to the fact that the double-shell structure increases the polarization source of the material, resulting in excellent microwave absorption properties; and on the other, the in situ generation of ·OH and O_2^- active species under microwave radiation and the synergistic coupling effect of metal plasma greatly improved the degradation efficiency of methylene blue. The findings of this study could provide a valuable reference for the green degradation of industrial dye wastewater and its sustainable development process.

Keywords: double-shell catalyst; electrospinning; microwave irradiation; degradation; active species; plasma

1. Introduction

With the rapid development of the economy and the increasing growth in population, the quantity of wastewater produced and its overall pollution load is continuously increasing worldwide [1–3]. According to reports from the United Nations World Water Development, over 80 percent of wastewater worldwide is released into the environment without treatment. Untreated wastewater not only jeopardizes human health but also seriously pollutes the ecological environment and adversely affects various economic activities [4,5]. Organic dyes are widely used in various industries (e.g., textile, pharmaceutical, food, cosmetic, plastic, photographic, and paper industries) and are an important source of water pollution; how to reduce the water pollution caused by organic dyes has attracted widespread attention from many different sectors [6–8].

Currently, adsorption, coagulation, advanced oxidation, and membrane separation are widely used for the removal of dyes from wastewater [9,10]. In recent years, advanced oxidation processes (AOPs) have made remarkable progress in the field of wastewater treatment [11]. The degradation of dyes by AOPs contributes to the in situ generation of free radicals with high oxidizing activity (·OH, O_2^-, SO_4^-) [12]. These radicals react non-selectively with most organic matter, resulting in the extensive degradation or mineralization of pollutants [13–15].

Microwave-enhanced AOPs can greatly improve the degradation rate and shorten the reaction time, since they have the ability to induce a hot spot effect, plasma discharge effect, and photoelectric effect in the microwave field. (i) The hot spots are a microplasma region on the catalyst surface where temperatures can reach over 1200 °C, promoting faster production of active species and oxidation of organic matter [16]. Qi et al., Hu et al., Garcia-Costa et al., Lei et al., and Shen et al. utilized the microwave hot spot effect to efficiently degrade different organic pollutants [17–21]. (ii) When microwaves act on certain tips or metals with special shapes or sizes, a charge builds up at the tip, causing the nearby electric field to be stronger than the air field, and thus a plasma discharge phenomenon occurs [22]. Qian et al., Zhao et al., Xu et al., and Zhang et al. used the plasma discharge effect generated under a microwave for the efficient treatment of refractory organic matter as well as white pollution [22–25]. (iii) Catalysts with semiconducting properties (e.g., transition metal oxides, semiconductors, and ferromagnetic metals) generate hole–electron pairs upon MW excitation, and the excited h^+ and e^- can migrate to the catalyst surface to participate in redox reactions [17,26,27]. Nowadays, the catalytic activity of many materials can be induced by microwave irradiation. When the MW irradiates these materials, the interaction between them can rapidly generate heat to produce a large number of active sites for the enhanced absorption and conversion of MW energy. Thus, it is a critical issue to develop microwave-active catalysts with strong microwave absorption ability and efficient microwave energy transfer ability [28–30].

Carbon nanomaterials are rich in hole structure, which can increase the electron transport path and improve the conductive loss. Meanwhile, the non-uniform structure of the phase interface can generate abundant interfacial polarization and multiple relaxation to enhance microwave attenuation [31,32]. However, due to poor impedance matching and single dielectric loss, the microwave absorption performance of carbon materials is not very ideal [33]. "Carbon-magnetic" composites are a good way to solve this problem. By introducing magnetic metals and microwave absorbers with high dielectric loss into carbon matrix materials, which can be enriched with a single microwave loss mechanism while obtaining ideal magnetic loss, impedance matching can be improved and microwave attenuation capability can be enhanced [34]. Magnetic materials (such as Fe, Co, Ni, and corresponding alloys) are ideal complementary materials because of their higher permeability, saturation magnetization, and snoek limit [35–37]. When the microwave irradiates the metal, a unique discharge phenomenon occurs, which can realize the coupling of the thermal effect and the plasma effect. Li [33] et al. synthesized the rod-like Ni@C composite, which has good microwave absorption and high reuse efficiency. SiC is an ideal microwave absorber because of its high microwave absorption characteristics in the gigahertz frequency. The superior thermal conductivity means it can still be used as a catalyst carrier at high temperatures. However, due to its inherent low conductivity and single polarization characteristics, SiC needs to be combined with different magnetic materials or carbon-based materials to enhance electrical conductivity and enrich its polarization types to improve its microwave absorption performance [38]. Samarjit Singh et al. prepared a Ni/SiC/graphene composite material with a minimum reflection loss of −59.15 dB, showing good microwave absorption performance [39].

Based on the above ideas, we propose a novel Ni@SiC/CNFs catalyst for MW-AOP. Carbon fibers, as supporting materials, in addition to excellent microwave absorption properties, also have a high electron mobility, which can promote charge separation/electron transfer and improve catalytic oxidation performance. SiC and Ni nanoparticles generate hole–electron pairs and strong plasma discharge under microwave irradiation. At the same time, the reaction of hole–electron pairs with O_2 and H_2O in the air generates strong oxidizing groups, which enhances the chemical reaction in degradation and improves the degradation efficiency. The hot spots and plasma generated by this discharge phenomenon can increase the degradation rate by reducing the presence of contaminants on the material and decreasing their participation in the chemical reaction time.

In this work, the Ni@SiC/CNF composites as MV-AOP catalysts were synthesized by the electrostatic spinning method. The morphology, microstructure, and phase composition of the Ni@SiC/CNFs were characterized through complementary analytical ways. The microwave catalytic performance of the Ni@SiC/CNFs was evaluated with MB as the target containment. The catalytic mechanism was proposed to explain the synergistic effect of the Ni@SiC/CNF catalyst under microwave irradiation.

2. Materials and Methods

2.1. Synthesis of Ni@SiC/CNFs

The Ni@SiC/CNF composites were prepared by electrostatic spinning combined with high-temperature carbonization (CNFs are carbon nanofibers, and SiC is silicon carbide). First of all, the precursor solution containing 10 mL of N, N-Dimethylformamide (DMF), 1.0 g of polyvinylpyrrolidone (PAN), 0.16 g of polycarbosilane (PCS), 1.7 mol of ($Ni(CH_3COO)_2$), and 2 mol of thiourea was configured. Subsequently, the prepared precursor solution was loaded into a 10 mL plastic syringe and injected at a flow rate of 7 μL/min, setting a distance of 17 cm between the needle and the receiver, a voltage of 17 kV, and a rotational speed of 600 r/min, to ultimately obtain a primed polymer fiber membrane. To maintain fiber morphology, the as-spun polymer nanofibers were subsequently pre-sintered at 280 °C for 2 h in the air followed by further carbonization at 1400 °C for 1 h under Ar atmosphere. Finally, the Ni@SiC/CNF (NSC) composites were obtained. By changing the content of PCS(0 g) in the precursor solution, the C/Ni (NC) composite material was prepared. By changing the content of ($Ni(CH_3COO)_2$) (0 mol) and thiourea (0 mol) in the precursor solution, the C/SiC (SC) composite material was prepared. Figure 1 is the schematic diagram of the preparation of the different composite materials.

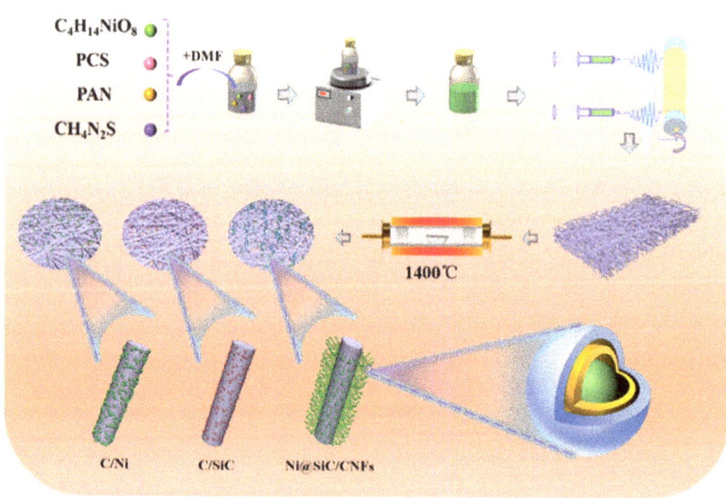

Figure 1. The flow diagram of preparing Ni@SiC/CNF, C/Ni, and C/SiC composites.

2.2. Characterization

The crystal structure of the sample was investigated using a powder X-ray diffractometer (XRD, Rigaku D/max-2200PC) with Cu Kα (λ = 0.15418 nm) radiation.

The morphology of the samples was observed using a field-emission scanning electron microscope (Chiyoda City, Japan, Hitachi, FE-SEM, S-4800).

XPS can be used for qualitative and quantitative analysis of the elements contained in the sample, and can also characterize the surface composition and chemical state of the sample.

High-resolution transmission electron microscopy (HRTEM) combined with energy dispersive spectroscopy (EDS) observation was performed on a Thermos Fisher Scientific, USA FEI Tecnai G2 F20S-TWIN system at 200 kV.

The molecular structure of the prepared samples was analyzed using a Renishaw-invia Raman instrument manufactured by Renishaw (London, UK), with lasers: 532 nm; 785 nm, wavelength range: 100–4000 cm^{-1}, optical plate size: 1–2 μm, and power: 0.005%–100% adjustable.

The UV–vis absorption spectra of the degradation were recorded on a UV/vis/NI R Spectrophotometer (LAMBDA950, PerkinElmer, Shanghai, China).

The dielectric property and magnetic permeability of the catalysts were analyzed by a vector network analyzer (Agilent 85071 E, Santa Clara, CA, USA). The block samples made in this study were tested by the waveguide method. The test range is X-band (2–18 GHz), and the samples are cut into cubic samples of 22.86 mm in length, 10.16 mm in width, and 1–3 mm in height.

2.3. Degradation Experiment Using Microwave

To examine the catalytic activity of the catalysts, degradation experiments were carried out using a controllable MW oven (WD750B, Galanz Company, Foshan, China) equipped with a self-made glass reactor and a condensing tube. The procedure was as follows: a methylene blue solution was prepared at a concentration of 20 mg/L; 20 mL of methylene blue solution and 20 mg of catalyst were added to a 50 mL reaction vessel, and degradation experiments were carried out.

The Teflon reaction vessel was followed by microwave irradiation (450 W, 2450 MHz). Upon the desired reaction time, the mixture was cooled to room temperature. The reaction mixture was then analyzed to evaluate the degradation efficiency by an ultraviolet spectrophotometer (UV–vis, UV-2450, SHIMADZU, Kyoto City, Japan). The removal rate of MB was calculated using the following formula:

$$\textit{Removal rate}\ (\%) = \frac{C_0 - C_t}{C_0} \times 100\%$$

C_0 and C_t are the initial concentration and the concentration after treatment time t of MB solution, respectively. Figure S1 is a schematic diagram of a microwave-assisted catalytic reactor.

3. Results

3.1. Microstructure and Morphology

As depicted in Figure 2a, the NSC catalyst shows three clear and sharp diffraction peaks at $2\theta = 26°$, 35.6°, and 45.5°, corresponding to the (002) crystal plane of C, (111) crystal plane of SiC, and (121) crystal plane of Ni, respectively. The corresponding crystal sizes are 23.2 nm, 29.6 nm, and 27.7 nm. The NC and SC catalysts only have the carbon diffraction peak and the corresponding Ni diffraction peak and SiC diffraction peak. The diffraction peak intensity of SiC is lower for the SC catalysts compared to the NSC catalysts. This is because the intervention of Ni can provide a more catalytic effect for the generation of SiC. At high temperatures, C and Si in the composites are solidly dissolved in the molten Ni particles, and under the catalysis of Ni, the reaction between C and Si occurs to generate SiC. Therefore, the introduction of Ni in the NSC catalysts can provide a more catalytic effect for the generation of SiC, and generate more SiC. In addition, the SC catalyst corresponds to the Mantou peak of carbon at 26°, while NC and NSC both show a sharp diffraction peak of C, indicating that Ni particles play a major role in the catalytic graphitization of composites at high temperatures. To prove the above conjecture, the graphitization of carbon was further confirmed using Raman spectroscopy (Figure 2b). Two typical carbon peaks at 1345 cm^{-1} and 1592 cm^{-1} were observed, corresponding to the D band and G band of carbon [40]. It can be found that the R-value (ID/IG) of SC was the highest at 0.924, while that of NSC was the lowest at 0.754 (Figure 2c), showing an obvious downward trend. It is well known that the D-band reflects the structural defects and disorder of carbon, while the G-band is related to graphitic carbon [41,42]. Therefore, a higher R-value indicates more lattice defects of the C atom and a higher graphitization degree on the contrary. Herein, the SC has the lowest graphitization, and the NSC has the highest graphitization, which is consistent with the XRD results.

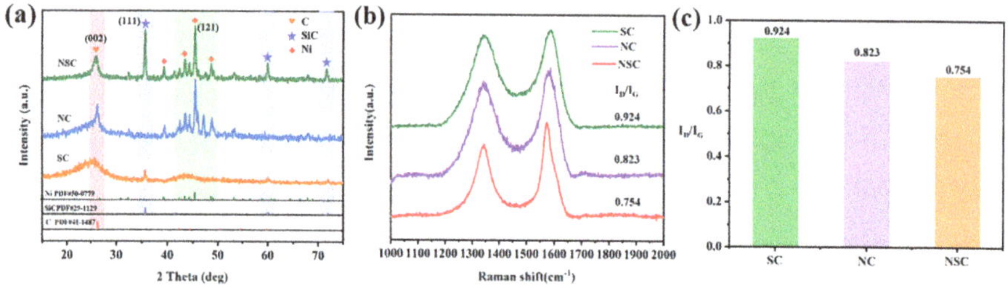

Figure 2. (**a**) XRD patterns; (**b**) Raman spectra; and (**c**) ID/IG graph of all the samples.

The chemical compositions and surface states of the NSC were analyzed by XPS. The XPS full-survey scan spectrum of NSC shown in Figure 3a confirms C, O, Si, and Ni elements. The low strength of the Ni element may be because most of the Ni particles on the surface are coated with C. Figure 3b reveals the high-resolution C 1s spectra, which are deconvoluted into three stripping peaks, where 283.5, 284.6, 285.8 eV can be indexed to the C-Si, C-C, C-O peaks, respectively [43,44]. The Si 2p peak can be fitted into three peaks with typical binding energy of 101.4 and 103.4 eV, which are aligned with the Si-C bond and the Si-O bond [45,46], respectively. Furthermore, the peak with a binding energy of 864 eV is attributed to Ni 2P, and the peaks at 853.1 eV and 873.2 eV are assigned to Ni 2p3/2 and Ni 2p1/2 [47], respectively. Furthermore, the survey spectra of NC and SC are shown in Figure S2. The XPS results are consistent with the XRD results, which proved that the three catalyst materials were successfully synthesized.

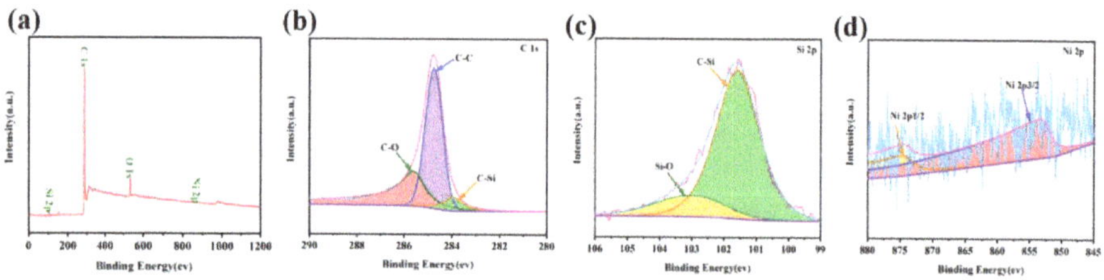

Figure 3. (**a**) XPS full spectrum; (**b**) XPS high-resolution spectrum of C 1s; (**c**) Si 2p; and (**d**) Ni 2p for the NSC composite materials.

The microscopic morphology of the NSC catalyst is revealed by the scanning electron microscopy (SEM) image shown in Figure 4a–d. SEM of NC and SC samples are shown in Figures S3 and S4 in the Supplementary Materials. These nanofibers are highly interconnected with random orientations to form 3D continuous networked structures with excellent mechanical stability [48]. As can be seen in Figure 4a, a large number of highly flexible nanowires grow on the surface of the carbon nanofibers, resulting in a one-dimensional hierarchical structure. Such a microscopic level network not only enhances mechanical connection between fibers but also promotes electron transport and multiple reflections of microwaves in the network, which could benefit the dielectric properties of the hybrid fibers [31].

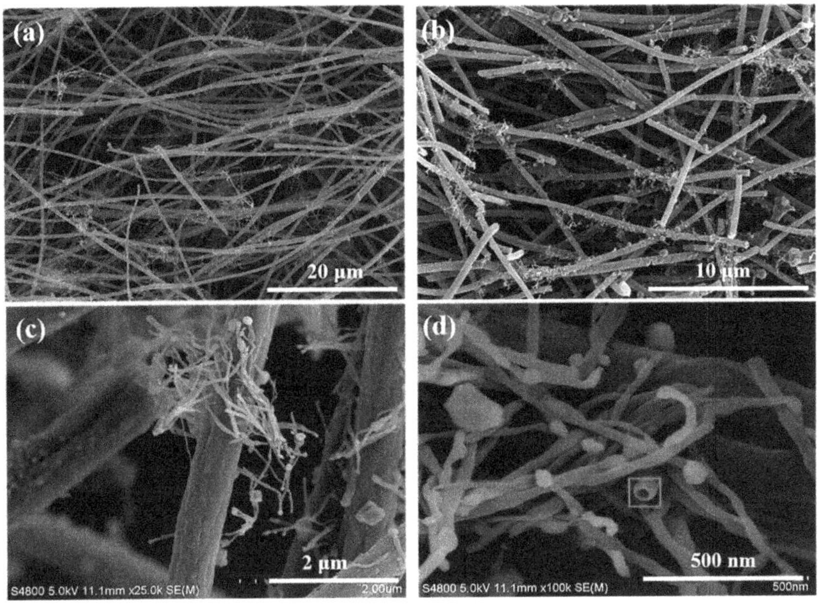

Figure 4. (**a–d**) SEM images of NSC catalyst at different magnifications.

TEM and HRTEM analyses were performed to characterize the microstructure in more detail (Figure 5a–h). It is not difficult to see that Ni particles are covered and carbon nanotubes with a diameter of 15 nm grow from the surface of the cover layer, which is caused by Ni nanoparticles catalyzing the graphitization of amorphous carbon at high temperatures [49–51]. Surprisingly, a thin layer of SiC was precipitated on the surface of Ni particles, followed by a graphite carbon layer in the high-resolution HRTEM image in Figure 5d. This occurs because C and Si are solidly dissolved in the molten Ni particles at high temperatures due to the precipitation temperatures of C and SiC being different; C is preferentially precipitated at low temperatures, while SiC is precipitated after nucleation at high temperatures of 1300 °C. Therefore, a SiC, C bilayer structure is formed on the surface of Ni particles. However, due to the lack of silicon sources, SiC can only exist as thin layers [52,53]. Furthermore, such an encapsulated structure can not only effectively protect the metal particles against corrosion in a harsh environment, but also can increase the interface layer between materials, so that the materials can produce more interface polarization under the action of the microwave, and improve the absorption capacity of materials to the microwave [54,55]. HRTEM images show that the periodic lattice fringes with good resolution are about 0.18 nm, 0.25 nm, and 0.35 nm, corresponding to the crystal planes of Ni (002), SiC (111), and C (002), respectively. In Figure 5c, we can see that the surface C layer is bent into a discontinuous sheet; defects will naturally occur at these positions, and a large number of dipoles are easily gathered at the defects, which is conducive to the dipole polarization effect of the absorber, which may lead to electromagnetic wave loss. The high-angle annular dark-field scanning TEM (HAADF-STEM) image and elemental mappings disclose the uniform distribution of C, Ni, and Si throughout the fibrous nanohybrids. It can be seen from Figures S3 and S4 that no nanostructure is generated on the surface of the NC and SC catalysts. Ni nanoparticles of the NC catalyst are uniformly anchored on the surface of the carbon nanofibers, while the SC surface is rough and no nanostructure can be seen from SEM. However, many fine SiC grains were found on the surface and the fibers showed a core–shell structure. There are many interfaces between different catalyst structures. Under the action of a microwave, the electrons move and gather at the interface, which makes it easy to generate interface polarization and improve the microwave absorption capacity [43].

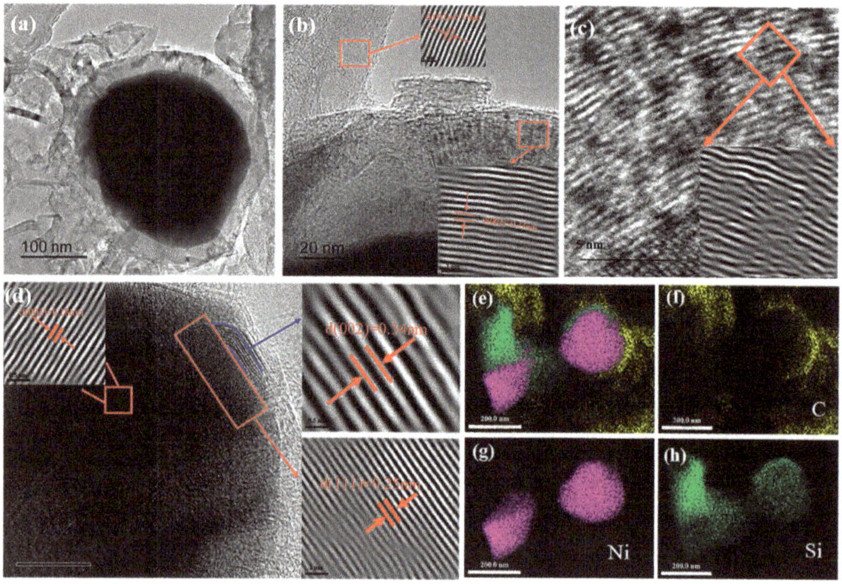

Figure 5. (**a**,**b**) TEM images of NSC catalyst; (**c**,**d**) HRTEM image; (**e**–**h**) EDS Element Mapping Diagram.

3.2. Microwave-Induced Catalytic Degradation of MB

Figure 6 shows the degradation curves of different catalytic materials for the MW catalytic oxidation of methylene blue at a given power in a MW (450 W, 2450 MHz). Before the microwave-assisted degradation of methylene blue, we first soaked these samples with methylene blue solution in the dark for 30 min to achieve adsorption equilibrium. Samples were taken every 30 s until either of the two visually visible samples faded and the test was stopped. The absorption peak of the MB chromophore and the parent ring structure is at 664 nm, so its degradation rate can be determined by the maximum absorption wavelength of 664 nm. By comparing the absorption peaks of the three catalysts at 664 nm, the UV–vis spectra of the NSC, SC, and NC catalysts are shown in Figure S5. It can be found that with the increase in time, the absorption peaks of the NSC catalyst at 664 nm decreased significantly, while the absorption peaks of NC and SC to MB decreased slowly. Figure 6a shows the time-degradation rates of the different catalysts. It can be seen from the figure that the NSC catalyst almost completely degraded after 90 s, with a degradation rate close to 100%, while the degradation rates of SC and NC were about 45% and 19%, respectively. Meanwhile, we tested the degradation performance of the three catalysts without microwave irradiation in order to verify the excitation effect of microwave irradiation on the materials. As can be seen from Figure 6b, the degradation rates of the three catalysts without microwave irradiation were basically around 3%. The results showed that the three catalysts could be excited under microwave irradiation to have certain degradation ability, among which the NSC catalyst could achieve efficient degradation of methylene blue under microwave irradiation. The degradation kinetics results of MB in 90 s is shown in Figure 6c, which indicates that the degradation kinetics of MB in 90 s by the three catalysts under microwave conditions belong to first-order reaction kinetics. The microwave-induced degradation rate constants of MB by different catalysts are shown in Figure 6d. The reaction rates k of NSC, SC, and NC are 0.1951, 0.0100, and 0.0051, respectively. In other words, the degradation rate of MB by NSC is about 19.51 times that of SC and 38.25 times that of NC. By comparing the SC and NC samples, the degradation efficiency of the SC system is found to be higher than that of NC. It is also confirmed that the presence of SiC in the ternary NSC catalyst contributes more to the degradation than Ni. The above experimental results show that NSC has better MW catalytic activity and higher MB degradation catalytic activity compared with the other two catalysts. In addition, we performed free radical scavenging tests and cyclic degradation tests on the NSC catalysts. As can be seen in Figure 6e, tert-butanol and sodium oxalate were added to the reaction solution as ·OH scavenger, O_2^- scavenger, and h^+ scavenger, respectively, to compare with the degradation of methylene blue by the NSC samples. The addition of tert-butanol and sodium oxalate reduced the degradation rate to 57.8% and 70.4%, respectively, and the degradation of methylene blue was inhibited, which indicated that ·OH, h^+, and O_2^- are important substances leading to the degradation of methylene blue. Through Figure 6f, we found that the NSC catalyst can maintain its original degradation ability after 10 cycles of testing, which indicates that the prepared new catalyst has better stability under microwave action.

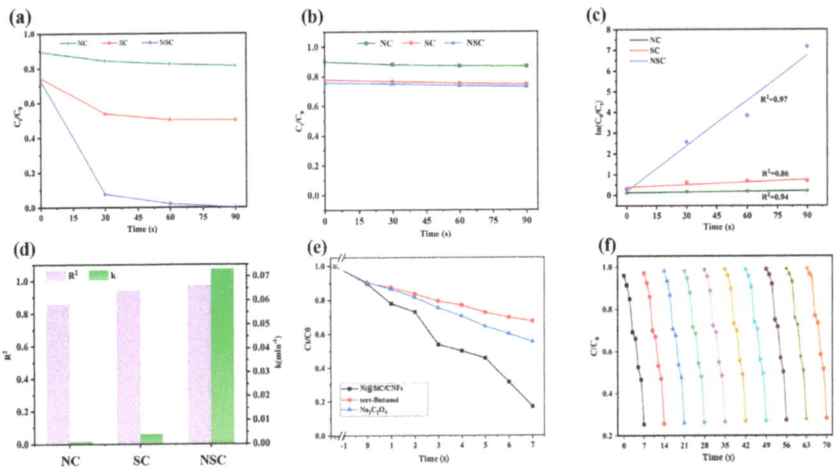

Figure 6. (**a**) Degradation curves of MB by different catalysts under microwave irradiation; (**b**) degradation curves of MB by different catalysts (without microwave irradiation); (**c**) the corresponding degradation rate curve; (**d**) the corresponding degradation slope and R^2 of composites under microwave catalysis; (**e**) free radical scavenging test; (**f**) the cyclic degradation performance of the NSC catalysts.

3.3. Electromagnetic Wave Absorption Characteristics and Catalytic Degradation Mechanism of NSC Composite Materials

To explore the reason why the NSC catalyst has a good degradation efficiency for MB under the action of microwave, we tested the electromagnetic wave absorption performance of the catalyst. Figure 7a–f plotted the electromagnetic parameters of the NSC catalyst to study its electromagnetic wave absorption performance. It is generally true that the −10 dB is equivalent to 90% of the total absorbed energy of the EM wave, and the −20 dB is equivalent to 99% of the total energy absorbed by the EM wave [56]. Reflection loss (RL) data about NSC for a frequency range of 2–18 GHz are illustrated by the curves in Figure 7a. The prepared NSC catalyst exhibits excellent microwave absorption performance, with a maximum reflection loss (RL_{max}) of 51.40 dB at 12.5 GHz. The corresponding matching thickness is 1.418 mm, and the effective RL bandwidth (RL > −10 dB) is 11.4–14.1 GHz.

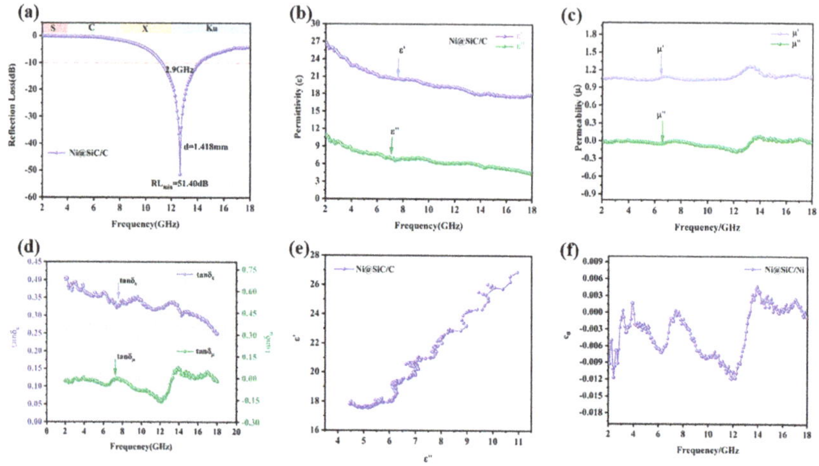

Figure 7. (**a**) RL curves at a thickness of 1.418 mm; (**b**) real/imaginary permittivity of NSC; (**c**) real/imaginary permeability of NSC; (**d**) frequency dependence of dielectric loss and magnetic loss of sample; (**e**) typical Cole–Cole semicircles (ε'' versus ε') for NSC; (**f**) c_0 values of NSC.

As shown in Figure 7b, the permittivity (ε', ε'') of the catalyst decreases gradually with the increase in frequency, which may be related to the polarization and interface relaxation at the interface between the Ni, SiC, and C layer. According to the value of $\tan\delta_\varepsilon$ ($\tan\delta_\varepsilon = \varepsilon''/\varepsilon'$), the dielectric loss capacity is further estimated. Interestingly, the $\tan\delta_\varepsilon$ values of the NSC catalysts peak at 9 GHz and 12.7 GHz, as shown in Figure 7d. The reason is that the dielectric loss ability is affected simultaneously by the polarization effect and by the conductive loss [57].

In addition to dielectric loss, magnetic loss also plays an important role. As shown in Figure 7d, the $\tan\delta_\mu$ value of the catalyst shows a similar trend to the imaginary part (μ'') of the permeability, indicating that the magnetic energy dissipation capacity is the key factor of the magnetic loss [58–60]. It is worth noting that in Figure 7d, the peak of the magnetic loss ($\tan\delta_\mu$) around 13.8 GHz comes from both natural and exchange resonance. It is clearly observed from Figure 7d that the value of $\tan\delta_\varepsilon$ is significantly larger than $\tan\delta_\mu$. This indicates that dielectric loss is the main contributor to excellent microwave absorption performance.

To further investigate the mechanism of dielectric and magnetic loss, Figure 7e–f illustrates the Cole–Cole plots and c_0 curves of the samples. Figure 7e shows the Cole–Cole semicircles of the sample, with each semicircle corresponding to Debye relaxation [61]. Obviously, there are multiple decomposable semicircles in the ε' and ε'' curves, indicating that the Debye relaxation process caused by interfacial polarization and dipole polarization occurs in the samples. In addition, there are almost no straight lines at the ends of the Cole–Cole diagram, indicating that there is almost no conduction loss during the whole process.

The magnetic loss mechanism mainly consists of three parts: exchange resonance, natural resonance, and eddy current loss. In general, the eddy current losses of absorbing materials can be described by the value of c_0 ($c_0 = \mu'' (\mu')^{-2} f^{-1}$). When the value of c_0 is constant with increasing frequency, eddy current loss is dominant in magnetic loss. If c_0 changes with f, eddy current losses are suppressed, and natural resonance and exchange resonance are the main sources of magnetic losses [62]. As shown in Figure 7f, the c_0 curve of the sample fluctuates significantly at the frequency of 2–18 GHz, indicating that exchange resonance and natural resonance play a crucial role in the magnetic loss of the catalyst.

Based on the above analysis, the mechanism of degradation of methylene blue by the NSC catalyst under microwave irradiation was further studied, as shown in Figure 8, and the efficient degradation rate of MB by the binuclear NSC catalyst composed of C, Ni, and SiC is mainly attributed to the following points. (1) The presence of many one-dimensional CNTs in the prepared catalysts is conducive to increasing the electron transport paths during microwave action and increasing the multiple reflections of MW, which gives the materials excellent microwave absorption properties and ensures that the MB can fully absorb microwave energy and convert it into thermal energy during the microwave degradation process. (2) SiC as a polar molecule, under microwave irradiation, will produce dipole polarization. In addition, there are a small number of defects in the catalyst, where the electron aggregation experiences dipole polarization. (3) There are abundant phase boundaries in the composites, which can cause more interfacial polarization and multiple relaxation. The polarization mechanism occurring in processes (2) and (3) will cause the material to generate a lot of heat, putting electrons in an active and unstable state, and the transition will generate hole–electron pairs. At the same time, one-dimensional carbon nanotubes can provide more conductive paths to facilitate the transport of excited hopping electrons. These hole–electron pairs react with O_2 and H_2O in water to form various active substances such as $\cdot OH$, O_2^-, and h^+ active substances. (4) Due to the characteristics of microwave selective heating, the dielectric constants of C and SiC in the composite material are different, and there is an obvious difference in the absorption ability of the microwave, which leads to the temperature around the strong wave-absorbing material being significantly higher than that of the weakly absorbing material, and there is a temperature gradient to form a hot spot, and the hot spot generates a high temperature conducive to the generation of the active substances ($\cdot OH$, O_2^-, and h^+). (5) A unique discharge phenomenon will be produced when the metal is irradiated by microwave. The discharge process will be accompanied by the release of plasma, which not only can shape local high temperatures to strengthen microwave heating, but also can generate plasma which can significantly promote a chemical reaction, which can generate $\cdot OH$, O_2^-, and h^+ active substances. Under the action of a microwave, it is the synergistic effect between various elements that makes the degradation of MB (methylene blue) easier to occur.

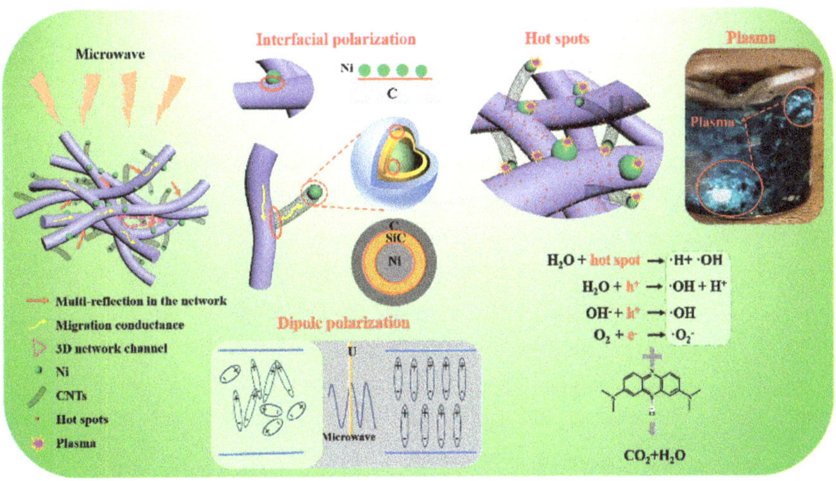

Figure 8. Degradation mechanism of methylene blue by Ni@SiC/CNFs under microwave irradiation.

Overall, the one-dimensional CNT structure extends the conductive path of electrons to enhance microwave dissipation; the polar SiC molecules are more likely to generate dipole polarization under microwave irradiation to enhance the consumption of electromagnetic waves; and the introduction of magnetic Ni particles enriches the depletion mechanism of the material system and generates plasma discharges to generate reactive groups to promote the reaction under microwave irradiation. The synergistic effect among the three components contributed to the excellent and efficient degradation capability of this catalyst.

4. Conclusions

In this study, catalysts with different structures were successfully prepared by adjusting the raw materials in a precursor solution by the electrostatic spinning method and were then used for the degradation of a high-concentration MB solution under microwave irradiation. The characterization results of the different catalysts suggest that the microstructure of the materials plays an important role in the degradation rate of MB. Under the action of a microwave, the NSC catalyst showed excellent catalytic activity, with MB degradation in 1.5 min. Compared with NC and SC, the reaction rates were 38.25 and 19.51 times higher, which was mainly attributed to the synergistic effect of the thermal effect, non-thermal effect, and plasma effect of the NSC catalyst under the action of the microwave. Under these effects, a large number of $\cdot OH$, h^+, and O_2^- active substances were produced, and these active substances could rapidly oxidize methylene blue dye, so that it was rapidly degraded into non-toxic and harmless small molecules. It is expected that Ni@SiC/CNFs with high microwave-assisted catalytic activity will greatly promote a practical application for the removal of organic pollutants.

Supplementary Materials: The following supporting information can be downloaded at: https://www.mdpi.com/article/10.3390/c10030072/s1, Figure S1: Diagram of microwave-assisted catalytic reactor; Figure S2: XPS of the NC and SC composite materials: (a) full spectrum, (b) C 1s, and (c) Ni 2p for the NC composite materials, respectively; (d) full spectrum, (e) C 1s, and (f) Si 2p for the SC composite materials, respectively; Figure S3: (a,b) SEM images of NC catalyst at different magnifications; (c) TEM images of NC catalyst; (d) HRTEM image; Figure S4: (a,b) SEM images of SC catalyst at different magnifications; (c) TEM images of NSC catalyst; (f) HRTEM image; (d–g) HAADF-STEM image and corresponding elemental mappings of (f) C, (g) SiC; Figure S5: Ultraviolet absorption spectra of MB by different composites: (a) C/Ni, (b) C/SiC, (c) Ni@SiC/CNFs.

Author Contributions: Conceptualization, H.O. and J.L.; validation, J.L and L.B.; formal analysis, T.S. and Y.L.; writing—original draft preparation, J.L.; writing—review and editing, C.L. All authors have read and agreed to the published version of the manuscript.

Funding: This research was supported by the National Nature Science Foundation of China (Grant No. 52173299, 52372087).

Data Availability Statement: All original data from the study have been fully included in the article/Supplementary Material and further enquiries can be directed to the corresponding author.

Acknowledgments: The authors would like to thank Jinfan Liu for his help in experimental section.

Conflicts of Interest: The authors declare no conflicts of interest.

References

1. Nasir, A.M.; Awang, N.; Jaafar, J.; Ismail, A.F.; Othman, M.H.D.A.; Rahman, M.; Aziz, F.; Mat Yajid, M.A. Recent progress on fabrication and application of electrospun nanofibrous photocatalytic membranes for wastewater treatment: A review. *J. Water Process Eng.* **2021**, *40*, 101878. [CrossRef]
2. Xia, H.; Li, C.; Yang, G.; Shi, Z.; Jin, C.; He, W.; Xu, J.; Li, G. A review of microwave-assisted advanced oxidation processes for wastewater treatment. *Chemosphere* **2022**, *287*, 131981. [CrossRef]
3. Liu, Z.; Zhang, W.; Liang, Q.; Huang, J.; Shao, B.; Liu, Y.; Liu, Y.; He, Q.; Wu, T.; Gong, J.; et al. Microwave-assisted high-efficiency degradation of methyl orange by using $CuFe_2O_4$/CNT catalysts and insight into degradation mechanism. *Environ. Sci. Pollut. Res. Int.* **2021**, *28*, 42683–42693. [CrossRef] [PubMed]
4. Khoshnam, M.; Farahbakhsh, J.; Zargar, M.; Mohammad, A.W.; Benamor, A.; Ang, W.L.; Mahmoudi, E. α-Fe_2O_3/graphene oxide powder and thin film nanocomposites as peculiar photocatalysts for dye removal from wastewater. *Sci. Rep.* **2021**, *11*, 20378. [CrossRef]
5. Zhang, Y.; Su, P.; Weathersby, D.; Zhang, Q.; Zheng, J.; Fan, R.; Zhang, J.; Dai, Q. Synthesis of γ-Fe_2O_3-ZnO-biochar nanocomposites for Rhodamine B removal. *Appl. Surf. Sci.* **2020**, *501*, 144217. [CrossRef]

6. Wong, S.; Ghafar, N.A.; Ngadi, N.; Razmi, F.A.; Inuwa, I.M.; Mat, R.; Amin, N.A.S. Effective removal of anionic textile dyes using adsorbent synthesized from coffee waste. *Sci. Rep.* **2020**, *10*, 2928. [CrossRef]
7. Liu, Y.; Song, L.; Du, L.; Gao, P.; Liang, N.; Wu, S.; Minami, T.; Zang, L.; Yu, C.; Xu, X. Preparation of Polyaniline/Emulsion Microsphere Composite for Efficient Adsorption of Organic Dyes. *Polymers* **2020**, *12*, 167. [CrossRef] [PubMed]
8. Chen, B.; Long, F.; Chen, S.; Cao, Y.; Pan, X. Magnetic chitosan biopolymer as a versatile adsorbent for simultaneous and synergistic removal of different sorts of dyestuffs from simulated wastewater. *Chem. Eng. J.* **2020**, *385*, 123926. [CrossRef]
9. Xue, C.; Mao, Y.; Wang, W.; Song, Z.; Zhao, X.; Sun, J.; Wang, Y. Current status of applying microwave-associated catalysis for the degradation of organics in aqueous phase—A review. *J. Environ. Sci.* **2019**, *81*, 119–135. [CrossRef]
10. Kubra, K.T.; Salman, M.S.; Hasan, M.N. Enhanced toxic dye removal from wastewater using biodegradable polymeric natural adsorbent. *J. Mol. Liq.* **2021**, *328*, 115468. [CrossRef]
11. Gągol, M.; Przyjazny, A.; Boczkaj, G. Wastewater treatment by means of advanced oxidation processes based on cavitation—A review. *Chem. Eng. J.* **2018**, *338*, 599–627. [CrossRef]
12. Tian, K.; Hu, L.; Li, L.; Zheng, Q.; Xin, Y.; Zhang, G. Recent advances in persulfate-based advanced oxidation processes for organic wastewater treatment. *Chin. Chem. Lett.* **2022**, *33*, 4461–4477. [CrossRef]
13. Liang, C.; Niu, H.-Y.; Guo, H.; Niu, C.-G.; Yang, Y.-Y.; Liu, H.-Y.; Tang, W.-W.; Feng, H.-P. Efficient photocatalytic nitrogen fixation to ammonia over bismuth monoxide quantum dots-modified defective ultrathin graphitic carbon nitride. *Chem. Eng. J.* **2021**, *406*, 126868. [CrossRef]
14. Peng, Y.; Li, Y.; Liu, L.; Hao, X.; Cai, K.; Xiong, J.; Hong, W.; Tao, J. New optimization approach for amphoteric/magnetic ramie biosorbent in dyestuff adsorption. *Biochem. Eng. J.* **2022**, *181*, 108379. [CrossRef]
15. Wang, H.; Zhao, Z.; Zhang, X.; Dong, W.; Cao, Z.; He, L.; Wang, X. Rapid decomplexation of Ni-EDTA by microwave-assisted Fenton reaction. *Chem. Eng. J.* **2020**, *381*, 122703. [CrossRef]
16. Garcia-Costa, A.L.; Zazo, J.A.; Casas, J.A. Microwave-assisted catalytic wet peroxide oxidation: Energy optimization. *Sep. Purif. Technol.* **2019**, *215*, 62–69. [CrossRef]
17. Qi, Y.; Mei, Y.; Li, J.; Yao, T.; Yang, Y.; Jia, W.; Tong, X.; Wu, J.; Xin, B. Highly efficient microwave-assisted Fenton degradation of metacycline using pine-needle-like $CuCo_2O_4$ nanocatalyst. *Chem. Eng. J.* **2019**, *373*, 1158–1167. [CrossRef]
18. Hu, L.; Wang, P.; Liu, G.; Zheng, Q.; Zhang, G. Catalytic degradation of p-nitrophenol by magnetically recoverable Fe_3O_4 as a persulfate activator under microwave irradiation. *Chemosphere* **2020**, *240*, 124977. [CrossRef]
19. Garcia-Costa, A.L.; Zazo, J.A.; Rodriguez, J.J.; Casas, J.A. Microwave-assisted catalytic wet peroxide oxidation. Comparison of Fe catalysts supported on activated carbon and γ-alumina. *Appl. Catal. B Environ.* **2017**, *218*, 637–642. [CrossRef]
20. Lei, Y.; Lin, X.; Liao, H. New insights on microwave induced rapid degradation of methyl orange based on the joint reaction with acceleration effect between electron hopping and Fe^{2+}-H_2O_2 reaction of $NiFeMnO_4$ nanocomposites. *Sep. Purif. Technol.* **2018**, *192*, 220–229. [CrossRef]
21. Shen, M.; Fu, L.; Tang, J.; Liu, M.; Song, Y.; Tian, F.; Zhao, Z.; Zhang, Z.; Dionysiou, D.D. Microwave hydrothermal-assisted preparation of novel spinel-$NiFe_2O_4$/natural mineral composites as microwave catalysts for degradation of aquatic organic pollutants. *J. Hazard. Mater.* **2018**, *350*, 1–9. [CrossRef]
22. Qian, C.; Dai, J.; Tian, Y.; Duan, Y.; Li, Y. Efficient degradation of Fipronil in water by microwave-induced argon plasma: Mechanism and degradation pathways. *Sci. Total Environ.* **2020**, *725*, 138487. [CrossRef] [PubMed]
23. Zhao, C.; Xue, L.; Zhou, Y.; Zhang, Y.; Huang, K. A microwave atmospheric plasma strategy for fast and efficient degradation of aqueous p-nitrophenol. *J. Hazard. Mater.* **2021**, *409*, 124473. [CrossRef] [PubMed]
24. Xu, G.; Jiang, H.; Stapelberg, M.; Zhou, J.; Liu, M.; Li, Q.J.; Cao, Y.; Gao, R.; Cai, M.; Qiao, J.; et al. Self-Perpetuating Carbon Foam Microwave Plasma Conversion of Hydrocarbon Wastes into Useful Fuels and Chemicals. *Environ. Sci. Technol.* **2021**, *55*, 6239–6247. [CrossRef] [PubMed]
25. Zhang, P.; Liang, C.; Wu, M.; Chen, X.; Liu, D.; Ma, J. High-efficient microwave plasma discharging initiated conversion of waste plastics into hydrogen and carbon nanotubes. *Energy Convers. Manag.* **2022**, *268*, 116017. [CrossRef]
26. Pang, Y.; Lei, H. Degradation of p-nitrophenol through microwave-assisted heterogeneous activation of peroxymonosulfate by manganese ferrite. *Chem. Eng. J.* **2016**, *287*, 585–592. [CrossRef]
27. Zhou, J.; You, Z.; Xu, W.; Su, Z.; Qiu, Y.; Gao, L.; Yin, C.; Lan, L. Microwave irradiation directly excites semiconductor catalyst to produce electric current or electron-holes pairs. *Sci. Rep.* **2019**, *9*, 5470. [CrossRef]
28. Pawar, S.P.; Gandi, M.; Bose, S. High performance electromagnetic wave absorbers derived from PC/SAN blends containing multiwall carbon nanotubes and Fe_3O_4 decorated onto graphene oxide sheets. *RSC Adv.* **2016**, *6*, 37633–37645. [CrossRef]
29. Menéndez, J.A.; Arenillas, A.; Fidalgo, B.; Fernández, Y.; Zubizarreta, L.; Calvo, E.G.; Bermúdez, J.M. Microwave heating processes involving carbon materials. *Fuel Process. Technol.* **2010**, *91*, 1–8. [CrossRef]
30. Li, C.; Xia, H.; Zhang, L.; Wang, S.; Peng, J.; Cheng, S.; Shu, J.; Jiang, X.; Zhang, Q. Analysis of dielectric characterization and microwave adsorbing properties in organism-contained spent carbon: An efficient regeneration method via microwave-assisted ultrasound. *Chem. Eng. Process. Process Intensif.* **2018**, *125*, 74–86. [CrossRef]
31. Yuan, X.; Wang, R.; Huang, W.; Kong, L.; Guo, S.; Cheng, L. Morphology Design of Co-electrospinning MnO-VN/C Nanofibers for Enhancing the Microwave Absorption Performances. *ACS Appl. Mater. Interfaces* **2020**, *12*, 13208–13216. [CrossRef]
32. Chen, C.; Xi, J.; Zhou, E.; Peng, L.; Chen, Z.; Gao, C. Porous Graphene Microflowers for High-Performance Microwave Absorption. *Nano-Micro Lett.* **2018**, *10*, 26. [CrossRef]

33. Li, J.; Zhang, F.; Lu, H.; Guo, W.; He, X.; Yuan, Y. Heterogeneous rod-like Ni@C composites toward strong and stable microwave absorption performance. *Carbon* **2021**, *181*, 358–369. [CrossRef]
34. Wang, B.; Wu, Q.; Fu, Y.; Liu, T. A review on carbon/magnetic metal composites for microwave absorption. *J. Mater. Sci. Technol.* **2021**, *86*, 91–109. [CrossRef]
35. Hidaka, H.; Saitou, A.; Honjou, H.; Hosoda, K.; Moriya, M.; Serpone, N. Microwave-assisted dechlorination of polychlorobenzenes by hypophosphite anions in aqueous alkaline media in the presence of Pd-loaded active carbon. *J. Hazard. Mater.* **2007**, *148*, 22–28. [CrossRef]
36. Sun, G.; Dong, B.; Cao, M.; Wei, B.; Hu, C. Hierarchical Dendrite-Like Magnetic Materials of Fe_3O_4, γ-Fe_2O_3, and Fe with High Performance of Microwave Absorption. *Chem. Mater.* **2011**, *23*, 1587–1593. [CrossRef]
37. Ren, Y.; Zhu, C.; Zhang, S.; Li, C.; Chen, Y.; Gao, P.; Yang, P.; Ouyang, Q. Three-dimensional SiO_2@Fe_3O_4 core/shell nanorod array/graphene architecture: Synthesis and electromagnetic absorption properties. *Nanoscale* **2013**, *5*, 12296–12303. [CrossRef]
38. Singh, S.; Bhaskar, R.; Narayanan, K.B.; Kumar, A.; Debnath, K. Development of silicon carbide (SiC)-based composites as microwave-absorbing materials (MAMs): A review. *J. Eur. Ceram. Soc.* **2024**, *44*, 7411–7431. [CrossRef]
39. Singh, S.; Maurya, A.K.; Gupta, R.; Kumar, A.; Singh, D. Improved microwave absorption behavioral response of Ni/SiC and Ni/SiC/graphene composites: A comparative insight. *J. Alloys Compd.* **2020**, *823*, 153780. [CrossRef]
40. Wang, Y.; Di, X.; Chen, J.; She, L.; Pan, H.; Zhao, B.; Che, R. Multi-dimensional C@NiCo-LDHs@Ni aerogel: Structural and componential engineering towards efficient microwave absorption, anti-corrosion and thermal-insulation. *Carbon* **2022**, *191*, 625–635. [CrossRef]
41. Hu, P.; Dong, S.; Li, X.; Chen, J.; Hu, P. Flower-like $NiCo_2S_4$ Microspheres Based on Nanosheet Self-Assembly Anchored on 3D Biomass-Derived Carbon for Efficient Microwave Absorption. *ACS Sustain. Chem. Eng.* **2020**, *8*, 10230–10241. [CrossRef]
42. Liu, L.; Yang, S.; Hu, H.; Zhang, T.; Yuan, Y.; Li, Y.; He, X. Lightweight and Efficient Microwave-Absorbing Materials Based on Loofah-Sponge-Derived Hierarchically Porous Carbons. *ACS Sustain. Chem. Eng.* **2018**, *7*, 1228–1238. [CrossRef]
43. Li, H.; Gao, S.; Tong, H.; Liu, Y.; Wu, A.; Hao, H. The capacitive loss of microwave energy in Ni@SiC@C core/bi-shell nanoparticles. *Chem. Eng. J.* **2022**, *434*, 134655. [CrossRef]
44. Chan, C.K.; Patel, R.N.; O'Connell, M.J.; Korgel, B.A.; Cui, Y. Solution-Grown Silicon Nanowires for Lithium-Ion Battery Anodes. *ACS Nano* **2010**, *4*, 1443–1450. [CrossRef]
45. Liu, Y.; Liu, Y.; Choi, W.C.; Chae, S.; Lee, J.; Kim, B.-S.; Park, M.; Kim, H.Y. Highly flexible, erosion resistant and nitrogen doped hollow SiC fibrous mats for high temperature thermal insulators. *J. Mater. Chem. A* **2017**, *5*, 2664–2672. [CrossRef]
46. Wang, B.; Wang, Y.; Lei, Y.; Wu, N.; Gou, Y.; Han, C.; Fang, D. Hierarchically porous SiC ultrathin fibers mat with enhanced mass transport, amphipathic property and high-temperature erosion resistance. *J. Mater. Chem. A* **2014**, *2*, 20873–20881. [CrossRef]
47. Huo, Y.; Tan, Y.; Zhao, K.; Lu, Z.; Zhong, L.; Tang, Y. Enhanced electromagnetic wave absorption properties of Ni magnetic coating-functionalized SiC/C nanofibers synthesized by electrospinning and magnetron sputtering technology. *Chem. Phys. Lett.* **2021**, *763*, 138230. [CrossRef]
48. Li, T.; Luo, G.; Liu, K.; Li, X.; Sun, D.; Xu, L.; Li, Y.; Tang, Y. Encapsulation of Ni_3Fe Nanoparticles in N-Doped Carbon Nanotube–Grafted Carbon Nanofibers as High-Efficiency Hydrogen Evolution Electrocatalysts. *Adv. Funct. Mater.* **2018**, *28*, 1805828. [CrossRef]
49. Liu, S.; Wang, Z.; Zhou, S.; Yu, F.; Yu, M.; Chiang, C.Y.; Zhou, W.; Zhao, J.; Qiu, J. Metal-Organic-Framework-Derived Hybrid Carbon Nanocages as a Bifunctional Electrocatalyst for Oxygen Reduction and Evolution. *Adv. Mater.* **2017**, *29*, 1700874. [CrossRef] [PubMed]
50. Wu, R.; Wang, D.P.; Rui, X.; Liu, B.; Zhou, K.; Law, A.W.; Yan, Q.; Wei, J.; Chen, Z. In-situ formation of hollow hybrids composed of cobalt sulfides embedded within porous carbon polyhedra/carbon nanotubes for high-performance lithium-ion batteries. *Adv. Mater.* **2015**, *27*, 3038–3044. [CrossRef]
51. Tang, J.; Salunkhe, R.R.; Liu, J.; Torad, N.L.; Imura, M.; Furukawa, S.; Yamauchi, Y. Thermal conversion of core-shell metal-organic frameworks: A new method for selectively functionalized nanoporous hybrid carbon. *J. Am. Chem. Soc.* **2015**, *137*, 1572–1580. [CrossRef] [PubMed]
52. Li, D.; Liao, H.; Kikuchi, H.; Liu, T. Microporous Co@C Nanoparticles Prepared by Dealloying CoAl@C Precursors: Achieving Strong Wideband Microwave Absorption via Controlling Carbon Shell Thickness. *ACS Appl. Mater. Interfaces* **2017**, *9*, 44704–44714. [CrossRef] [PubMed]
53. Huang, F.; Fan, S.; Li, X.; Qu, X.; Tian, Y.; Zhang, X.; Zhang, Z.; Dong, X.; Cao, T. Enhanced dielectric and conductivity properties of carbon-coated SiC nanocomposites in the terahertz frequency range. *Nanotechnology* **2021**, *32*, 265705. [CrossRef] [PubMed]
54. Deng, J.; Ren, P.; Deng, D.; Yu, L.; Yang, F.; Bao, X. Highly active and durable non-precious-metal catalysts encapsulated in carbon nanotubes for hydrogen evolution reaction. *Energy Environ. Sci.* **2014**, *7*, 1919–1923. [CrossRef]
55. Cui, X.; Ren, P.; Deng, D.; Deng, J.; Bao, X. Single layer graphene encapsulating non-precious metals as high-performance electrocatalysts for water oxidation. *Energy Environ. Sci.* **2016**, *9*, 123–129. [CrossRef]
56. Cai, Z.; Su, L.; Wang, H.; Xie, Q.; Gao, H.; Niu, M.; Lu, D. Hierarchically assembled carbon microtube@SiC nanowire/Ni nanoparticle aerogel for highly efficient electromagnetic wave absorption and multifunction. *Carbon* **2022**, *191*, 227–235. [CrossRef]
57. Di, X.; Wang, Y.; Lu, Z.; Cheng, R.; Yang, L.; Wu, X. Heterostructure design of Ni/C/porous carbon nanosheet composite for enhancing the electromagnetic wave absorption. *Carbon* **2021**, *179*, 566–578. [CrossRef]

58. Li, D.; Guo, K.; Wang, F.; Wu, Z.; Zhong, B.; Zuo, S.; Tang, J.; Feng, J.; Zhuo, R.; Yan, D.; et al. Enhanced microwave absorption properties in C band of Ni/C porous nanofibers prepared by electrospinning. *J. Alloys Compd.* **2019**, *800*, 294–304. [CrossRef]
59. Ye, X.; Zhang, J.; Chen, Z.; Xiang, J.; Jiang, Y.; Xie, F.; Ma, X. Microwave absorption properties of Ni/C@SiC composites prepared by precursor impregnation and pyrolysis processes. *Def. Technol.* **2023**, *21*, 94–102. [CrossRef]
60. Wu, F.; Liu, Z.; Wang, J.; Shah, T.; Liu, P.; Zhang, Q.; Zhang, B. Template-free self-assembly of MXene and CoNi-bimetal MOF into intertwined one-dimensional heterostructure and its microwave absorbing properties. *Chem. Eng. J.* **2021**, *422*, 130591. [CrossRef]
61. Huang, B.; Wang, Z.; Hu, H.; Xiu, Z.; Huang, X.; Yue, J.; Wang, Y. Enhancement of the microwave absorption properties of PyC-SiCf/SiC composites by electrophoretic deposition of SiC nanowires on SiC fibers. *Ceram. Int.* **2020**, *46*, 9303–9310. [CrossRef]
62. Li, W.; Guo, F.; Wei, X.; Du, Y.; Chen, Y. Preparation of Ni/C porous fibers derived from jute fibers for high-performance microwave absorption. *RSC Adv.* **2020**, *10*, 36644–36653. [CrossRef] [PubMed]

C Journal of Carbon Research

MDPI

Article

Enhanced Adsorption of Arsenate from Contaminated Waters by Magnesium-, Zinc- or Calcium-Modified Biochar—Modeling and Mechanisms

Despina Vamvuka *, Elena Sdoukou, Antonios Stratakis and Despina Pentari

School of Mineral Resources Engineering, Technical University of Crete, 73100 Chania, Crete, Greece; esdoukou@tuc.gr (E.S.); astratakis@tuc.gr (A.S.); dpentari@tuc.gr (D.P.)
* Correspondence: dvamvouka@tuc.gr

Abstract: The adsorption of arsenate from wastewaters was investigated by applying Mg-, Zn- or Ca-modified nut residue biochar activated by nitrogen/steam. The parameters studied were the contact time, adsorbent dose, initial arsenate concentration and solution pH. The adsorption mechanism was investigated. Various analyses of the material before and after arsenate adsorption were carried out, and experimental data were simulated by applying two isotherm models. The results indicated that the maximum removal efficiency of arsenate was 29.4% at an initial concentration of 10 mg/L. The modification of biochar by Mg, Zn or Ca oxides increased the removal rate significantly, from 49.4% at 100 mg/L As^{5+} up to 8%, 97% and 97%, respectively. Zn-modified biochar presented an excellent performance for both low and high As^{5+} concentrations. All experimental data were accurately fitted by the Freundlich isotherm model (R^2 = 0.94–0.97), confirming a multilayer adsorption mechanism. For a biochar dose of 2 g/L, the maximum capacity of adsorption was enhanced after Mg-, Zn- or Ca-modification from 12.4 mg/g to 35 mg/g, 50 mg/g and 49 mg/g, respectively. The potential mechanisms of adsorption were ligand exchange, chemical complexation, surface precipitation and electron coordination.

Keywords: arsenate; adsorption; wastewaters; modified biochar; steam

1. Introduction

Arsenic is a highly toxic and carcinogenic heavy metal to living organisms. It can be naturally derived from lithogenic sources and mobilized by geothermal or biological environments. But most importantly, it can be derived from anthropogenic sources, such as uncontrolled mining activities, the combustion of fossil fuels, petroleum refineries and industries of pesticides, herbicides, fertilizers, crop desiccants and glass manufacturing [1,2]. The most prevalent species are arsenite and arsenate, the latter being less soluble and toxic [3]. The discharge of these species to the soil, surface or groundwater can seriously threaten human health and the whole environment. Thus, the World Health Organization has set the maximum allowable limit in drinking water to 10 μg/L [4,5].

Several technologies have been developed for removing arsenic from contaminated water, including ion-exchange, reverse osmosis, membrane filtration, chemical oxidation, coagulation, precipitation and adsorption [1–3,6–8]. Among these technologies, adsorption is considered more efficient, less costly and easier to operate, with minimum by-products generation [1,2].

Traditionally, active carbon has been used as an efficient adsorbent of heavy metals from wastewaters. However, its high cost arising from the availability and processing of raw materials and disposal and recycling issues [9] limit its large-scale application. On the other hand, biochar, the solid product of the pyrolysis of carbonaceous biomass materials, widely applied as a soil amendment and carbon sink in soils for thousands of years [10], has attracted increased attention recently as an environmentally friendly adsorbent for organic or inorganic pollutants [1]. Although less efficient than active carbon, biochar can be produced at a low cost from a variety of waste materials [11], it is rich in carbon, it shows great thermal stability and it presents favorable physicochemical properties, such as a large surface area and porosity [3,12,13].

However, biochars possess functional groups that are negatively charged on their surface, limiting the adsorption ability of anionic pollutants such as arsenate, which mainly exists in the environment in the form of the oxyanions AsO_4^{3-} or AsO_3^{2-} [13]. Several studies reported a low adsorption capacity of agricultural-derived biochars, such as plant residues, empty fruit bunches, rice husks, corn cobs and corn straws for arsenate, ranging between 0.8 mg/g and 7.1 mg/g [7,8,14]. Therefore, the physicochemical properties of biochars should be modified in order to immobilize As^{5+} ions from aqueous solutions.

The adsorption of As^{5+} ion- on iron oxide-modified biochars has been found to be an effective option for their removal from wastewaters. A Fe_3O_4-loaded activated carbon, using waste biomass, showed an excellent capability for arsenate adsorption, with a maximum capacity of 204.2 mg/g [6]. However, nanosized γ-Fe_2O_3 particles embedded in cotton wood [10], walnut shell [5] and rice husk [12] biochars exhibited a much lower adsorption capacity of 3.1 mg/g, 1.91 mg/g and up to 0.7 mg/g, respectively. In the case of a MnO-modified corn straw biochar, the maximum adsorption capacity for As^{5+} was reported to be 14.4 mg/g [1]; for bismuth impregnated wheat straw biochar, it was 16.2 mg/g [4]; and for KOH-modified corn cobs, it was 31 mg/g [11]. Mg-modified materials for arsenate adsorption have been scarcely used. In one study [13], Mg-Al modified corn stalk was synthesized to immobilize As, Pb and Cd simultaneously. The removal efficiency of arsenate was found to increase from −62% to 52%. Also, there is limited knowledge on the impregnation of ZnO on biochars and its effect on improving solid properties. The results obtained by corn cob-derived ZnO-impregnated biochar showed a maximum equilibrium adsorption capacity of 25.9 mg/g of As^{5+} [14]. However, the modification by ZnO has been principally used for organic substances and dyes in photocatalytic experiments and, to a lesser extent, in heavy metals adsorption [15]. Furthermore, to the authors' knowledge, the use of Ca oxides, as impregnation agent for the production of engineered biochar, has been investigated only for rice husks, with a low adsorption capacity of about 0.5 mg/g [12].

Based on the above discussion, the innovation of the present study lies in the investigation of the potential of a nut residue not previously investigated, an almond shell, for the efficient adsorption of As^{5+} from contaminated water solutions after its physical activation by nitrogen/steam and its modification by $MgCl_2.6H_2O$, $Zn(NO_3)_2.4H_2O$ and quarry dust, a waste of mortars production as a novel modifier. The objective was to explore the performance of the material by varying experimental parameters, as well as the mechanism of adsorption. Valuable information was provided by carrying out advanced specific analyses of the material before and after As^{5+} adsorption and by applying two isotherm models to simulate experimental data.

2. Materials and Methods

2.1. Raw Material and Pretreatment

The agrowaste used in this work was almond shell (AS), provided by a nuts production industry located in North Greece. The material was ground in a cutting mill, sieved to obtain a particle size less than 500 μm and homogenized by a Jones riffler in order to obtain representative samples for the tests.

Four different feedstocks were prepared for the adsorption experiments: non-modified almond shell biochar (ASB), Mg-modified biochar (ASB-Mg), Zn-modified biochar (ASB-Zn) and Ca-modified biochar (ASB-Ca). For Mg modification, raw AS material was dispersed in $MgCl_2.6H_2O$ solution at a solid-to-liquid ratio of 1/10 and stirred for 24 h at 200 rpm and room temperature. After filtration, the material was dried at 110 °C and then activated as described below. The Mg-modified biochar produced was washed to constant pH with distilled water and dried in the oven at 110 °C before use. For the preparation of Zn-modified feedstock, biochar was mixed with NaOH 0.5 M at a solid-to-liquid ratio of

1:5 and stirred at 200 rpm, while being heated up to 60 °C [14]. Zinc nitrate tetrahydrate $Zn(NO_3)_2.4H_2O$, purchased by Sigma-Aldrich Chemical Co. (St. Louis, MO, USA), was then added slowly to the solution, which was heated up to 80 °C and maintained at this temperature for 2 h, to ensure $Zn(OH)_2$ and ZnO precipitation. Subsequently, the slurry was introduced in a supersonic bath for about 1 h, filtered, washed with distilled water and dried in the oven overnight at 60 °C. Finally, for the modification of the AS with CaO, the raw material was mixed with quarry dust, a waste of mortars production provided by the Finomix AE company, consisting mainly of calcite and calcite magnesian (Figure 1h [16]), at a mass ratio of 1:1. The mixture was impregnated with distilled water and stirred for 12 h at 120 rpm. The slurry was further stirred at 80 °C for 3 h, filtered, washed with distilled water until neutrality and dried at 110 °C overnight. The modified material was activated by nitrogen/steam, as described below.

2.2. Production of the Adsorbent and Characterization

All modified or non-modified materials were physically activated by nitrogen and steam in order to increase their specific surface area and pore volume for the adsorption of As^{5+}. The equipment used was a fixed bed reactor unit, presented in detail in a previous investigation [16]. The sample was first pyrolyzed in nitrogen (150 mL/min) up to 700 °C, a temperature which was attained with a furnace heating rate of 10 °C/min and kept for 30 min. Nitrogen gas was then replaced by steam by injecting continuously de-ionized water through a syringe pump, with a flow rate of 0.5 mL/min. A uniform flow of steam was achieved by forcing water through a 2 m tube surrounding the reactor within the furnace before contacting the sample bed. The activation lasted 1 h and was followed by reactor cooling with nitrogen. In the case of quarry dust-modified material, the pyrolysis step was conducted at 750 °C for 2 h to calcine the carbonate of calcium. The content of the quarry dust in CaO was 94.5%, as reported previously by the authors [16].

The biochar was characterized following the European CEN/TC335 standards and employing programmable furnaces and the CHNS Thermoscientific analyzer, model Flash 2000 (Thermo Fisher Scientific Inc., Waltham, MA, USA). The specific surface area and pore volume were measured by the BET method, using the Autosorb 1Q-C-MP equipment of Quantachrome, for relative pressures of 0.015–0.32. Before the test, the sample was out-gassed under vacuum at 200 °C for 1 h. Fourier Transform Infrared Spectroscopy (FTIR) was adopted to identify the chemical functional groups on the biochar's surface before and after arsenate adsorption, using a Spectrum 100 spectrophotometer of Perkin Elmer. FTIR spectra were obtained for the 400–4000 cm^{-1} range of wave numbers, with a resolution of 4 cm^{-1}. Finally, any crystalline mineral phases of biochar before and after arsenate adsorption were detected by an X-ray diffractometer (XRD), type S2 Ranger EPS (Bruker AXS, Karlsruhe, Germany). The evaluation of XRD spectra was performed using the COD (Crystallography Open Database) database and DIFFRAC-Plus Evaluation software (v. 12.0., Karlsruhe, Germany).

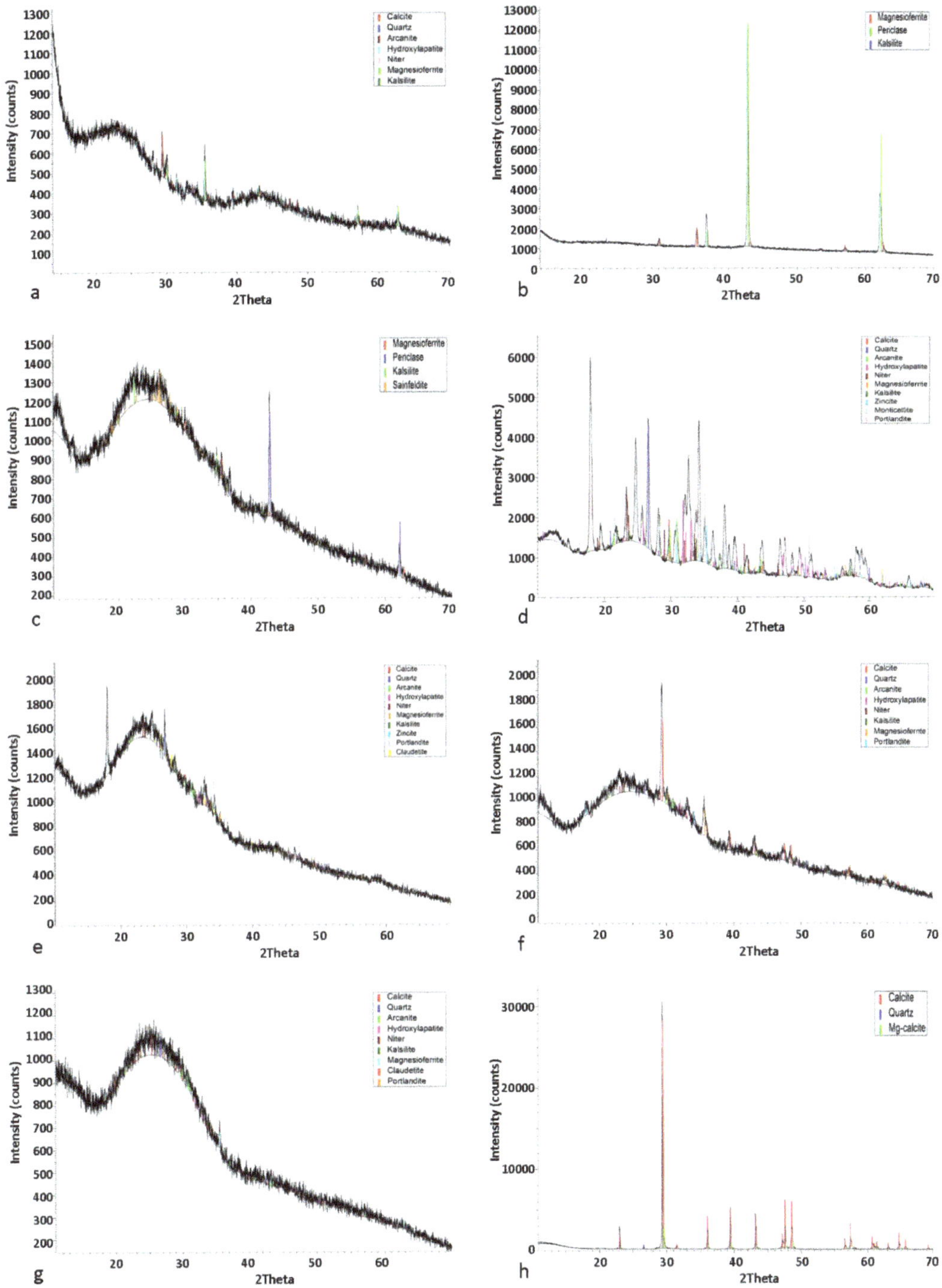

Figure 1. XRD spectra of (**a**) original biochar modified by (**b**) Mg^{2+} prior to adsorption, (**c**) Mg^{2+} after adsorption, (**d**) Zn^{2+} prior to adsorption, (**e**) Zn^{2+} after adsorption, (**f**) Ca^{2+} prior to adsorption, (**g**) Ca^{2+} after adsorption and (**h**) quarry dust.

2.3. Arsenate Adsorption Experiments

The reagent used in the current work was sodium arsenate dibasic heptahydrate ($Na_2HAsO_4.7H_2O$) of analytical grade, purchased by Sigma Aldrich Chemical Co. Arsenate concentrations 5, 10, 25, 50, 75 and 100 mg/L were prepared in de-ionized water. The pH of each solution was adjusted to 6 or 7, using 0.1 M NaOH or 0.1 M HCl.

For the kinetic experiments, the adsorbent was immersed in a 10 mg/L standard arsenate solution at a load of 4 g/L and stirred at 200 rpm on a mechanical shaker for 24 h at room temperature. Liquid samples were withdrawn and filtered after specific contact times up to 24 h. The aliquot pH was measured again and then analyzed by an inductively coupled plasma mass spectrometer ICP-MS, model 7500× of Agilent Technologies (Santa Clara, CA, USA), assisted by an Anton Paar Multiwave 3000 oven (Anton Paar GmbH, Graz, Austria) for the digestion of the samples, to determine the amount of residual As^{5+} in each solution. The heavy metal concentration adsorbed by the biochar was calculated as a difference of the initial and final aqueous concentration values.

Once the equilibrium contact time was established, the As^{5+} adsorption experiments were carried out following the same experimental conditions as before for all arsenate solution concentrations prepared. Duplicate tests were performed, and the data are reported as average values.

2.4. Equilibrium and Adsorption Modelling

The pseudo-first-order model (Equation (1)) and the pseudo-second-order model (Equation (2)) were used to simulate the kinetic tests [4,8]:

$$\log(q_e - q_t) = \log q_e - \frac{k_1}{2.303}t \quad (1)$$

$$\frac{t}{q_t} = \frac{1}{k_2 {q_e}^2} + \frac{t}{q_e} \quad (2)$$

The modelling of the adsorption data and the determination of the adsorption capability of the biochar under study were performed by adopting the Langmuir (Equation (3)) and Freundlich (Equation (4)) isotherm models [8,11]:

$$\frac{C_e}{q_e} = \frac{1}{bC_e} + \frac{C_e}{Q} \quad (3)$$

$$\log q_e = \log k + \frac{1}{n}\log C_e \quad (4)$$

3. Results and Discussion

3.1. Properties of Non-Modified and Modified Adsorbent Materials

As Table 1 shows, the biochar of AS generated after nitrogen/steam activation was rich in organic matter and carbon, had a low content of minerals and hydrogen, and was free of nitrogen and sulfur, implying no toxic emissions during processing. However, the oxygen concentration was high, making the material more hydrophilic, with increased polarity [17,18]. The specific surface of the non-modified biochar was enhanced after the nitrogen/steam thermal treatment, and it was about 260 times higher than that of the raw material (2.5 m^2/g). The microporosity was also increased. The elemental concentration of AS biochar is within the range of values reported for hard wood [18]. The specific surface area and pore volume are higher than most reported data for various biochar materials, ranging between 61 and 349 m^2/g and 10 and 29 × 10^{-2} cm^3/g, respectively [1,2,4,6,13]. On the other hand, all modified samples presented lower values of the specific surface area (ASB-Mg: 108 m^2/g, ASB-Zn: 453 m^2/g, ASB-Ca: 120 m^2/g), which could be assigned to the coverage of the solid surface by the oxides of MgO, ZnO and CaO, respectively, or the dilution effect.

Table 1. Chemical and structural characteristics of biochar (% dry) *.

Fixed Carbon	Ash	C	H	N	O	S	Specific Surface Area (m^2/g)	Micropore Volume × 10^2 (cm^3/g)	Average Pore Size(Å)
97.5	2.5	66.0	1.1	-	30.4	-	657.0	32.9	24.0

* "-" not applicable or below detection limits.

The XRD spectra of all adsorbent materials studied before and after the adsorption of arsenate ions are illustrated in Figure 1. The mineral matter of the original AS material was rich in Ca and K phases, mainly as calcite and arcanite, respectively. P was incorporated in hydroxylapatite, whereas niter, quartz, kalsilite and magnesioferrite were detected in low amounts.

After biochar was modified by MgO, Figure 1b shows that periclase was formed on its surface. Following arsenate adsorption, a new crystalline phase was identified in the spectra of Figure 1c, sainfeldite ($Ca_5(AsO_4)_2(AsO_3OH)_2.4H_2O$), indicating that under alkaline conditions, As^{5+} was removed by precipitation. To our knowledge, this mineral phase has not been reported previously for the adsorption of arsenate. The XRD of AS biochar modified by ZnO is represented in Figure 1d. Apart from the minerals detected for the raw material, zincite (ZnO) was identified after modification. Following arsenate adsorption, from Figure 1e, it can be seen that a new mineral phase claudetite (As_2O_3) was formed. This mineral was also detected in the XRD spectra of biochar modified by quarry dust after the adsorption of As^{5+} (Figure 1g). Figure 1f,g also show the CaO of quarry dust after the adsorption of As^{5+} (Figure 1g). Figure 1f,g also show that the CaO of quarry dust was transformed to portlandite ($Ca(OH)_2$) after the activation of the biochar with steam. There is very limited information from the literature data on XRD results after arsenate adsorption by biochar materials. As^{5+} precipitation was evidenced only in two reports, for calcium oxide-rich rice husk biochar [12] and dolomite-modified material [19].

The chemical functional groups on modified AS biochar, as detected by the FTIR, are indicated in Figure 2 and Table 2. In the case of Mg-modified material, the low-frequency peaks at wave numbers 434 cm^{-1} and 538 cm^{-1} are characteristic of the Mg-O composite [13,20]. The broad bands, shown in Figure 2a, at 1268 cm^{-1} and 1580 cm^{-1} are attributed to the C-O stretching vibration of alkyl/aryl ethers or aromatic esters and cycloalkenes, respectively. After As^{5+} adsorption, the FTIR spectrum presented in Figure 2b was quite different. Strong intensities at 434 cm^{-1} and 538 cm^{-1} wave numbers disappeared, indicating that new bonding occurred between the Mg-O and As^{5+} ion. Also, the new peaks, which appeared at wave numbers 640–1050 cm^{-1}, represent the stretching vibration of As-O or As-OH bonds [4,7], confirming the adsorption of As^{5+} on the biochar's surface. On the other hand, the disappearance of the 1268 cm^{-1} peak after arsenate adsorption implies some interaction between C-O groups with As^{5+} [1]. The new strong peak at 1442 cm^{-1} and the weak one at 2372 cm^{-1} are attributed to the C-H stretching of alkanes and the vibrational stretching of O=C=O, respectively. Furthermore, the new intense band at 3448 cm^{-1}, shown in Figure 2b, which is attributed to hydroxyl group O-H vibrations, suggests that the O-H of water molecules were also adsorbed on the biochar's surface during the process [4].

When ASB was modified by $ZnO/Zn(OH)_2$, Figure 2c shows that the introduction of this inorganic substance resulted in quite strong adsorption peaks between 450 cm^{-1} and 950 cm^{-1}, assigned to Zn-O vibrations [13] or bending vibrations of the hydroxyl groups in Zn-OH [3]. By comparing Figure 2c,d, corresponding to the Zn-modified spectra of the biochar before and after arsenate adsorption, it can be observed that several peaks are common in these graphs. Thus, the peaks at wave numbers 1384 cm^{-1}, 1594/1642 cm^{-1}, 2368 cm^{-1} and 3494 cm^{-1} are linked with stretching vibrations of O-H from alcohols/phenols, C=C from cycloalkenes and O=C=O and O-H from alcohol groups, respectively. As^{5+} adsorption (Figure 2d) is indicated by the characteristic bands centered at 652 cm^{-1} and 884 cm^{-1}, representing As-O or As-OH bonds [4,7].

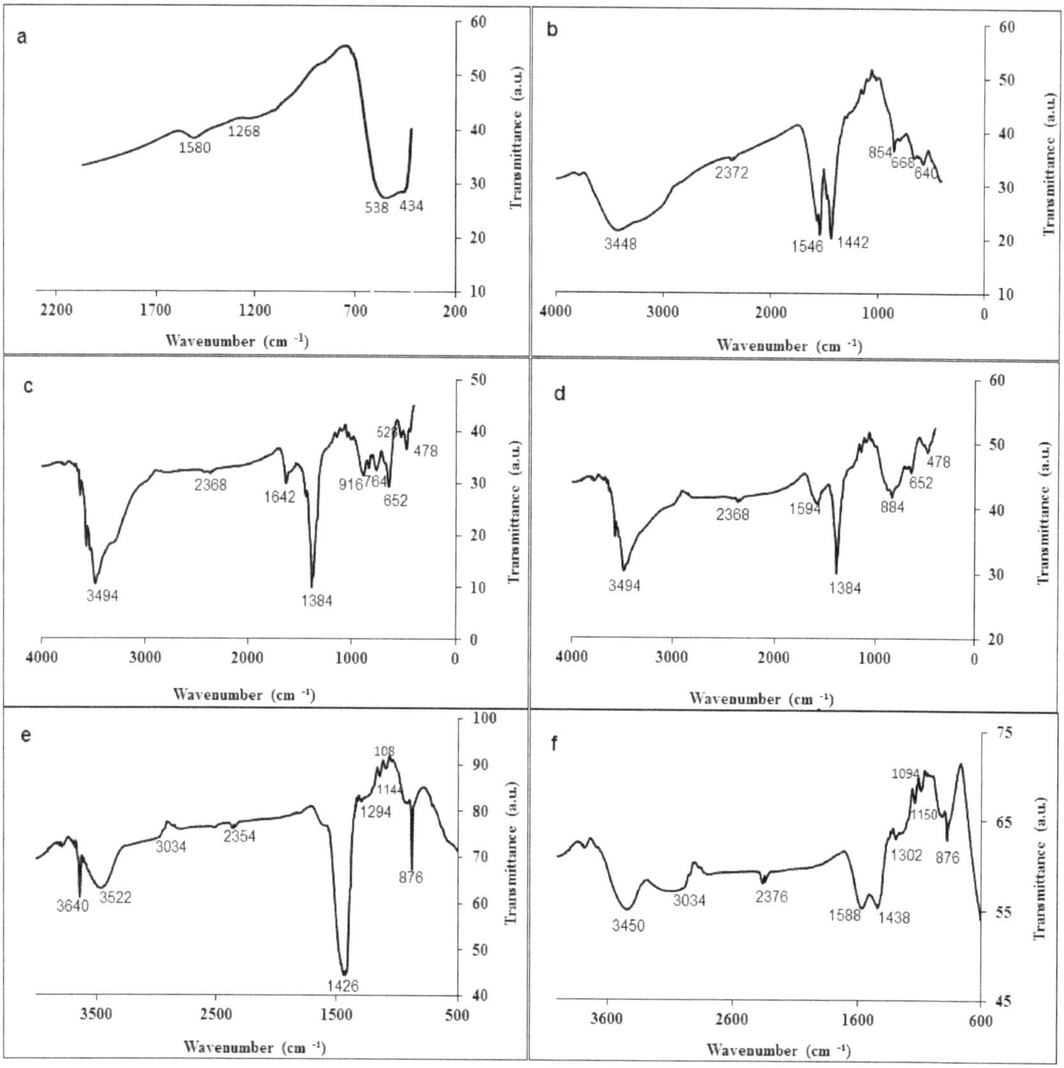

Figure 2. FTIR spectra of biochar modified by (**a**) Mg^{2+} prior to adsorption, (**b**) Mg^{2+} after adsorption, (**c**) Zn^{2+} prior to adsorption, (**d**) Zn^{2+} after adsorption, (**e**) Ca^{2+} prior to adsorption and (**f**) Ca^{2+} after adsorption.

The FTIR of ASB modified by quarry dust is illustrated in Figure 2e,f. At the 876 cm^{-1} wave number, a strong vibration characteristic of Ca-O groups [21] of calcined quarry dust was shown in Figure 2e. Several small peaks in the region of 1084–1294 cm^{-1} are attributed to the C-O stretching of alcohols, aliphatic ethers and aromatic esters. The intense absorbance band at 1426 cm^{-1} represents the O-H bending of carboxylic acid, while the small peak at 2354 cm^{-1} corresponds to the vibrational stretching of O=C=O. Broad bands between 3034 cm^{-1} and 3640 cm^{-1} are all representative of O-H groups from alcohols. After arsenate adsorption, from Figure 2f, it can be noticed that the intensity of the peak corresponding to Ca-O bonds was considerably decreased, revealing some interaction of quarry dust with As^{5+} and possibly the overlapping of As-O, As-OH bonds in this region. Also, the peaks appearing in the 1084–1426 cm^{-1} domain before adsorption were shifted to higher wave numbers after As^{5+} adsorption, 1094–1438 cm^{-1}, implying π–π interaction between the biochar matrix and As^{5+} [22]. The new broad band at 1588 cm^{-1}, in Figure 2f, is assigned to cycloalkenes. All peaks at higher wave numbers (above 2000 cm^{-1}) found on quarry dust-modified material (Figure 2e), representing O=C=O and O-H stretching from alcohol groups, were also present after the adsorption of arsenate (Figure 2f).

Table 2. Chemical functional groups of modified biochar.

Modified Biochar	Wave Number (cm^{-1})	Functional Groups
ASB-Mg	434/538	Mg-O
	640/668/854	As-O/As-OH
	1268	C-O
	1442	C-H
	1546/1580	C=C
	2372	O=C=O
	3448	O-H
ASB-Zn	478/528/652/764/884/916	Zn-O/Zn-OH
	652/884	As-O/As-OH
	1384	O-H
	1594/1642	C=C
	2368	O=C=O
	3494	O-H
ASB-Ca	876	Ca-O/As-O/As-OH
	1084/1144/1150/1294/1302	C-O
	1426/1438	O-H
	1588	C=C
	2354/2376	O=C=O
	3034/3450/3522/3640	O-H

3.2. Kinetics of Arsenate Adsorption

The effect of the contact time on the uptake of As^{5+} ions by ASB from the aqueous solution was studied within a time range of 0.5–24 h for an initial arsenate concentration of 10 mg/L. Figure 3 shows that arsenate delayed to adhere on the biochar surface and the adsorption rate was low, removing about 12% of arsenate from the solution in 3 h and 27% in 6 h. Then, it gradually decreased, finally achieving equilibrium in 12 h. The maximum removal of arsenate ions was 29.4% for an initial pH = 7 (at pH = 6, the efficiency was lower, so pH = 7 was selected for subsequent tests). This behavior implies a low bonding energy between the adsorbate and adsorbent, resulting in higher contact times, and this has also been observed in previous studies [8,11].

Two classical dynamic models (Equations (1) and (2)) were employed for the experimental data, and the characteristic parameters are displayed in Table 3. As can be seen, the model that was found to describe the experimental results more accurately was the pseudo-second-order model. According to this model, the reaction rate is proportional to the number of active sites on the surface of the adsorbent, and the rate-limiting step is chemisorption reaction. These findings greatly agree with the literature data [3,5,8,11,13,14].

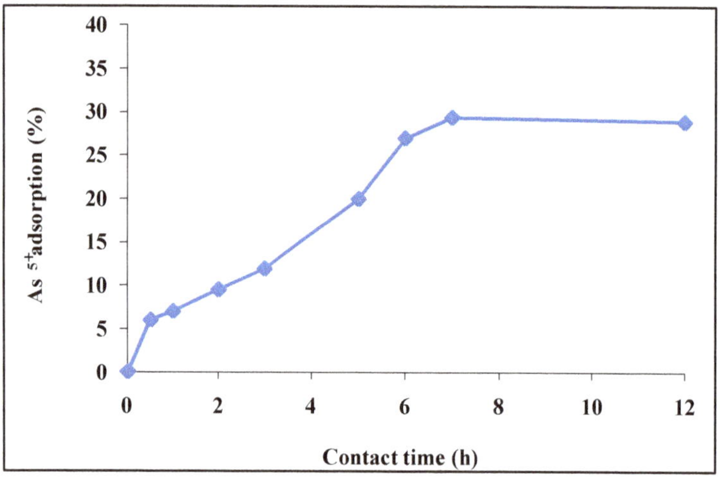

Figure 3. Kinetics of arsenate adsorption.

Table 3. Kinetic results of arsenate adsorption by ASB.

Kinetic Model Parameters					
Pseudo-First-Order			Pseudo-Second-Order		
qe (mg/g)	k_1 (1/h)	R^2	qe (mg/g)	k_2 (1/h)	R^2
0.733	0.009	0.427	0.733	0.973	0.759

3.3. Arsenate Adsorption Capacity of AS Biochar

The adsorption capacity of AS biochar, at various initial As^{5+} concentrations in the solution and adsorbent dose, is compared with that of Mg-, Zn- and Ca-modified biochar in Table 4. As can be seen, at an adsorbent dose of 4 mg/L, the non-modified material showed a poor adsorption capacity for As^{5+}, especially at low initial concentrations. The removal efficiency of As^{5+} increased gradually with an increasing ion flux density, attaining a maximum of 49.4% at 100 mg/L As^{5+}. This trend of favoring access to biochar's adsorption sites with a higher As^{5+} concentration gradient, was also observed when the material was modified by quarry dust. However, the removal rate significantly increased in this case, and the maximum value was 97% at 100 mg/L As^{5+}. For Mg- and Zn-modified biochar, it can be noticed that at low ion concentrations, up to 10 mg/L, arsenate was practically completely adsorbed by the solid, with the uptake rate declining thereafter at higher concentrations. For Mg-modified material, the removal efficiency was 86% at 100 mg/L As^{5+}, while Zn-modified biochar presented an excellent performance for both low and high As^{5+} concentrations, achieving the lowest value of 97% at 100 mg/L.

The effect of the adsorbent dosage plays a significant role in the removal efficiency of heavy metals from aqueous solutions. Table 4 shows that when the amount of biochar was lowered to 2 g/L, the adsorption of arsenate was improved (with the exception of Mg-modified material), agreeing with earlier studies [1,13].

Another critical parameter of the process, governing the surface charge of the biochar and the ionization of arsenate in the solution, is the pH. As arsenic is an anionic heavy metal, electrostatic attraction between the ion and the adsorbent would be favored under acidic conditions and weakened under alkaline conditions. At a basic pH, As^{5+} species exist as $HAsO_4^{2-}$ [7,13,14]. From Table 4, presenting the solution pH before and after As^{5+} adsorption at equilibrium, it can be observed that the pH was alkaline throughout the process and the adsorption efficiency increased with the pH, reaching maximum values at pH values between 9.7 and 11.5. As AS biochar was found to be alkaline (pH = 10.2) and abundant in negatively charged functional groups (identified by FTIR), this implies that As^{5+} ions interacted with –O-H groups on the solid surface through ligand exchange, releasing them into the solution. Some previous studies [3,14] also reported that adsorption between negatively charged oxyanions of As^{5+} and ZnO might occur through the ligand exchange of the hydroxyl group from Zn-OH, as the removal percentage of As^{5+} was enhanced with increasing pH values. Furthermore, others reported that for Mg- or Ca-modified materials, MgO [17,20] or CaO [12] were positively charged on the biochar's surface, efficiently removing As^{5+} ions from alkaline solutions. Possible mechanisms for the current results will be discussed below.

Table 4. Removal efficiency of arsenate for various initial concentrations and adsorbent doses.

Initial Ion Concentration (mg/L)	Adsorbent Dose 4 g/L							
	Removal Efficiency (%)				pH after Adsorption			
	ASB	ASB-Mg	ASB-Zn	ASB-Ca	ASB	ASB-Mg	ASB-Zn	ASB-Ca
5	2.2	99.9	100.0	89.0	8.02	9.97	9.58	11.19
10	29.3	99.1	100.0	94.0	7.89	9.65	9.34	11.85
25	24.4	96.0	98.5	95.8	7.90	9.60	9.17	11.77
50	41.2	77.0	98.4	96.2	7.96	9.56	8.90	11.63
75	45.2	85.0	98.0	96.7	7.99	9.46	8.78	11.58
100	49.4	86.0	97.0	97.0	8.04	9.33	8.71	11.48
	Adsorbent Dose 2 g/L							
5	2.9	94.0	100.0	92.0	7.90	10.05	9.65	11.2
10	30.0	93.3	100.0	95.0	7.81	9.81	9.32	11.9
25	31.2	82.2	99.7	97.0	7.83	9.75	9.15	11.8
50	45.1	64.3	98.8	97.3	7.88	9.65	8.92	11.6
75	46.9	63.8	99.1	97.7	7.91	9.55	8.87	11.6
100	50.6	62.7	99.9	98.0	7.97	9.47	8.80	11.5

3.4. Equilibrium of Arsenate Adsorption

The arsenate uptake of the ASB adsorbent was simulated by the Langmuir and Freundlich models. The adsorption isotherms, which were defined in the experimental section at equilibrium conditions, are illustrated in Figures 4–7, and calculated parameters are included in Table 5. All arsenate equilibrium test data were best fitted with the Freundlich model, conforming the theory that arsenate molecules migrated to heterogeneous surface sites on the biochar, following a multilayer adsorption mechanism, instead of a monolayer mechanism assumed by the Langmuir model. The Freundlich correlation coefficient ranged between 0.94 and 0.97, as opposed to the Langmuir correlation coefficient, which varied between 0.42 and 0.83 for the biochar-modified material.

Table 5. Model results of arsenate adsorption by non-modified or modified ASB.

	Adsorbent Dose 4 g/L					
	Langmuir Model			Freundlich Model		
	Q (mg/g)	b (L/mg)	R^2	k (L/g)	1/n	R^2
ASB	12.469	0.009	0.205	0.126	1.080	0.888
ASB-Mg	15.528	0.945	0.773	6.092	0.331	0.941
ASB-Zn	26.666	1.974	0.835	12.981	0.325	0.943
ASB-Ca	20.533	0.188	0.819	5.375	1.325	0.993
	Adsorbent Dose 2 g/L					
ASB	81.301	0.585	0.508	27.327	0.698	0.810
ASB-Mg	33.333	0.126	0.838	4.795	0.487	0.977
ASB-Zn	39.062	9.143	0.714	66.512	0.637	0.889
ASB-Ca	16.0	0.416	0.602	15.170	1.800	0.973

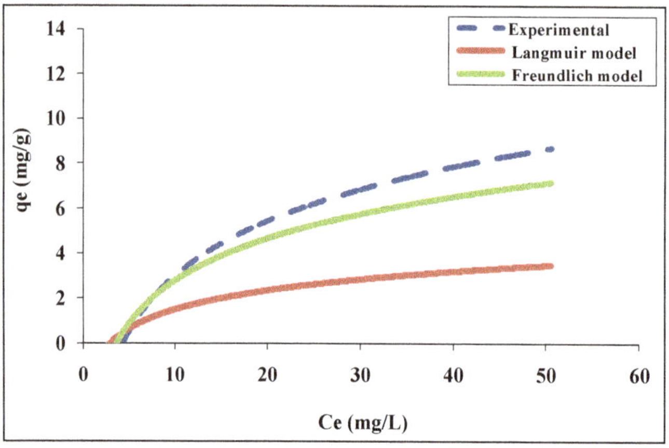

Figure 4. Isotherms of the original biochar at an adsorption dose of 4 g/L.

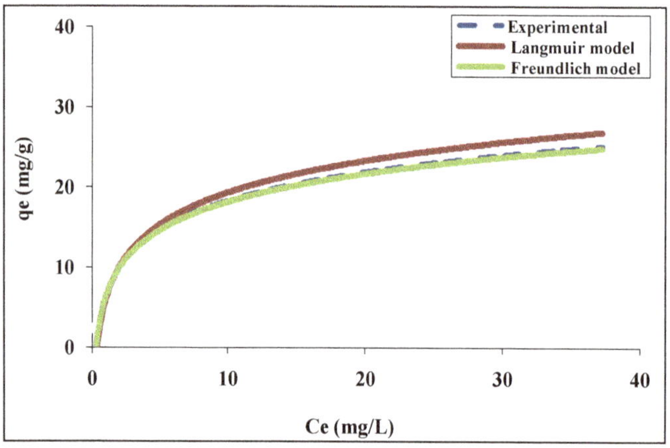

Figure 5. Isotherms of the Mg-modified biochar at an adsorption dose of 2 g/L.

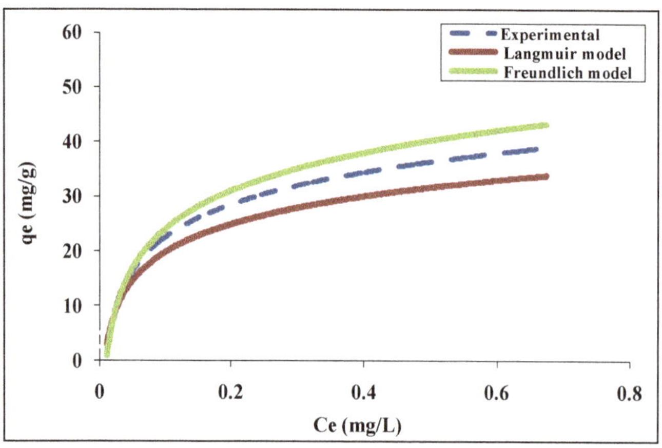

Figure 6. Isotherms of the Zn-modified biochar at an adsorption dose of 2 g/L.

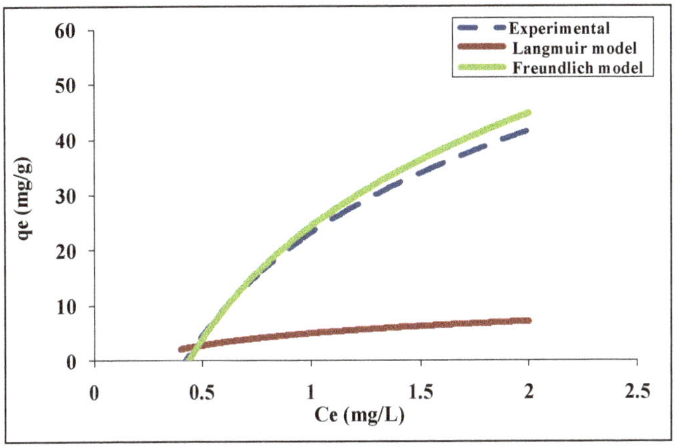

Figure 7. Isotherms of the Ca-modified biochar at an adsorption dose of 2 g/L.

Past investigations using a variety of materials and modifiers reported that either the Freundlich [3,12,13] or Langmuir [4,5,7,10,14] models best described the equilibrium As^{5+} concentration and sorption capacity, pointing to the complexity of the adsorption mechanism. The present results showed that for non-modified ASB, the maximum adsorption capacity was 12.4 mg/g, which is higher than the values determined (0.8–7.1 mg/g) for similar materials [1,7,8,11,14]. However, for modified material by Mg, Zn or Ca oxides, the maximum value was increased to 21.5 mg/g, 24.3 mg/g and 24.3 mg/g, respectively. On the other hand, the maximum capacity of ASB for As^{5+} adsorption could be enhanced up to 35 mg/g, 50 mg/g and 49 mg/g for Mg-, Zn- and Ca -modified material, respectively, in case the adsorbent dose is 2 g/L (Figures 5–7). Previous studies for MnO-modified corn straw biochar reported a maximum adsorption capacity for As^{5+} of 14.4 mg/g [1]; for KOH-modified municipal solid wastes, a value of 31 mg/g was reported [11]; and for ZnO-modified corn cobs, a value of 25.9 mg/g was reported [14]. On the other hand, for Fe-coated biochars of rice husks, walnut shells, cotton wood and pine sawdust, the maximum As^{5+} capacities were found to range between 0.1 mg/g and 43.7 mg/g [5–7,10,12]. A comparison with the current results is made in Table 6.

Table 6. Comparison of the literature data with the current results.

Feedstock	Activation Method	Adsorbent Dose (g/L)	Arsenate Adsorption Capacity (mg/g)	Reference
Empty fruit bunch Rice husk Fe-coated rice husk	Commercial biochar	5	5.5 7.1 16	[7]
Roots of C. esculenta	Nitrogen/steam 700 °C	0.6	2.2	[8]
Corncob ZnO-corncob	Nitrogen 600 °C	4	5 25.9	[14]
MnO-corn straw	Nitrogen 600 °C	0.5–4	14.4	[1]
KOH-MSW	Nitrogen 600 °C	2	31	[11]
Fe-coated walnut shell	Nitrogen microwave	1	1.9	[5]
Fe-coated pine sawdust	Nitrogen 600 °C	0.2	43.7	[6]
Fe-coated cotton wood	Nitrogen 600 °C	2	3.1	[10]
Fe-coated rice husk	Nitrogen 300 °C	16	0.7	[12]
Almond shell Mg/Zn/Ca-almond shell	Nitrogen/steam 700 °C	2	12.4 35/50/49	This study

3.5. Mechanisms of Arsenate Adsorption

An improved adsorption of As^{5+} on biochar can be achieved by enlarging its surface area and optimizing its porosity. However, this is not the only criterion for the efficient removal of As^{5+} and does not always increase the adsorption capacity [9,11,12]. Other mechanisms also play important roles such as electrostatic attraction, anion exchange, chemical complexation, electron coordination, and surface precipitation [2,3,7,11–14].

As previously presented by the ultimate analysis and the FTIR spectra, the biochar under study was enriched in oxygen and incorporated in negatively charged functional groups. Also, it was shown that the pH was alkaline throughout the adsorption process and As^{5+} species existed as $HAsO_4{}^{2-}$ in the solution. Furthermore, for all modified materials, it was shown that the adsorption efficiency increased with the pH, reaching maximum values at pH = 9.7–11.5. Therefore, the ligand exchange between As^{5+} ions and hydroxyl groups from Ca-OH, Mg-OH or Zn-OH on a solid surface appeared to be the mechanism of As^{5+} adsorption, while electrostatic attraction was not possible. These findings are supported by previous studies [3,8,14].

The type and concentration of biochar's surface functional groups are known to significantly control the adsorption capacity of organic or inorganic sorbents [7,11,13]. Through FTIR analysis, it was shown that, after arsenate adsorption As-O and As-OH bonds appeared on biochar's surface for all modified materials. Moreover, the decrease in Ca-O and the disappearance of Mg-O peaks after adsorption imply some interaction of quarry dust and MgO with As^{5+}, respectively. Consequently, another mechanism responsible for the adsorption of As^{5+} could be chemical complexation between As^{5+} and the functional groups of the adsorbent. Also, for quarry dust-modified biochar, the shifting of peaks to higher wave numbers after adsorption reveals electron coordination between the As^{5+} ion and adsorbent matrix as a potential adsorption mechanism. These results are in agreement with the majority of previous investigations reporting such mechanisms [4,7,9,11,13,22].

Furthermore, according to the XRD spectra presented above and analyzed, arsenate was found to be removed from the solution by precipitation with inorganic or organic species of biochar, as new mineral phases were detected after adsorption from MgO-, ZnO- and CaO-modified material, sainfeldite and claudetite, respectively. Therefore, surface precipitation was also identified as an important As^{5+} adsorption mechanism, in agreement with some other studies [12,17]. The sorption mechanisms are shown in Figure 8.

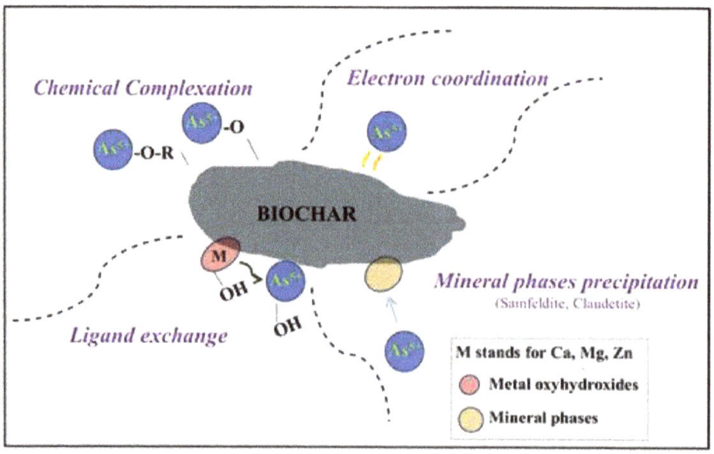

Figure 8. Mechanisms of arsenate sorption.

4. Conclusions

Almond shell biochar presented a negatively charged surface. The adsorption of arsenate occurred by chemisorption and attained equilibrium in 12 h, with a maximum removal efficiency of 29.4%, at an initial concentration of 10 mg/L. The modification of biochar by Mg, Zn or Ca oxides increased the removal rate significantly, from 49.4% at 100 mg/L As^{5+} up to 86%, 97% and 97%, respectively. Zn-modified biochar presented an excellent performance for both low and high As^{5+} concentrations.

All experimental data were accurately fitted by the Freundlich isotherm model (R^2 = 0.94–0.97), indicating a multilayer adsorption mechanism. When the adsorbent dose was 4 g/L, the arsenate adsorption capacity of biochar increased after Mg-, Zn- or Ca-modification from 12.4 mg/g to 21.5 mg/g, 24.3 mg/g and 24.3 mg/g, respectively. For an adsorbent dose of 2 g/L, these values were raised up to 35 mg/g, 50 mg/g and 49 mg/g, respectively.

Structural analyses, pH measurements, X-ray diffraction and Fourier Trasform Infrared analyses, before and after the adsorption of the As^{5+} ion, showed that ligand exchange between As^{5+} ions and hydroxyl groups on the biochar's surface, chemical complexation between the As^{5+} ion and the adsorbent as well as surface precipitation with modified biochar species were the potential mechanisms of adsorption. For quarry dust-modified material, electron coordination was also a possible mechanism.

Author Contributions: Conceptualization, D.V.; methodology, D.V. and D.P.; software, E.S. and A.S.; validation, D.V. and D.P.; investigation, D.V. and D.P.; data curation, D.P., E.S. and A.S.; writing—original draft preparation, D.V.; writing—review and editing, D.V.; visualization, D.V.; supervision, D.V. All authors have read and agreed to the published version of the manuscript.

Funding: This research received no external funding.

Data Availability Statement: The original contributions presented in the study are included in the article, further inquiries can be directed to the corresponding author/s.

Acknowledgments: The authors kindly thank the laboratories of Hydrocarbons Chemistry and Ore Processing and Beneficiation of the Technical University of Crete for the ultimate and FTIR analysis and T. Ioannides and M. Smyrnioti from the Inst. of Chemical Engineering Sciences in Patra for the BET measurements of the samples.

Conflicts of Interest: The authors declare no conflicts of interest.

References

1. Yu, Z.; Zhou, L.; Huang, Y.; Song, Z.; Qiu, W. Effects of a manganese-oxide-modified biochar composite on adsorption of arsenic in red soil. *J. Environ. Manag.* **2015**, *163*, 155–162. [CrossRef] [PubMed]
2. Braghiroli, F.L.; Calugaru, I.L.; Merchan, C.G.; Neculita, C.M.; Bouafif, H.; Koubaa, A. Efficiency of eight modified materials for As (V) removal from synthetic and real mine effluents. *Miner. Eng.* **2020**, *151*, 106310. [CrossRef]
3. Xia, D.; Tan, F.; Zhang, C.; Jiang, X.; Chen, Z.; Li, H.; Zheng, Y.; Li, Q.; Wang, Y. $ZnCl_2$-activated biochar from biogas residue facilitates aqueous As (III) removal. *Appl. Surf. Sci.* **2016**, *377*, 361–369. [CrossRef]
4. Zhu, N.; Yan, T.; Qiao, J.; Cao, H. Adsorption of arsenic, phosphorous and chromium by bismuth impregnated biochar: Adsorption mechanism and depleted adsorbent utilization. *Chemosphere* **2016**, *164*, 32–40. [CrossRef] [PubMed]
5. Duan, X.; Zhang, C.; Srinivasakannan, C.; Wang, X. Waste walnut shell valorization to iron loaded biochar and its application to arsenic removal. *Res. Effic. Technol.* **2017**, *3*, 29–36.
6. Liu, Z.; Zhang, F.; Sasai, R. Arsenate removal from water using Fe_3O_4-loaded activated carbon prepared from waste biomass. *Chem. Eng. J.* **2010**, *160*, 57–62. [CrossRef]
7. Samsuri, A.W.; Zadeh, F.S.; Bardan, B.J.S. Adsorption of As (III) and As (V) by Fe coated biochars and biochars produced from empty fruit bunch and rice husk. *J. Environ. Chem. Eng.* **2013**, *1*, 981–988. [CrossRef]
8. Banerjee, S.; Mukherjee, S.; LaminKa-Ot, A.; Joshi, S.R.; Mandal, T.; Halder, G. Biosorptive uptake of Fe^{2+}, Cu^{2+} and As^{5+} by activated biochar derived from Colocasia esculenta: Isotherm, kinetics, thermodynamics and cost estimation. *J. Adv. Res.* **2016**, *7*, 597–610. [CrossRef]
9. Alatalo, S.; Repo, E.; Makila, E.; Salonen, J.; Vakkilainen, E.; Sillanpaa, M. Adsorption behaviour of hydrothermally treated municipal sludge and pulp and paper industry sludge. *Biores. Technol.* **2013**, *147*, 71–76. [CrossRef]
10. Zhang, M.; Gao, B.; Varnoosfaderani, S.; Hebard, A.; Yao, Y.; Inyang, M. Preparation and characterization of a novel magnetic biochar for arsenic removal. *Biores. Technol.* **2013**, *130*, 457–462. [CrossRef]

11. Jin, H.; Capareda, S.; Chang, Z.; Gao, J.; Xu, Y.; Zhang, J. Biochar pyrolytically produced from municipal solid wastes for aqueous As (V) removal: Adsorption property and its improvement with KOH. *Biores. Technol.* **2014**, *169*, 622–629. [CrossRef] [PubMed]
12. Agrafioti, E.; Kalderis, D.; Diamadopoulos, E.J. Ca and Fe modified biochars as adsorbents of arsenic and chromium in aqueous solutions. *Environ. Manag.* **2014**, *146*, 444–450. [CrossRef] [PubMed]
13. Li, Q.; Liang, W.; Liu, F.; Wang, G.; Wan, J.; Zhang, W.; Peng, C.; Yang, J. Simultaneous immobilization of arsenic, lead and cadmium by magnesium-aluminum modified biochar in mining soil. *J. Environ. Manag.* **2022**, *310*, 114792. [CrossRef] [PubMed]
14. Cruz, G.J.F.; Mondal, D.; Rimaycuna, J.; Soukup, K.; Gomez, M.M.; Solis, J.L.; Lang, J. Agrowaste derived biochars impregnated with ZnO for removal of arsenic and lead in water. *J. Environ. Chem. Eng.* **2020**, *8*, 103800. [CrossRef]
15. Chen, M.; Bao, C.; Hu, D.; Jin, X.; Huang, Q. Facile and low-cost fabrication of ZnO/biochar nanocomposites from jute fibers for efficient and stable photodegradation of methylene blue dye. *J. Anal. Appl. Pyrolysis* **2019**, *139*, 319–332. [CrossRef]
16. Vamvuka, D.; Elmazaj, J.; Berkis, M. Enhanced H_2 gas production from steam gasification of a winery waste through CO_2 capture by waste concrete fines and use of alkali catalysts. *Renew. Energy* **2023**, *219*, 119428. [CrossRef]
17. Jiang, M.; Yang, Y.; Lei, T.; Ye, Z.; Huang, S.; Fu, X.; Liu, P.; Li, H. Removal of phosphate by a novel activated sewage sludge biochar: Equilibrium, kinetic and mechanism studies. *Applic. Ener. Comb. Sci.* **2022**, *9*, 100056. [CrossRef]
18. Hassan, M.; Liu, Y.; Naidu, R.; Parikh, S.J.; Du, J.; Qi, F.; Willett, I.R. Influences of feedstock sources and pyrolysis temperature on the properties of biochar and functionality as adsorbents: A meta-analysis. *Sci. Tot. Environ.* **2020**, *744*, 140714. [CrossRef]
19. Salameh, Y.; Albadarin, A.B.; Allen, S.; Walker, G.; Ahmad, M.N.M. Arsenic (III, V) adsorption onto charred dolomite: Charring optimization and batch studies. *Chem. Eng. J.* **2015**, *259*, 663–671. [CrossRef]
20. Shin, H.; Tiwari, D.; Kim, D.J. Phosphate adsorption/desorption kinetics and P bioavailability of Mg-biochar from ground coffee waste. *J. Water Process Eng.* **2020**, *37*, 101484. [CrossRef]
21. Li, J.; Li, B.; Huang, H.; Lv, X.; Zhao, N.; Guo, G.; Zhang, D. Removal of phosphate from aqueous solution by dolomite-modified biochar derived from urban dewatered sewage sludge. *Sci. Tot. Environ.* **2019**, *687*, 460–469. [CrossRef] [PubMed]
22. Li, R.; Wang, J.J.; Zhou, B.; Zhang, Z.; Liu, S.; Lei, S.; Xiao, R. Simultaneous capture removal of phosphate, ammonium and organic substances by MgO impregnated biochar and its potential use in swine wastewater treatment. *J. Clean. Prod.* **2017**, *147*, 96–107. [CrossRef]

C Journal of Carbon Research

MDPI

Article

Highly Porous Cellulose-Based Carbon Fibers as Effective Adsorbents for Chlorpyrifos Removal: Insights and Applications

Tamara Tasić [1], Vedran Milanković [1], Christoph Unterweger [2], Christian Fürst [2], Stefan Breitenbach [2,3], Igor A. Pašti [4] and Tamara Lazarević-Pašti [1,*]

[1] VINCA Institute of Nuclear Sciences—National Institute of the Republic of Serbia, University of Belgrade, Mike Petrovica Alasa 12–14, 11000 Belgrade, Serbia; tamara.tasic@vin.bg.ac.rs (T.T.); vedran.milankovic@vin.bg.ac.rs (V.M.)

[2] Wood K Plus—Kompetenzzentrum Holz GmbH, Altenberger Strasse 69, 4040 Linz, Austria; c.unterweger@wood-kplus.at (C.U.); c.fuerst@wood-kplus.at (C.F.); s.breitenbach@wood-kplus.at (S.B.)

[3] Institute of Chemical Technology of Inorganic Materials (TIM), Johannes Kepler University Linz, Altenberger Strasse 69, 4040 Linz, Austria

[4] Faculty of Physical Chemistry, University of Belgrade, Studentski Trg 12-16, 11158 Belgrade, Serbia; igor@ffh.bg.ac.rs

* Correspondence: tamara@vin.bg.ac.rs

Abstract: The extensive utilization of the organophosphate pesticide chlorpyrifos, combined with its acute neurotoxicity, necessitates the development of effective strategies for its environmental removal. While numerous methods have been explored for chlorpyrifos removal from water, adsorption is the most promising. We investigated the potential of two cellulose-derived porous carbons as adsorbents for chlorpyrifos removal from water, prepared by either CO_2 or H_2O activation, resulting in similar morphologies and porosities but different amounts of heteroatom functionalities. The kinetics of batch adsorption removal from water fits well with the pseudo-first-order and pseudo-second-order kinetic models for both materials. The Freundlich, Langmuir, Dubinin–Radushkevich, and Sips isotherm models described the process of chlorpyrifos adsorption very well in all investigated cases. The maximum adsorption capacity determined from the Sips isotherm model gave values of 80.8 ± 0.1 mg g^{-1} and 132 ± 3 mg g^{-1} for the H_2O and CO_2 activated samples, respectively, reflecting the samples' differences in heteroatom functionalities. Additionally, the application of either adsorbent led to reduced toxicity levels in all tested samples, implying that no harmful by-products were generated during adsorption. Comparative analysis with the existing literature further validates the study's findings, suggesting the efficacy and applicability of cellulose-based porous carbons for sustainable chlorpyrifos remediation.

Keywords: organophosphate; pesticide remediation; cellulose; adsorption; neurotoxicity

1. Introduction

Chlorpyrifos (CHP) is a widely used organophosphate pesticide in agriculture and is well known for its effectiveness against a broad spectrum of pests [1]. Its significance stems from several key factors. Firstly, CHP is valued for its efficacy in controlling various agricultural pests, including aphids, caterpillars, beetles, and mites, making it a versatile tool for farmers across various crops [2]. Moreover, its ability to persist on treated crops and soil provides prolonged protection against pests, particularly in regions with continuous pest pressure throughout the growing season [3]. Additionally, CHP is considered cost-effective compared to other alternatives, making it an attractive choice for farmers, especially in pest-prone areas. Its versatility in various formulations like liquids, granules, and dusts allows for flexible application methods tailored to different crop types, pest species, and environmental conditions [4]. Furthermore, with a history dating back to the 1960s [5],

CHP has earned farmers' trust due to its reliability and proven effectiveness over decades, solidifying its widespread adoption in agriculture as a go-to pesticide for protecting crops.

However, it was proven that CHP poses significant environmental and health concerns due to its persistence and potential toxicity to non-target organisms, including humans [6,7]. Its extended persistence in the environment increases the risk of chronic exposure to both target and non-target organisms, with residues leaching into groundwater and contaminating aquatic ecosystems [8]. Acute exposure to CHP can result in symptoms ranging from nausea and dizziness to convulsions or death [9,10], while chronic low-level exposure, especially during critical developmental stages, has been associated with neurological developmental delays and cognitive deficits in children [11,12]. Additionally, CHP's non-selective toxicity poses risks to beneficial insects [13,14], birds [13,14], and aquatic life [14,15], disrupting ecological balance and agricultural productivity.

Adsorption is a crucial and efficient remediation strategy for mitigating CHP contamination in soil and water environments due to its effectiveness, simplicity, and environmental compatibility [16,17]. By utilizing adsorbent materials, such as carbon-based substances, CHP molecules can be selectively captured and immobilized onto the surface of the adsorbent, effectively removing them from the contaminated media. This process achieves high removal efficiencies and reduces CHP concentrations to safe levels, minimizing the risks associated with CHP exposure to human health and the environment [16,17]. Adsorption offers several advantages, including cost-effectiveness, minimal energy requirements, and optimizing adsorption capacities by manipulating material properties [18].

Carbon-based porous materials offer several advantages for adsorption applications, making them highly effective at removing contaminants like CHP from soil and water environments. Firstly, carbon-based materials can possess exceptionally high surface areas due to their porous structures, providing plenty of adsorption sites for molecules' adsorption. This increased surface area enhances the adsorption capacity of the material, allowing for the efficient removal of contaminants [19,20]. Additionally, the porous structure of carbon-based materials facilitates rapid diffusion of contaminants into the material's interior, further enhancing adsorption efficiency [19]. Moreover, the surface chemistry of carbon-based materials can be easily modified or tailored to enhance adsorption selectivity towards specific contaminants. This tunable surface chemistry optimizes adsorption properties, making carbon-based porous materials versatile and adaptable for various adsorption applications [21].

Cellulose-derived porous carbons represent a promising class of carbon-based materials with diverse applications in various fields, including environmental remediation, energy storage, and catalysis [22,23]. The synthesis of cellulose-derived porous carbons typically involves a multi-step process, beginning with the dissolution of cellulose from natural sources in a suitable solvent, followed by spinning or casting to form precursor fibers or films. Subsequent thermal treatment, such as carbonization and activation, transforms the precursor materials into carbonaceous structures while removing volatile components, forming porous carbon materials [17,22]. Additionally, cellulose is easily chemically modifiable, enhancing its versatility for specific uses and enabling the preparation of diverse macroporous structures like textiles, papers, or aerogels [24–26]. The usually small and constant diameter of cellulose fibers allows narrow pore size distributions to be achieved in contrast to granular precursor materials. Due to their small dimensions, activated carbon fibers have pores mostly situated at the fiber surface and thus provide good accessibility to their active sites [27]. In our previous study, we showed that cellulose-based activated carbon fibers contain pores in the micropore and mesopore ranges [22]. In combination with the above-mentioned macroporous structures or macropores being formed between the single fibers, cellulose-derived porous carbons can exhibit a hierarchical porous structure comprising micropores, mesopores, and macropores, which has been described as advantageous for various applications, including enhanced adsorption capacities [28,29]. These materials' specific surface area and pore size distribution can be tailored through control over precursor composition, processing conditions, and activation methods. Additionally, the surface chemistry of cellulose-derived carbon materials can be modified through functionalization or doping with heteroatoms, further enhancing their adsorption selectivity and catalytic properties [16,17].

In environmental remediation, these materials demonstrate excellent adsorption capabilities for various pollutants, including organic contaminants, heavy metals, and emerging contaminants like CHP. Moreover, their high surface area and porous structure enable the efficient capture and removal of contaminants from water and soil matrices [22].

The existing literature on the adsorption of CHP onto carbon-based materials reveals a significant research gap concerning the limited exploration of biomass-derived carbon materials in this context. While numerous studies have investigated the adsorption behavior of CHP on various carbonaceous substrates, including activated carbons, carbon nanotubes, and graphene-based materials, there is a notable insufficiency of research specifically focusing on biomass-derived carbon materials. This research gap is particularly noteworthy given the unique properties of cellulose-derived porous carbons, as already mentioned above, which offer potential advantages for CHP adsorption applications. The limited exploration of cellulose-derived carbon materials in CHP adsorption studies represents an untapped opportunity to advance our understanding of their adsorption mechanisms, optimize their adsorption capacities, and expand their practical applications in environmental remediation.

This study investigates the potential of using cellulose-derived porous carbons as an adsorbent for CHP removal from water. Initially, the manuscript presents the results of the physicochemical characterization of these materials. Following this, CHP adsorption onto cellulose-based carbon materials in water is conducted. The kinetics and thermodynamics of CHP adsorption on these materials are thoroughly investigated and discussed. Furthermore, the neurotoxicity of the water samples is closely monitored throughout the remediation process, aiming to identify the potential formation of more toxic by-products, such as oxo-forms of pesticides. Finally, the results obtained from this study were compared with those from the literature to assess their comparability.

2. Materials and Methods

2.1. Adsorbent Preparation and Characterization

In order to prepare adsorbents, cellulose fibers (Viscose 1.7 dtex, 38 mm, Lenzing AG, Lenzing, Austria) were impregnated in a 7.4 wt% diammonium hydrogen phosphate (DAHP) solution and carbonized in a nitrogen atmosphere at a heating rate of 1 °C min^{-1} until 850 °C followed by an isothermal step for 30 min. The subsequent activation was performed with a carbon dioxide flow of 80 L h^{-1} (adsorbent VF_{CO_2}) or water vapor (adsorbent VF_{H_2O}), enabled by the injection of water (1.0 mL min^{-1}) with a peristaltic pump. In both cases, the activation was performed at 870 °C for 165 min.

The examination of samples' structure and elemental composition utilized the Phenom-ProX Scanning Electron Microscope (SEM) provided by Thermo Fisher Scientific (Waltham, MA, USA) combined with Energy Dispersive X-ray Analysis (EDX).

N_2 adsorption at an isothermal temperature of −196.15 °C was used in a gas sorption system (Autosorb-iQ, Anton Paar QuantaTec Inc., Graz, Austria) to examine the specific surface area and total pore volume of the samples. The samples were outgassed for 2 h at 200 °C under vacuum before analysis.

2.2. Adsorption Experiments

The adsorption experiments were conducted under batch (stationary) conditions. Initially, the adsorbents were dispersed in double-distilled water. To achieve the desired concentration of both the adsorbent and CHP, an appropriate amount of the organophosphate stock solution (Pestanal, Sigma Aldrich, Søborg, Denmark) was added. The mixture of adsorbent and CHP underwent shaking and incubation at 25 °C for specified durations using a laboratory shaker (Orbital Shaker-Incubator ES-20, Grant-Bio, Cambridgeshire, UK). Following incubation, the mixture underwent centrifugation at 14,500 rpm for 10 min. Supernatant filtration was performed using a nylon filter (pore size 220 nm KX Syringe Filter, Kinesis, Cole Parmer, St. Neots, UK). The resulting filtrate underwent ultra-performance liquid chromatography (UPLC) analysis to determine the concentration of organophosphates after adsorption.

The adsorption efficiency, denoted as uptake, was calculated using the following formula: uptake = 100% × $(C_0 - C_{eq})/C_0$, where C_0 represents the initial concentration of organophosphates. The concentrations of CHP in filtrates after adsorption (C_{eq}) were determined using UPLC. Control experiments were conducted under the same conditions as batch experiments but without the presence of adsorbents.

CHP analysis was conducted using the Waters ACQUITY UPLC system with a photodiode array (PDA) detector controlled by the Empower 3 version software. The ACQUITY UPLC™ BEH C18 column (1.7 µm, 100 mm × 2.1 mm, Waters GmbH, Eschborn, Germany) was utilized under isocratic conditions, with 10% acetonitrile in water (*v/v*) as mobile phase A and pure acetonitrile as mobile phase B. The eluent flow rate was 0.25 mL min^{-1}, and the injection volume was consistent at 5 µL. The analysis employed a mobile phase composition of 20% A and 80% B. The retention time was (2.70 ± 0.05) min, and it was detected at 205 nm.

2.3. Neurotoxicity of Chlorpyrifos

The neurotoxicity of CHP solutions was evaluated through measurements of AChE inhibition, aiming to track and quantify changes in toxicity before and after adsorption. This approach also enabled an investigation into potential transformations of CHP into more toxic forms during hydrolysis or oxidation, which could exert harmful effects at concentrations below the detection limits of UPLC. A modified Ellman's procedure [30,31] was employed to assay AChE activity, with in vitro experiments conducted by exposing 1 U/mL AChE to CHP solutions pre- and post-adsorption at 37 °C in 50 mmol dm^{-3} phosphate buffer at pH 8.0 (final volume 0.650 cm^3). The enzymatic reaction was initiated by adding acetylcholine-iodide along with 5,5′-dithio-bis-(2-nitrobenzoic acid) (DTNB) as a chromogenic reagent and allowed to proceed for 8 min before being halted by 10% sodium dodecyl sulfate (SDS). The resultant enzymatic reaction product, thiocholine, reacted with DTNB to form 5-thio-2-nitrobenzoate, whose optical absorption was measured at 412 nm. Physiological effects were quantified as AChE inhibition, expressed as a percentage according to the formula: AChE inhibition (%) = 100 × $(A_0 - A)/A_0$, where A_0 represents the AChE activity in the absence of CHP and A represents the activity measured after exposure to CHP. Acetylcholinesterase from electric eel, acetylthiocholine iodide (ASChI), and DTNB were procured from Sigma-Aldrich St. Louis, MO, USA, while potassium-hydrogen phosphate ($K_2HPO_4 \cdot 3H_2O$) and acetonitrile were obtained from Merck KgaA, Darmstadt, Germany.

3. Results and Discussion

3.1. SEM and EDX of the Synthesized Adsorbents

SEM was utilized to assess the morphology of the samples, with SEM micrographs of samples VF_{H_2O} and VF_{CO_2} presented in Figure 1(a1–a4). The micrographs show that the morphology of these samples remains consistent and reflects that of the precursor cellulose fibers (besides shrinkage of approx. 30%). This observation aligns with our previous research findings [17,22,32]. Additionally, the chemical composition of the investigated activated carbon fibers was analyzed using EDX (see mapping in Figure 1(b1–b5)). For both materials, carbon was identified as the dominant element in all samples with 91.3% and 84.7%, followed by oxygen with 8.5% and 13.6% and phosphorus with 0% and 1.6%, for VF_{H_2O} and VF_{CO_2}, respectively. The phosphorus found in the VF_{CO_2} sample originates from the impregnation with DAHP, and it was effectively removed by H_2O steam as well as a significant amount of surface oxygen, suggesting that P is in the form of phosphates on the VF surface. Minor traces of sodium at 0.13% and 0.11% and sulfur at 0.04% and 0% were also found in VF_{H_2O} and VF_{CO_2}, in that order, likely originating from the precursor material. The distributions of the elements found in the samples were found to be uniform, without apparent separation of phases with higher concentrations of particular elements.

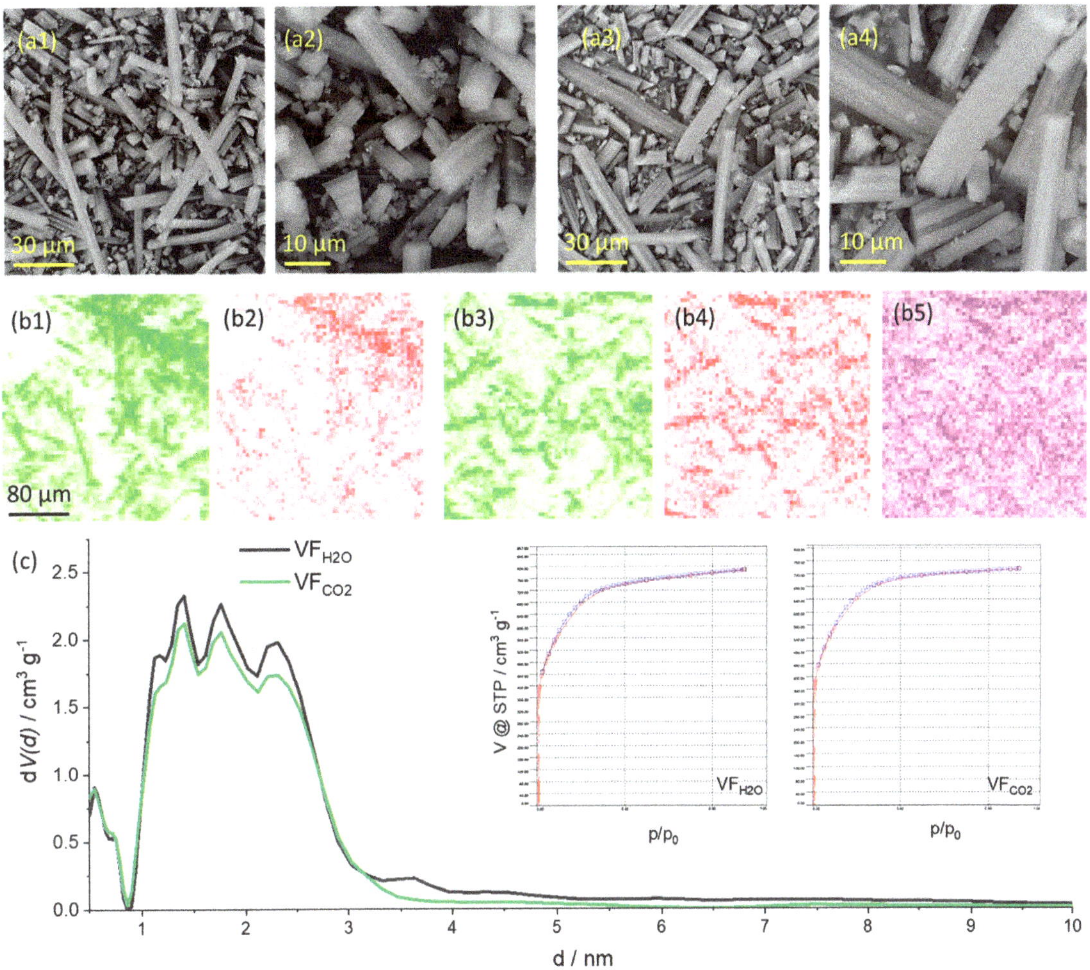

Figure 1. SEM micrographs of VF_{H2O} (**a1**,**a2**) and VF_{CO2} (**a3**,**a4**) at magnification ×2000 and ×5000. The middle row shows EDX maps of VF_{H2O} (**b1**—carbon; **b2**—oxygen) and VF_{CO2} (**b3**—carbon; **b4**—oxygen; **b5**—phosphorus), bottom row (**c**) shows pore size distributions of the two samples, with insets giving adsorption/desorption isotherms.

Considering the specific surface areas and pore volumes of the studied samples, there is another observation that is very important for the analysis of materials' performance as adsorbents. Namely, the specific surfaces (S_{BET}) are very similar and amount to 2443 $m^2\ g^{-1}$ for the VF_{H2O} sample and 2323 $m^2\ g^{-1}$ for the VF_{CO2} sample. The difference is only 5%, which is close to the uncertainty of the applied BET method, suggesting that the surface areas of these two samples are practically the same. The identical situation is found for total pore volumes (V_{tot}), where we found a V_{tot} of 1.22 $cm^3\ g^{-1}$ and 1.14 $cm^3\ g^{-1}$ for VF_{H_2O} and VF_{CO_2}, respectively, i.e., a 6.6% difference. This small difference is visible in the pore size distribution (PSD) curves (Figure 1c), but overall, the distributions are very similar.

3.2. Adsorption Kinetics

The kinetics of CHP adsorption onto VF_{H_2O} and VF_{CO_2} were examined utilizing various models, including the non-linear pseudo-first (PFO) and pseudo-second-order (PSO) models, the Elovich kinetic model, and the intraparticle diffusion model. These models' formulations are detailed in Table 1 as follows:

Table 1. Equations for kinetic models.

Kinetic Model	Equation
Pseudo-first-order model	$q_t = q_e\left(1 - e^{-k_1 t}\right)$
Pseudo-second-order model	$q_t = \frac{q_e^2 k_2 t}{1+q_e k_2 t}$
Elovich kinetic model	$q_t = \frac{1}{\beta}(1 + \alpha\beta t)$
Intraparticle diffusion model	$q_t = k_{id}t^{0.5} + C$

In these equations, q_t denotes the quantity of adsorbed adsorbate at time t (mg g^{-1}), while q_e signifies the equilibrium adsorption capacity (mg g^{-1}). The rate constants k_1 (min^{-1}) and k_2 (g mg^{-1} min^{-1}) describe the adsorption rate for the PFO and PSO models. Within the Elovich model, α denotes the initial adsorption rate (mg g^{-1} min^{-1}), and β represents the desorption constant (g mg^{-1}). The parameter k_{id} in the intraparticle diffusion model characterizes the adsorption rate constant (mg g^{-1} $min^{-0.5}$), while C denotes the boundary layer (mg g^{-1}).

The experimental data, alongside their corresponding fits, are presented in Figures 2 and 3. The derived equilibrium adsorption capacities and rate constants are tabulated in Tables 2 and 3, respectively.

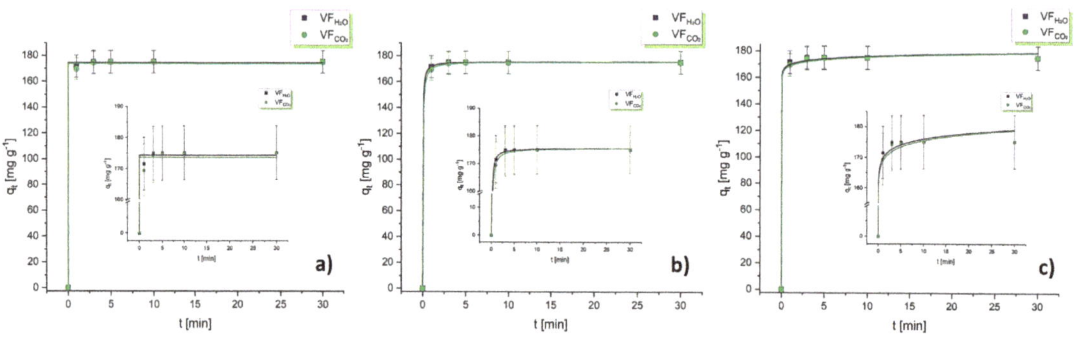

Figure 2. Graphical representation of kinetic models for CHP adsorption (concentration 5×10^{-4} mol dm^{-3}) onto carbon materials VF_{H_2O} and VF_{CO2} (concentration 1 mg mL^{-1}) at 25 °C using non-linear (**a**) PFO, (**b**) PSO, and (**c**) Elovich model.

From Figure 2, it can be seen that the adsorption equilibrium for removing CHP is achieved in 10 min, indicating rapid adsorption of CHP on both materials. Also, the experimental data aligned very well with the PSO and PFO models, as indicated by the high R^2 and low χ^2 values. The q_e values from both kinetic models are similar, confirming both fit robustness. It can be unambiguously concluded that the adsorption of CHP onto VF_{CO2} materials is slower compared to the VF_{H2O} material, according to k_1 and k_2 values and their corresponding uncertainties. By observing α values obtained from the Elovich kinetic model, which are very high, and β, which are very low, it can be concluded that the initial CHP adsorption rate onto the investigated materials is very high. These results confirmed previous statements regarding PFO and PSO.

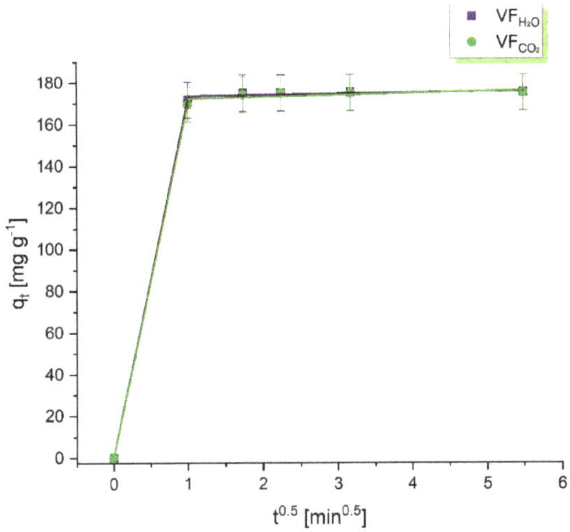

Figure 3. Intraparticle diffusion model for CHP adsorption (concentration 5×10^{-4} mol dm^{-3}) onto carbon materials VF_{H2O} and VF_{CO2} (concentration 1 mg mL^{-1}).

Table 2. Kinetic parameters for CHP adsorption (concentration 5×10^{-4} mol dm^{-3}) onto carbon materials VF_{H2O} and VF_{CO2} (concentration 1 mg mL^{-1}) at 25 °C.

Material	**Pseudo-First-Order Kinetics**			
	q_e (mg g^{-1})	**k_1 (min^{-1})**	**χ^2**	**R^2**
VF_{H2O}	174 ± 1	$(5.71 \pm 0.01) \times 10^4$	2.443	0.999
VF_{CO2}	174 ± 1	$(4.99 \pm 0.01) \times 10^4$	5.799	0.999
	Pseudo-second-order kinetics			
	q_e (mg g^{-1})	k_2 (min^{-1})	χ^2	R^2
VF_{H2O}	176 ± 1	0.251 ± 0.001	0.147	0.999
VF_{CO2}	176 ± 1	0.158 ± 0.001	0.150	0.999
	Elovich model			
	α (mg g^{-1} min^{-1})	β (g mg^{-1})	χ^2	R^2
VF_{H2O}	$(1.74 \pm 0.01) \times 10^{27}$	0.364 ± 0.001	6.317	0.999
VF_{CO2}	$(8.93 \pm 0.01) \times 10^{24}$	0.334 ± 0.001	5.366	0.999

Table 3. Kinetic parameters for intraparticle diffusion model of CHP (concentration 5×10^{-4} mol dm^{-3}) adsorption onto investigated adsorbents (concentration 1 mg mL^{-1}).

Material	**Intraparticle Diffusion Model**		
VF_{H2O}	I part	C (mg g^{-1})	0
		k_{id} (mg g^{-1} min$^{-0.5}$)	172 ± 1
	II part	C (mg g^{-1})	173 ± 4
		k_{id} (mg g^{-1} min$^{-0.5}$)	0.539 ± 0.006
VF_{CO2}	I part	C (mg g^{-1})	0
		k_{id} (mg g^{-1} min$^{-0.5}$)	169 ± 1
	II part	C (mg g^{-1})	171 ± 3
		k_{id} (mg g^{-1} min$^{-0.5}$)	0.904 ± 0.009

The intraparticle diffusion model showed two linear stages for CHP adsorption onto both materials. The first stage characterizes the diffusion of CHP molecules from the solution to the external surface of materials. In contrast, the second stage is associated with the final equilibrium stage. The k_{id} values decrease after the breakpoints, indicating a slower adsorption rate in each consecutive stage. The boundary layer increases throughout stages, reaching values comparable with q_e values obtained from PFO and PSO models. It suggests that the boundary layer in CHP's adsorption process onto materials is highly significant.

However, we should note that the kinetic analysis is performed using the data that point to an extremely fast reaching of the adsorption equilibrium, while the main information about the kinetics of adsorption is stored in the initial part of the q_t vs. t curve, which is largely inaccessible to accurate measurements due to technical limitations. For these reasons, we do not unambiguously ascribe the adsorption process to any of the applied kinetic models and note that the absolute values of kinetic parameters should be taken with care. On the other hand, it is safe to conclude that the adsorption process is indeed faster on the VF_{H2O} sample, which is also obvious from the measured data points.

3.3. Adsorption Isotherms

Adsorption isotherms describe the process of CHP adsorption onto synthesized carbon materials and enable obtaining information about the adsorption mechanism, as well as the surface characteristics and affinities of the investigated materials towards the investigated pesticide. Experimental data were analyzed using the non-linear Freundlich, Langmuir, Temkin, Dubinin–Radushkevich, and Sips adsorption isotherms. Their equations are presented in the following Table 4, where q_e (mg g^{-1}) represents equilibrium adsorption capacity, C_e (mg dm^{-3}) equilibrium adsorbate concentration, K_F ((dm^3 mg^{-1})$^{1/n}$) and n as Freundlich constants, K_L (dm^3 mg^{-1}) and q_{max} (mg g^{-1}) as Langmuir constant and theoretical maximum adsorption capacity of the monolayer, b_T (J g mol^{-1} mg^{-1}) and K_T (dm^3 mg^{-1}) for Temkin isotherm constants, q_{DR} for maximum adsorption capacity, K_{DR} (mol^2 J^{-2}) for the constant associated with the mean free adsorption energy per mole of adsorbent ($\varepsilon = RT \times \ln(1 + 1/C_e)$), K_s (dm^3 mg^{-1}) and b_s represent Sips isotherm constants.

Table 4. Adsorption isotherm models' equations.

Adsorption Isotherm Model	Equation
Freundlich model	$q_e = K_F C_e^{\frac{1}{n}}$
Langmuir model	$q_e = \frac{q_{max} K_L C_e}{1+K_L C_e}$
Temkin model	$q_e = \frac{RT}{b_T} lnK_T C_e$
Dubinin–Radushkevic model	$q_e = q_{DR} e^{-K_{DR}\varepsilon^2}$
Sips model	$q_e = \frac{q_{max}(K_s C_e)^{b_s}}{1+(K_s C_e)^{b_s}}$

The obtained isotherms are depicted in Figure 4, and the parameters of the adsorption isotherms are shown in Table 5. Figure 4 illustrates the non-linear fitting forms of all investigated isotherms for CHP adsorption onto VF_{H_2O} and VF_{CO_2} materials. Based on the parameters presented in Table 5, all applied models fit reasonably well with the experimental results, but the Freundlich, Langmuir, Dubinin–Radushkevich, and Sips isotherm models best described the experimental data, while the Temkin isotherm gave relatively large χ^2 values. As already known, the Sips isotherm combines Langmuir and Freundlich isotherm models [33,34]. It is a more flexible model for describing adsorption in systems exhibiting monolayer and multilayer adsorption. Also, it introduces an additional parameter known as the heterogeneity factor (denoted as b_s), which enables a smoother transition between Langmuir and Freundlich behaviors. The Sips isotherm yields better results in describing experimental data compared to the standalone use of Langmuir or Freundlich isotherms, particularly in cases where the adsorption behavior deviates from ideal monolayer or multilayer adsorption, which is not surprising in the case of heterogeneous surfaces with different possible adsorption sites or adsorption domains.

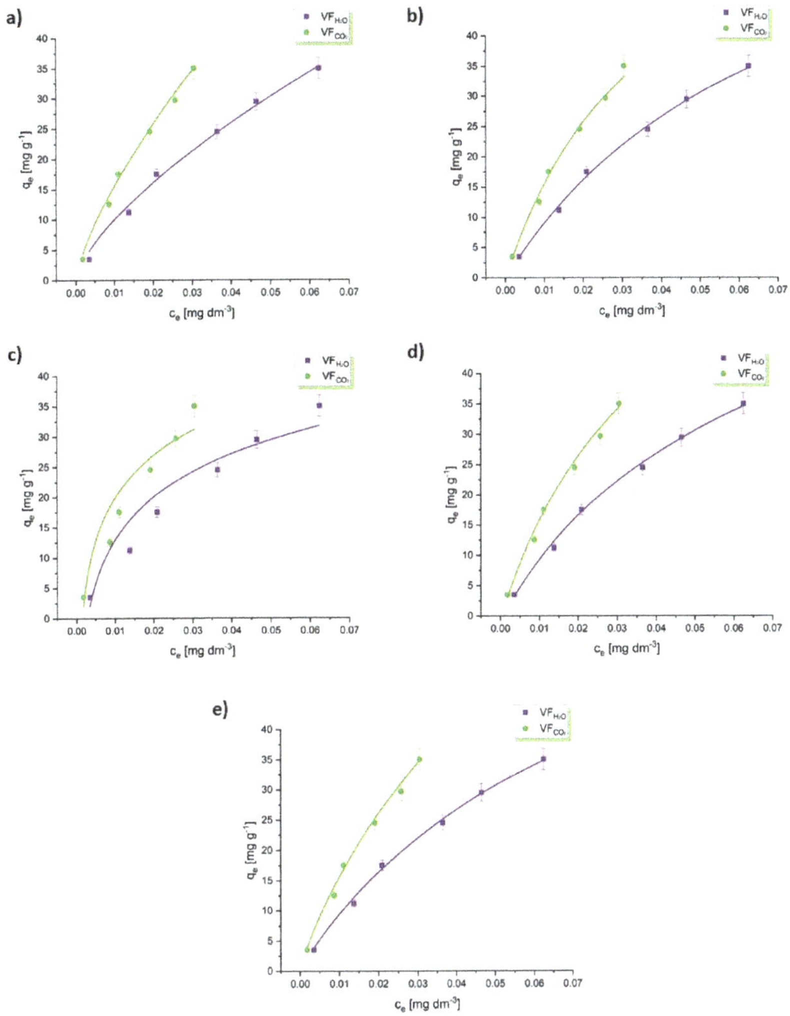

Figure 4. Non-linear adsorption isotherm models for CHP: (**a**) Freundlich, (**b**) Langmuir, (**c**) Temkin, (**d**) Dubinin–Raduskevic, and (**e**) Sips model, adsorbent dose 1 mg mL^{-1} at 25 °C.

The n value of the Freundlich isotherm model is higher than 1, indicating that adsorption is a favorable process. From the Langmuir model, we can see that the adsorption capacities of CHP are 73 mg g^{-1} and 75 mg g^{-1}. However, when observing the q_{max} values from the Sips isotherm, we can see that they are 80.8 ± 0.1 mg g^{-1} and 132 ± 3 mg g^{-1} for VF_{H2O} and VF_{CO2}, respectively, indicating that these investigated materials have a high potential to adsorb CHP. A similar conclusion can be derived from the results of the DR model fit, where q_{DR} was found to be 76.4 ± 0.1 and 101 ± 2 mg g^{-1} for VF_{H2O} and VF_{CO2}, respectively.

Given that the values for the parameter b_s are less than 1 for both materials and considering all the above-mentioned information, we can conclude that the adsorption process is most likely multilayer adsorption on the heterogeneous surface of carbon materials [35,36].

Table 5. Parameters for CHP adsorption using Freundlich, Langmuir, Temkin, Dubinin–Raduskevich, and Sips adsorption isotherm, adsorbent dose 1 mg mL^{-1}.

Material	**Freundlich Isotherm**				
	K_F ((dm^3 mg^{-1})$^{1/n}$)	n		χ^2	R^2
VF_{H2O}	239 ± 2	1.45 ± 0.04		1.341	0.992
VF_{CO2}	451 ± 1	1.37 ± 0.02		0.778	0.996
	Langmuir Isotherm				
	K_L (dm^3 mg^{-1})	q_{max} (mg g^{-1})		χ^2	R^2
VF_{H2O}	26.91 ± 0.4	73 ± 6		2.101	0.993
VF_{CO2}	13.77 ± 0.03	75 ± 4		0.684	0.998
	Temkin Isotherm				
	K_T (dm^3 mg^{-1})	b_T (J g mol^{-1} mg^{-1})		χ^2	R^2
VF_{H2O}	354 ± 6	242 ± 3		12.247	0.930
VF_{CO2}	705 ± 9	243 ± 7		15.829	0.909
	Dubinin–Radushkevich Isotherm				
	q_{DR} (mg g^{-1})	K_{DR} (mol^2 J^{-2})	E (J mol^{-1})	χ^2	R^2
VF_{H2O}	76.4 ± 0.1	(1.60 ± 0.02) × 10^{-8}	(5.58 ± 0.01) × 10^3	0.497	0.997
VF_{CO2}	101 ± 2	(1.41 ± 0.01) × 10^{-8}	(5.95 ± 0.01) × 10^3	0.952	0.995
	Sips Isotherm				
	K_s (dm^3 mg^{-1})^{bs}	b_s	q_{max} (mg g^{-1})	χ^2	R^2
VF_{H2O}	10.7 ± 0.2	0.917 ± 0.001	80.8 ± 0.1	0.834	0.995
VF_{CO2}	8.27 ± 0.03	0.828 ± 0.002	132 ± 3	1.224	0.993

Given that the adsorption energy per mole of adsorbent is <8 kJ mol^{-1}, we can conclude that it is a physisorption process, further supporting the findings based on the analysis of the Sips isotherm. Also, based on Temkin's constant b_T, we can conclude that the adsorption process is exothermic since its values are high and positive [37,38]. Based on the above-presented results and the good fit of the Sips and D-R isotherms, it is likely that the adsorption takes place on the energetically inhomogeneous surface in a combination of mono- and multilayer physisorption.

Linking adsorption kinetics and adsorption thermodynamics to carbon material properties reveals some interesting findings: The fact that the morphology is similar for both used carbons excludes it as a relevant parameter. Moreover, the values of S_{BET} and V_{tot}, which are similar for the two samples, also eliminate them as decisive parameters. In fact, we observe that the kinetics of CHP removal are quite similar, but the VF_{H2O} sample shows slightly higher kinetic constants (Table 2). It cannot be unambiguously claimed that the differences in kinetics are because of the slightly larger pore volume of the VF_{H2O} sample, but it could be a plausible explanation. On the other hand, adsorption thermodynamics differs in terms of adsorption capacities found by Dubinin–Raduskevich and Sips isotherms, suggesting that the VF_{CO2} sample has a higher adsorption capacity than the VF_{H2O} sample. If we recall the results of the EDX analysis, this sample has higher oxygen and phosphorus contents, which likely makes the surface of porous carbon fibers more oxophylic and strongly solvated. In such a scenario, the higher adsorption capacity of the VF_{CO2} sample can be understood on the basis of the combined experimental–theoretical work of Maliyekkal et al. [39]. Using graphene as a model of carbon surface, the authors have demonstrated that pesticide adsorption (including CHP) is mediated by water molecules, which solvate the carbon surface and pesticide molecules, bridging the interactions in that way. It should be noted that CHP can also interact with carbon surfaces with preserved sp^2 domains via aromatic moiety through π-π stacking. However, it is plausible that the interactions via solvent shells could lead to higher adsorption capacities.

3.4. Neurotoxicity of Chlorpyrifos

To assess the neurotoxicity of CHP both before and after treatment with our materials, we followed the procedure outlined in Section 2.3. Our primary objective was to determine if there was an increase in the formation of more toxic compounds when CHP interacted with the adsorbents. This potential increase could result from oxidation, as the thio-forms of organophosphate pesticides may transform into their respective oxo-forms. It is well established that oxo-forms of organophosphate compounds are more potent AChE inhibitors [40].

We tested the highest concentration of CHP investigated in our adsorption measurements (5×10^{-4} mol dm^{-3}). Before treatment, this concentration of CHP inhibited AChE by 90 $\pm$ 3% compared to the control. Following contact with both investigated adsorbents, AChE inhibition was reduced to only 5 $\pm$ 1%. The same values of the AChE inhibition for both samples after water treatment and a very small value of the AChE inhibition suggest that the CHP concentration remaining in the solution is shifted to the upper part of the (sigmoidal) AChE inhibition curve (Figure 5), where the inhibition is weakly dependent on the CHP concentration and the differences in adsorption capacities are not reflected in the AChE inhibition. Moreover, these findings suggest that no additional toxic products are formed during adsorption, indicating that all investigated adsorbents can effectively detoxify CHP.

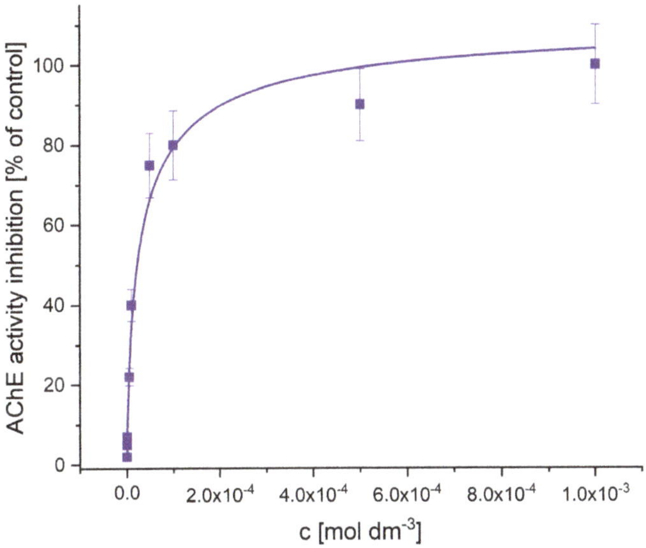

Figure 5. The dependence of AChE (1 U/mL) activity inhibition on CHP concentration. The results are given as a mean value from 3 measurements and expressed as % relative to the control value.

3.5. Where Do We Stand? Comparison with the Literature Data

Considering the detrimental effect of CHP on the human body and ecosystems, it is no surprise that many researchers have become interested in utilizing adsorption for its removal. Table 6 compares CHP adsorption using cellulose fibers with other reported adsorbents. Several studies have investigated different carbonaceous materials derived from agricultural residues, waste products, and synthesized compounds for their CHP adsorption capacities. The Langmuir and Freundlich isotherm models were usually employed to analyze the adsorption data in the reviewed studies, and the former one was used to extract adsorption capacity (as this quantity is not a parameter of the Freundlich equation, see Table 4). The choice of model depended on the nature of the adsorbent and the characteristics of the adsorption process. The Freundlich model, which describes adsorption on heterogeneous surfaces, was found to fit well in some cases, such as studies involving sugarcane bagasse biochar [41,42], cinnamon stick-derived materials [43], and *Moringa oleifera* seed waste [44]. On the other hand, the Langmuir model, which assumes monolayer adsorption on homogeneous surfaces, was utilized effectively in studies involving polyvinylamine-modified nanocellulose [45], walnut shell biochar [46], and various composite materials [47,48]. Efficiency-wise, the adsorption capacities varied significantly among the different carbon materials tested. For instance, nanocellulose modified with polyvinylamine exhibited a high adsorption capacity of 98.12 mg g^{-1} [45], whereas banana peels, orange peels, pomegranate peels, and date stone-derived carbon materials showed very low capacities ranging from 1.12×10^{-3} to 2.52×10^{-3} mg g^{-1} [49].

Interestingly, some studies explored functionalized carbon materials, such as magnetic graphene oxide, carboxymethyl cellulose, and aminoguanidine-modified magnetic graphene oxide, which demonstrated enhanced adsorption capacities compared to their non-functionalized counterparts. For instance, the adsorption capacity of magnetic graphene oxide and carboxymethyl cellulose was reported as 108.30 mg g^{-1} [50], while aminoguanidine-modified magnetic graphene oxide exhibited a capacity of 85.47 mg g^{-1} [51]. Moreover, the study on functionalized dextrin/graphene oxide composites reported an exceptionally high adsorption capacity of 769.23 mg g^{-1} [47], indicating the potential of such advanced materials for efficient CHP removal. The choice of carbon material significantly influenced the adsorption capacity for CHP, with functionalized and composite materials generally exhibiting higher efficiencies.

Table 6. Comparison of maximum adsorption capacity for CHP removal onto different materials.

Type of Material/Precursor	The Maximum Adsorption Capacity (mg g^{-1})	The Maximum Adsorption Capacity Calculated from	Best Fitting Isotherm Model	Reference
Sugarcane bagasse	3.20	Experimental results	Freundlich	[41]
Sugarcane bagasse-derived biochar	6.25	Experimental results	Freundlich	[42]
Cinnamon sticks	12.37	Langmuir	Langmuir	[43]
Polyvinylamine-modified nanocellulose	98.12	Langmuir		[45]
Moringa oleifera seed waste	25.00	Langmuir	Langmuir, Freundlich	[44]
Activated biochar from tobacco	0.68 using activated biochar 1.60 using chemically activated biochar	Langmuir	Langmuir, Freundlich	[52]
Walnut shell biochar	3.54	Langmuir	Langmuir, Freundlich	[46]
Nano-magnetized and carbonaceous adsorbents obtained from orange peels	68.00, 108.00, and 100.00 for the untreated orange peels, carbonaceous orange peels, and nano-magnetized orange peels	Langmuir	Langmuir, Freundlich	[53]
Nanostructured biochar (nPPAB) was prepared from *Punica granatum* peels	100.00	Langmuir	Langmuir	[54]
Banana peels, orange peels, pomegranate peels and date stones	1.12×10^{-3}, 2.52×10^{-3}, 1.64×10^{-3}, and 1.03×10^{-3} for banana peels, orange peels, pomegranate peels, and date stones			[49]

Table 6. *Cont.*

<table>
<tr><th>Type of Material/Precursor</th><th>The Maximum Adsorption Capacity (mg g^{-1})</th><th>The Maximum Adsorption Capacity Calculated from</th><th>Best Fitting Isotherm Model</th><th>Reference</th></tr>
<tr><td>Spent coffee grounds</td><td>7.00</td><td>Langmuir</td><td>Langmuir</td><td>[16]</td></tr>
<tr><td>Magnetic graphene oxide and carboxymethyl cellulose (MGOC)</td><td>108.30</td><td>Langmuir</td><td></td><td>[50]</td></tr>
<tr><td>Aminoguanidine modified magnetic graphene oxide as a robust nanoadsorbent</td><td>85.47</td><td>Langmuir</td><td>Freundlich</td><td>[51]</td></tr>
<tr><td>Functionalized dextrin/graphene oxide composite</td><td>769.23</td><td>Langmuir</td><td>Freundlich</td><td>[47]</td></tr>
<tr><td>Magnetic chitosan/graphene quantum dot/iron oxide nanocomposite hydrogel beads</td><td>39.95</td><td>Langmuir</td><td>Jossens</td><td>[48]</td></tr>
<tr><td>Superhydrophilic graphene oxide/electrospun cellulose nanofibre</td><td>3.97</td><td>Langmuir</td><td>Freundlich</td><td>[55]</td></tr>
<tr><td>Graphene-based materials</td><td>1.02×10^3 and 2.77×10^3 for industrial-quality graphene and graphene nanoplatelets</td><td>Langmuir</td><td>Langmuir, Freundlich</td><td>[56]</td></tr>
<tr><td>Graphene oxide</td><td>98.039</td><td>Langmuir</td><td>Langmuir</td><td>[57]</td></tr>
<tr><td rowspan="2">Cellulose fibers</td><td>73 ± 6 (VF_{H2O})
75 ± 4 (VF_{CO2})</td><td>Langmuir</td><td rowspan="2">Freundlich, Langmuir, and Sips</td><td rowspan="2">This study</td></tr>
<tr><td>80.8 ± 0.1 (VF_{H2O})
132 ± 3 (VF_{CO2})</td><td>Sips</td></tr>
</table>

When comparing the q_{max} values from this study obtained from the Langmuir isotherm model, which was most frequently used to extract adsorption capacities (q_{max} values are 73 ± 6 mg g^{-1} and 75 ± 4 mg g^{-1} for VF_{H2O} and VF_{CO2}, respectively), with the values from the literature presented in Table 6, it can be seen that the values reported here are comparable with previously published ones. This indicates the high adsorption capacity of the materials used in this study. However, it is essential to consider whether the obtained results are realistic and how justified it is to compare them with literature data, particularly considering non-standardized experimental conditions for accessing adsorption capacities.

The Langmuir model is often used in adsorption studies but is not always entirely realistic. The model assumes monolayer adsorption on a homogeneous surface, which is not always the case in real systems. In many situations, adsorption occurs on more complex surfaces that do not fulfill all the assumptions of the Langmuir model. When comparing the results with values from the literature, caution should be taken due to potential differences in experimental conditions, material characteristics, and analysis methods. For instance, different samples, sample preparations, particle sizes, and experimental conditions can significantly affect adsorption results. Therefore, direct comparison with literature values may be imprecise. Nevertheless, the fact that the q_{max} values from the Langmuir model in this study are comparable with others shown in Table 6 indicates a high adsorption capacity of the investigated porous carbons obtained from cellulose fibers.

4. Conclusions

The presented study highlights the potential of cellulose-based porous carbon fibers as efficient adsorbents for the removal of CHP from water. Physicochemical characterization revealed very high surface areas but similar porosity for both prepared porous carbon fiber samples. Differences could be observed in the chemical composition as the CO_2-activated sample contains more oxygen and, in contrast to the water vapor-activated counterpart, P residues from the DAHP pretreatment. The kinetics of CHP adsorption indicated rapid equilibrium attainment within 10 min, with excellent fitting to the PSO and PFO models. Although CHP adsorption onto VF_{H2O} materials occurred faster than VF_{CO2}, both exhibited high initial adsorption rates. The favorable nature of adsorption was supported by a satisfactory fit with both Sips and Dubinin–Radushkevich isotherm models, suggesting a multilayer adsorption process on heterogeneous surfaces. The determined q_{max} values from the Sips isotherm model are 80.8 ± 0.1 mg g^{-1} and 132 ± 3 mg g^{-1} for VF_{H_2O} and VF_{CO_2}, respectively. Higher adsorption capacities for the VF_{CO2} sample can be explained by solvent-mediated CHP adsorption due to the higher amount of heteroatom functionalities of this carbon. Furthermore, the significant reduction in acetylcholinesterase inhibition following contact with the adsorbents indicates effective detoxification of CHP without the formation of additional toxic by-products. A comparison of the obtained q_{max} values with literature data revealed favorable adsorption capacities of the materials used in this study, indicating their excellent performance. These results suggest that the cellulose-derived porous carbons presented in this study are very efficient CHP adsorbents, making them an attractive material for various applications. Their low cost, availability, and efficiency make them preferable to other materials, such as biomass-derived materials or commercial materials like graphene oxide.

Author Contributions: Conceptualization, I.A.P. and T.L.-P.; methodology, I.A.P., S.B., C.U., C.F. and T.L.-P.; software, I.A.P.; formal analysis, V.M., T.T., T.L.-P. and S.B.; investigation, V.M., T.T., S.B. and T.L.-P.; resources, I.A.P., C.F. and T.L.-P.; writing—original draft preparation, I.A.P., V.M., T.T. and T.L.-P.; writing—review and editing, I.A.P., S.B., C.U., C.F. and T.L.-P. All authors have read and agreed to the published version of the manuscript.

Funding: S.B., C.U., and C.F. gratefully acknowledge financial support through the COMET Programme (Competence Centers for Excellent Technologies) funded by the Austrian ministries BMK, BMAW, and the federal states of Upper Austria, Lower Austria, and Carinthia, operated by the Austrian Research Promotion Agency (FFG). Further funding was received from the European Regional Development Fund (EFRE) and the province of Upper Austria through the program IBW 2021–2027 (Project Sus2C). T.L.P. V.M., and T.T. acknowledge the support provided by the Serbian Ministry of Education, Science and Technological Development (contract number: 451-03-66/2024-03/200017). I.A.P. acknowledges the support provided by the Serbian Ministry of Education, Science, and Technological Development (contract number: 451-03-65/2024-03/200146).

Data Availability Statement: The data used to support the findings of this study can be made available by the corresponding author upon request.

Conflicts of Interest: The authors declare no conflicts of interest.

References

1. Tudi, M.; Yang, L.; Wang, L.; Lv, J.; Gu, L.; Li, H.; Peng, W.; Yu, Q.; Ruan, H.; Li, Q.; et al. Environmental and Human Health Hazards from Chlorpyrifos, Pymetrozine and Avermectin Application in China under a Climate Change Scenario: A Comprehensive Review. *Agriculture* **2023**, *13*, 1683. [CrossRef]
2. Madesh, K.; Chandraleka, R.; Komala, G.; Tripathi, P. *Distribution and Penetration of Insecticides*; P.K. Publishers & Distributors: Ghaziabad, India, 2024; pp. 219–254.
3. Foong, S.Y.; Ma, N.L.; Lam, S.S.; Peng, W.; Low, F.; Lee, B.H.K.; Alstrup, A.K.O.; Sonne, C. A recent global review of hazardous chlorpyrifos pesticide in fruit and vegetables: Prevalence, remediation and actions needed. *J. Hazard. Mater.* **2020**, *400*, 123006. [CrossRef] [PubMed]
4. Rakhimol, K.; Thomas, S.; Volova, T.; Jayachandran, K. *Controlled Release of Pesticides for Sustainable Agriculture*; Springer: Berlin/Heidelberg, Germany, 2020. [CrossRef]
5. EPA United States Environmental Protection Agency. Available online: https://www.epa.gov/ingredients-used-pesticide-products/chlorpyrifos (accessed on 15 April 2024).
6. Wołejko, E.; Łozowicka, B.; Jabłońska-Trypuć, A.; Pietruszyńska, M.; Wydro, U. Chlorpyrifos occurrence and toxicological risk assessment: A review. *Int. J. Environ. Res. Public Health* **2022**, *19*, 12209. [CrossRef] [PubMed]

7. Sud, D.; Kumar, J.; Kaur, P.; Bansal, P. Toxicity, natural and induced degradation of chlorpyrifos. *J. Chil. Chem. Soc.* **2020**, *65*, 4807–4816. [CrossRef]
8. Nandhini, A.R.; Harshiny, M.; Gummadi, S.N. Chlorpyrifos in environment and food: A critical review of detection methods and degradation pathways. *Environ. Sci. Process. Impacts* **2021**, *23*, 1255–1277. [CrossRef]
9. Reigart, J.R. *Recognition and Management of Pesticide Poisonings*; DIANE Publishing: Collingdale, PA, USA, 2009.
10. Marrs, T.C.; Ballantyne, B. *Pesticide Toxicology and International Regulation*; John Wiley & Sons: Hoboken, NJ, USA, 2004.
11. Richardson, R.J. Assessment of the neurotoxic potential of chlorpyrifos relative to other organophosphorus compounds: A critical review of the literature. *J. Toxicol. Environ. Health* **1995**, *44*, 135–165. [CrossRef] [PubMed]
12. Albers, J.W.; Cole, P.; Greenberg, R.S.; Mandel, J.S.; Monson, R.R.; Ross, J.H.; Snodgrass, W.R.; Spurgeon, A.; Van Gemert, M. Analysis of chlorpyrifos exposure and human health: Expert panel report. *J. Toxicol. Environ. Health Part B* **1999**, *2*, 301–324. [CrossRef]
13. U.S. Environmental Protection Agency. *Reregistration Eligibility Science Chapter for Chlorpyrifos, Fate and Environmental Risk Assessment Chapter*; U.S. Environmental Protection Agency: Washington, DC, USA, 2000.
14. Pesticides, O. *Reregistration Eligibility Decision for Chlorpyrifos*; U.S. Environmental Protection Agency: Washington, DC, USA, 2006.
15. Kamrin, M.A. *Pesticide Profiles: Toxicity, Environmental Impact, and Fate*; CRC Press: Boca Raton, FL, USA, 1997. [CrossRef]
16. Milanković, V.; Tasić, T.; Pejčić, M.; Pašti, I.; Lazarević-Pašti, T. Spent Coffee Grounds as an Adsorbent for Malathion and Chlorpyrifos—Kinetics, Thermodynamics, and Eco-Neurotoxicity. *Foods* **2023**, *12*, 2397. [CrossRef] [PubMed]
17. Tasić, T.; Milanković, V.; Batalović, K.; Breitenbach, S.; Unterweger, C.; Fürst, C.; Pašti, I.A.; Lazarević-Pašti, T. Application of Viscose-Based Porous Carbon Fibers in Food Processing—Malathion and Chlorpyrifos Removal. *Foods* **2023**, *12*, 2362. [CrossRef]
18. Vareda, J.P.; Valente, A.J.M.; Durães, L. Assessment of heavy metal pollution from anthropogenic activities and remediation strategies: A review. *J. Environ. Manag.* **2019**, *246*, 101–118. [CrossRef]
19. Soffian, M.S.; Halim, F.Z.A.; Aziz, F.; Rahman, M.A.; Amin, M.A.M.; Chee, D.N.A. Carbon-based material derived from biomass waste for wastewater treatment. *Environ. Adv.* **2022**, *9*, 100259. [CrossRef]
20. Sabzehmeidani, M.M.; Mahnaee, S.; Ghaedi, M.; Heidari, H.; Roy, V.A. Carbon based materials: A review of adsorbents for inorganic and organic compounds. *Mater. Adv.* **2021**, *2*, 598–627. [CrossRef]
21. Aydin, D.; Gübbük, İ.H.; Ersöz, M. Recent advances and applications of nanostructured membranes in water purification. *Turk. J. Chem.* **2024**, *48*, 2. [CrossRef] [PubMed]
22. Jocić, A.; Breitenbach, S.; Bajuk-Bogdanović, D.; Pašti, I.A.; Unterweger, C.; Fürst, C.; Lazarević-Pašti, T. Viscose-Derived Activated Carbons Fibers as Highly Efficient Adsorbents for Dimethoate Removal from Water. *Molecules* **2022**, *27*, 1477. [CrossRef] [PubMed]
23. Gavrilov, N.; Breitenbach, S.; Unterweger, C.; Fürst, C.; Pašti, I.A. Exploring the Impact of DAHP Impregnation on Activated Carbon Fibers for Efficient Charge Storage and Selective O_2 Reduction to Peroxide. *C* **2023**, *9*, 105. [CrossRef]
24. Pérez-Madrigal, M.M.; Edo, M.G.; Alemán, C. Powering the future: Application of cellulose-based materials for supercapacitors. *Green Chem.* **2016**, *18*, 5930–5956. [CrossRef]
25. Liebner, F.; Potthast, A.; Rosenau, T.; Haimer, E.; Wendland, M. Cellulose aerogels: Highly porous, ultra-lightweight materials. *Holzforschung* **2008**, *62*, 129–135. [CrossRef]
26. Magalhães, S.; Fernandes, C.; Pedrosa, J.F.S.; Alves, L.; Medronho, B.; Ferreira, P.J.T.; Rasteiro, M.D.G. Eco-Friendly Methods for Extraction and Modification of Cellulose: An Overview. *Polymers* **2023**, *15*, 3138. [CrossRef] [PubMed]
27. Pandolfo, A.G.; Hollenkamp, A.F. Carbon properties and their role in supercapacitors. *J. Power Sources* **2006**, *157*, 11–27. [CrossRef]
28. Cai, L.-F.; Zhan, J.-M.; Liang, J.; Yang, L.; Yin, J. Structural control of a novel hierarchical porous carbon material and its adsorption properties. *Sci. Rep.* **2022**, *12*, 3118. [CrossRef]
29. Li, W.-K.; Shi, Y.-P. Recent advances of carbon materials on pesticides removal and extraction based determination from polluted water. *TrAC Trends Anal. Chem.* **2024**, *171*, 117534. [CrossRef]
30. Ellman, G.L.; Courtney, K.D.; Andres, V.; Featherstone, R.M. A new and rapid colorimetric determination of acetylcholinesterase activity. *Biochem. Pharmacol.* **1961**, *7*, 88–95. [CrossRef]
31. Lazarević-Pašti, T.D.; Bondžić, A.M.; Pašti, I.A.; Vasić, V.M. Indirect electrochemical oxidation of organophosphorous pesticides for efficient detection via acetylcholinesterase test. *Pestic. Biochem. Physiol.* **2012**, *104*, 236–242. [CrossRef]
32. Breitenbach, S.; Gavrilov, N.; Pašti, I.; Unterweger, C.; Duchoslav, J.; Stifter, D.; Hassel, A.W.; Fürst, C. Biomass-Derived Carbons as Versatile Materials for Energy-Related Applications: Capacitive Properties vs. Oxygen Reduction Reaction Catalysis. *C* **2021**, *7*, 55. [CrossRef]
33. Foo, K.Y.; Hameed, B.H. Insights into the modeling of adsorption isotherm systems. *Chem. Eng. J.* **2010**, *156*, 2–10. [CrossRef]
34. Sips, R. On the Structure of a Catalyst Surface. *J. Chem. Phys.* **1948**, *16*, 490–495. [CrossRef]
35. Kumara, N.T.R.N.; Hamdan, N.; Petra, M.I.; Tennakoon, K.U.; Ekanayake, P. Equilibrium Isotherm Studies of Adsorption of Pigments Extracted from Kuduk-kuduk (*Melastoma malabathricum* L.) Pulp onto TiO_2 Nanoparticles. *J. Chem.* **2014**, *2014*, 468975. [CrossRef]
36. Keren, Y.; Borisover, M.; Bukhanovsky, N. Sorption interactions of organic compounds with soils affected by agricultural olive mill wastewater. *Chemosphere* **2015**, *138*, 462–468. [CrossRef]
37. Mabuza, M.; Premlall, K.; Daramola, M.O. Modelling and thermodynamic properties of pure CO2 and flue gas sorption data on South African coals using Langmuir, Freundlich, Temkin, and extended Langmuir isotherm models. *Int. J. Coal Sci. Technol.* **2022**, *9*, 45. [CrossRef]

38. Smječanin, N.; Bužo, D.; Mašić, E.; Nuhanović, M.; Sulejmanović, J.; Azhar, O.; Sher, F. Algae based green biocomposites for uranium removal from wastewater: Kinetic, equilibrium and thermodynamic studies. *Mater. Chem. Phys.* **2022**, *283*, 125998. [CrossRef]
39. Maliyekkal, S.M.; Sreeprasad, T.S.; Krishnan, D.; Kouser, S.; Mishra, A.K.; Waghmare, U.V.; Pradeep, T. Graphene: A Reusable Substrate for Unprecedented Adsorption of Pesticides. *Small* **2013**, *9*, 273–283. [CrossRef]
40. Colovic, M.B.; Krstic, D.Z.; Lazarevic-Pasti, T.D.; Bondzic, A.M.; Vasic, V.M. Acetylcholinesterase inhibitors: Pharmacology and toxicology. *Curr. Neuropharmacol.* **2013**, *11*, 315–335. [CrossRef]
41. Jacob, M.M.; Ponnuchamy, M.; Kapoor, A.; Sivaraman, P. Bagasse based biochar for the adsorptive removal of chlorpyrifos from contaminated water. *J. Environ. Chem. Eng.* **2020**, *8*, 103904. [CrossRef]
42. Jacob, M.M.; Ponnuchamy, M.; Kapoor, A.; Sivaraman, P. Adsorptive decontamination of organophosphate pesticide chlorpyrifos from aqueous systems using bagasse-derived biochar alginate beads: Thermodynamic, equilibrium, and kinetic studies. *Chem. Eng. Res. Des.* **2022**, *186*, 241–251. [CrossRef]
43. Ettish, M.N.; El-Sayyad, G.S.; Elsayed, M.A.; Abuzalat, O. Preparation and characterization of new adsorbent from Cinnamon waste by physical activation for removal of Chlorpyrifos. *Environ. Chall.* **2021**, *5*, 100208. [CrossRef]
44. Hamadeen, H.M.; Elkhatib, E.A.; Badawy, M.E.I.; Abdelgaleil, S.A.M. Green low cost nanomaterial produced from Moringa oleifera seed waste for enhanced removal of chlorpyrifos from wastewater: Mechanism and sorption studies. *J. Environ. Chem. Eng.* **2021**, *9*, 105376. [CrossRef]
45. Yang, J.; Ma, C.; Tao, J.; Li, J.; Du, K.; Wei, Z.; Chen, C.; Wang, Z.; Zhao, C.; Ma, M. Optimization of polyvinylamine-modified nanocellulose for chlorpyrifos adsorption by central composite design. *Carbohydr. Polym.* **2020**, *245*, 116542. [CrossRef]
46. Tulun, Ş.; Akgül, G.; Alver, A.; Çelebi, H. Adaptive neuro-fuzzy interference system modelling for chlorpyrifos removal with walnut shell biochar. *Arab. J. Chem.* **2021**, *14*, 103443. [CrossRef]
47. Hassanzadeh-Afruzi, F.; Esmailzadeh, F.; Taheri-Ledari, R.; Maleki, A. High adsorption capability of chlorpyrifos and Congo red in aqueous samples by a functionalized dextrin/graphene oxide composite. *Int. J. Environ. Sci. Technol.* **2023**, *20*, 10731–10750. [CrossRef]
48. Barzegarzadeh, M.; Amini-Fazl, M.S.; Sohrabi, N. Ultrasound-assisted adsorption of chlorpyrifos from aqueous solutions using magnetic chitosan/graphene quantum dot-iron oxide nanocomposite hydrogel beads in batch adsorption column and fixed bed. *Int. J. Biol. Macromol.* **2023**, *242*, 124587. [CrossRef]
49. Hussain, O.A.; Hathout, A.S.; Abdel-Mobdy, Y.E.; Rashed, M.M.; Abdel Rahim, E.A.; Fouzy, A.S.M. Preparation and characterization of activated carbon from agricultural wastes and their ability to remove chlorpyrifos from water. *Toxicol. Rep.* **2023**, *10*, 146–154. [CrossRef]
50. Dolatabadi, M.; Naidu, H.; Ahmadzadeh, S. Adsorption characteristics in the removal of chlorpyrifos from groundwater using magnetic graphene oxide and carboxy methyl cellulose composite. *Sep. Purif. Technol.* **2022**, *300*, 121919. [CrossRef]
51. Mahdavi, V.; Taghadosi, F.; Dashtestani, F.; Bahadorikhalili, S.; Farimani, M.M.; Ma'mani, L.; Mousavi Khaneghah, A. Aminoguanidine modified magnetic graphene oxide as a robust nanoadsorbent for efficient removal and extraction of chlorpyrifos residue from water. *J. Environ. Chem. Eng.* **2021**, *9*, 106117. [CrossRef]
52. Celso Gonçalves, A.; Zimmermann, J.; Schwantes, D.; Tarley, C.R.T.; Conradi Junior, E.; Henrique Dias de Oliveira, V.; Campagnolo, M.A.; Ziemer, G.L. Renewable Eco-Friendly Activated Biochar from Tobacco: Kinetic, Equilibrium and Thermodynamics Studies for Chlorpyrifos Removal. *Sep. Sci. Technol.* **2022**, *57*, 159–179. [CrossRef]
53. Asghar, A.; Mabarak, S.; Ashraf, B.; Rizwan, M.; Massey, S.; Asghar, B.H.; Shahid, B.; Rasheed, T. A sustainable approach for the removal of chlorpyrifos pesticide from aqueous phase using novel nano magnetized biochar. *Inorg. Chem. Commun.* **2024**, *159*, 111790. [CrossRef]
54. Hamadeen, H.M.; Elkhatib, E.A. Nanostructured modified biochar for effective elimination of chlorpyrifos from wastewater: Enhancement, mechanisms and performance. *J. Water Process Eng.* **2022**, *47*, 102703. [CrossRef]
55. Aris, N.I.F.; Rahman, N.A.; Wahid, M.H.; Yahaya, N.; Abdul Keyon, A.S.; Kamaruzaman, S. Superhydrophilic graphene oxide/electrospun cellulose nanofibre for efficient adsorption of organophosphorus pesticides from environmental samples. *R. Soc. Open Sci.* **2020**, *7*, 192050. [CrossRef]
56. Lazarević-Pašti, T.; Aničijević, V.; Baljozović, M.; Aničijević, D.V.; Gutić, S.; Vasić, V.; Skorodumova, N.V.; Pašti, I.A. The impact of the structure of graphene-based materials on the removal of organophosphorus pesticides from water. *Environ. Sci. Nano* **2018**, *5*, 1482–1494. [CrossRef]
57. Yadav, S.; Goel, N.; Kumar, V.; Singhal, S. Graphene Oxide as Proficient Adsorbent for the Removal of Harmful Pesticides: Comprehensive Experimental Cum DFT Investigations. *Anal. Chem. Lett.* **2019**, *9*, 291–310. [CrossRef]

C Journal of Carbon Research

MDPI

Review

Advanced Graphene-Based Technologies for Antibiotic Removal from Wastewater: A Review (2016–2024)

Joydip Sengupta [1] and Chaudhery Mustansar Hussain [2,*]

[1] Department of Electronic Science, Jogesh Chandra Chaudhuri College, Kolkata 700033, India; joydip@jcccollege.ac.in

[2] Department of Chemistry and Environmental Science, New Jersey Institute of Technology, Newark, NJ 07102, USA

* Correspondence: chaudhery.m.hussain@njit.edu

Abstract: The increasing presence of antibiotics in wastewater poses significant environmental risks, including the promotion of antibiotic resistance and harm to aquatic ecosystems. This study reviews advancements in graphene-based technologies for removing antibiotics from wastewater between 2016 and 2024. Graphene-based platforms, such as graphene oxide (GO), reduced graphene oxide (rGO), and graphene composites, have shown great promise in this field because of their exceptional adsorption capacities and rapid photocatalytic degradation capabilities. Functionalized graphene materials and graphene integrated with other substances, such as metal oxides and polymers, have enhanced performance in terms of antibiotic removal through mechanisms such as adsorption and photocatalysis. These technologies have been evaluated under various conditions, such as pH and temperature, demonstrating their practical applicability. Despite challenges related to scalability, cost-effectiveness, and environmental impact, the advancements in graphene-based technologies during this period highlight their significant potential for effective antibiotic removal, paving the way for safer and more sustainable environmental management practices.

Keywords: graphene; antibiotics; wastewater; photocatalytic degradation; adsorption

1. Introduction

The increasing contamination of water sources with pharmaceutical pollutants, particularly antibiotics, poses significant environmental and public health challenges [1]. Antibiotics in wastewater contribute to the development of antibiotic-resistant bacteria [2], disrupt aquatic ecosystems [3], and enter the food chain [4], making their removal an urgent global priority. Conventional wastewater treatment methods [5], while effective at removing many contaminants, often fail to eliminate antibiotics, necessitating the development of more advanced treatment technologies.

Graphene-based materials (GBMs) have emerged as promising solutions for the removal of a wide range of pollutants because of their exceptional physicochemical properties, including high surface area, chemical stability, and unique adsorption capabilities [6]. Since its discovery, graphene and its derivatives, such as GO and rGO, have been extensively studied for environmental applications, particularly in water treatment [7]. The use of graphene-based technologies for antibiotic removal from wastewater has gained considerable attention because of their efficiency in adsorbing various pharmaceutical compounds, including antibiotics, and their potential for large-scale implementation [8].

This review aims to provide a comprehensive overview of the advancements in graphene-based technologies for the removal of antibiotics from wastewater from 2016–2024. By critically analyzing recent literature following the Preferred Reporting Items for Systematic Reviews and Meta-Analyses (PRISMA) guidelines [9], this study highlights the various mechanisms involved in antibiotic adsorption, the role of graphene derivatives in enhancing removal efficiency, and the challenges associated with the practical implementation of these technologies. This review also identifies future research directions, emphasizing the need for more sustainable and economically viable solutions in wastewater treatment.

1.1. Graphene-Based Materials

Graphene, a two-dimensional (2D) form of carbon in which atoms are arranged in a hexagonal lattice, has emerged as one of the most promising materials in nanotechnology and materials science. As discovered in 2004 [10], graphene's unique combination of properties, such as exceptional electrical and thermal conductivity, mechanical strength, flexibility, and optical transparency, has led to its exploration in a wide range of applications, from electronics to energy storage and beyond [11]. These exceptional properties have spurred the development of various GBMs, each tailored to exploit specific aspects of graphene's capabilities. The diversity of these materials stems from modifications in the structure, composition, and synthesis processes of graphene, allowing for the customization of its properties to meet the demands of different technological applications [12].

1.1.1. Types of Graphene-Based Materials

GBMs can be classified into several categories based on their structure (Figure 1). Each type of graphene-based material offers unique properties that make it suitable for specific applications.

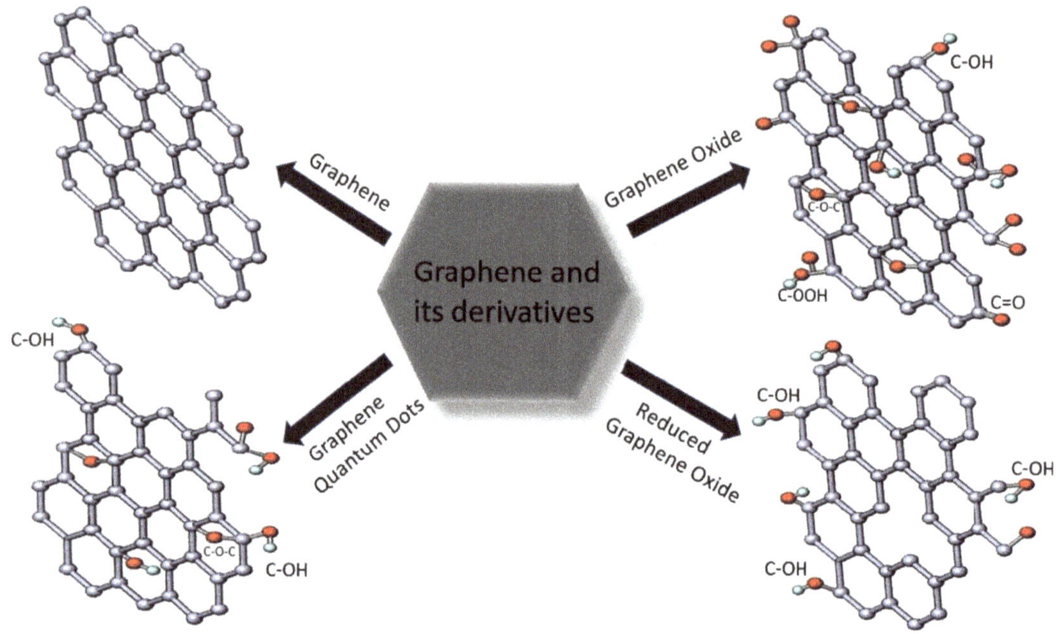

Figure 1. Graphene and its derivatives (reproduced with permission from [13]).

Pristine Graphene

Pristine graphene is the simplest form of graphene and consists of a single layer of carbon atoms arranged in a honeycomb lattice. This structure has exceptional properties, including high electrical conductivity [14], strength exceeding that of steel [15], and excellent thermal conductivity [16]. These traits make it ideal for electronic devices [17] and composite materials [18]. However, its large-scale use is limited by production challenges and difficulties in processing it in its pure form.

Graphene Oxide

GO is a graphene derivative with oxygen-containing functional groups such as hydroxyl, carboxyl, and epoxy groups, which disrupt its lattice and alter its properties [19]. These groups make GO more hydrophilic than pristine graphene, allowing it to disperse easily in water, which is ideal for liquid processing [20]. However, they also reduce the electrical conductivity of the material [21]. Despite this, GO is widely used in applications such as water purification [22] and as a precursor for other graphene-based materials such as rGO [23].

Reduced Graphene Oxide

rGO is produced by partially reducing GO [24] to restore some of the sp^2 carbon network, improving its electrical conductivity while retaining some oxygen groups [25]. This balance between hydrophilicity and conductivity makes rGO useful for applications such as supercapacitors [26] and batteries [27]. However, the reduction process is often incomplete, resulting in defects that affect its performance [28] and distinguish it from both pristine graphene and GO.

Graphene Quantum Dots (GQDs)

GQDs are nanoscale graphene fragments under 10 nanometers in size with unique electronic, optical, and chemical properties due to quantum confinement. Their tunable bandgap [29] makes them ideal for optoelectronics [30], bioimaging [31], sensing [32], and energy conversion [33]. Doping with elements such as nitrogen or sulfur further enhances their functionality [34].

1.1.2. Synthesis Methods for the Graphene-Based Materials

The synthesis of graphene and its derivatives is a critical aspect of their development, as the chosen method affects the material's properties and determines its suitability for different applications. Several synthesis methods have been developed (Figure 2), each with its own set of advantages, limitations, and areas of application.

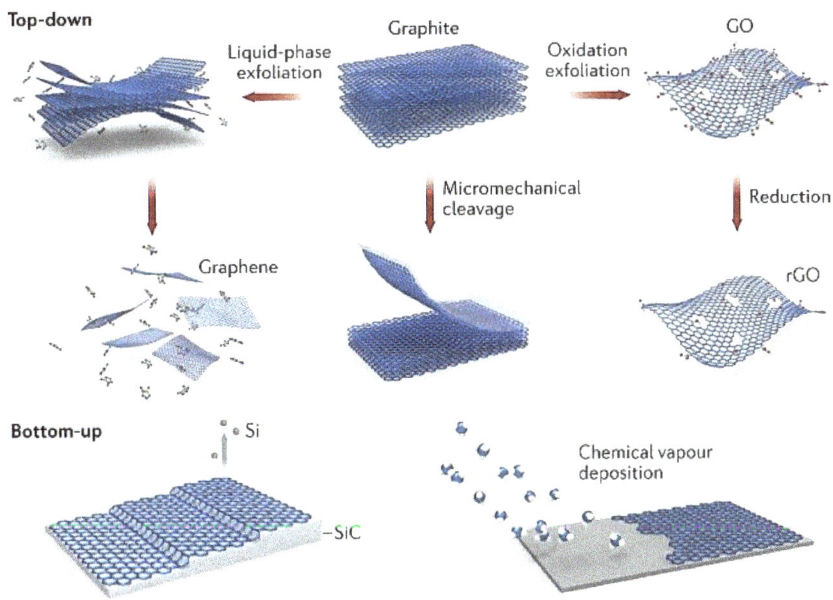

Figure 2. Different synthesis methods for graphene-based materials (reproduced with permission from [35]).

Liquid-Phase Exfoliation

Liquid-phase exfoliation (LPE) is a common method for synthesizing graphene from bulk graphite [36]. It involves dispersing graphite in a liquid medium with surfactants to prevent reaggregation. The process uses solvents such as N-methyl-2-pyrrolidone (NMP) [37] and applies ultrasonication to separate the graphene layers [38]. After sonication, exfoliated sheets are isolated from unexfoliated graphite through centrifugation [39]. While LPE is simple and scalable for large-scale production [40], it can introduce defects, contamination from surfactants, and variation in the thickness of the resulting graphene [41].

Oxidation Exfoliation for Graphene Oxide (GO) and Reduction for Reduced Graphene Oxide (rGO)

Oxidation exfoliation is a chemical method used to produce GO from graphite, which can be reduced to create rGO. The process begins by oxidizing graphite with strong agents such as sulfuric acid, nitric acid, and potassium permanganate [42], introducing oxygen-containing functional groups that increase the interlayer spacing for easier separation [43]. The oxidized graphite is then exfoliated in water to yield single- or few-layer GO, which can be reduced chemically [44], thermally [45], or electrochemically [46] to form rGO. While rGO has properties closer to those of pristine graphene, it still contains defects and residual oxygen groups [47]. This method is advantageous because of the high dispersibility of GO in solvents [48] and scalability [49] for producing graphene-based materials, although it has limitations such as lower electrical conductivity and environmental concerns related to the use of strong chemicals [50].

Micromechanical Cleavage

Micromechanical cleavage, or the "Scotch tape method", is the original technique for isolating graphene from graphite and remains essential in graphene research [51]. This involves physically exfoliating graphene layers via adhesive tape, which is pressed onto a graphite crystal and then peeled off to remove thin layers. This process is repeated, and the tape is then pressed onto a substrate such as silicon dioxide (SiO_2) to transfer the graphene [52]. While this method produces high-quality, defect-free graphene that is ideal for research, it is not suitable for industrial-scale production because of its labor-intensive nature, low yield, and limited capacity for large quantities [53].

Silicon Carbide (SiC) to Graphene

The SiC-to-graphene method involves thermally decomposing SiC to produce high-quality, large-area graphene on its surface [54], making it suitable for electronic applications. The process starts with heating SiC in an ultrahigh vacuum or argon atmosphere [55], causing silicon atoms to sublimate and leave behind a carbon-rich surface that forms graphene layers. The number of layers can be controlled by adjusting the heating duration and temperature [56]. This method produces high-quality graphene with few defects, which is crucial for electronic applications [57], but it is expensive because of the cost of SiC substrates and high-temperature processing, and it can be challenging to uniformly control the number of layers over large areas [58].

Chemical Vapor Deposition (CVD)

CVD is a widely used method for producing large-area graphene films, particularly on metal substrates such as copper [59] or nickel [60]. The process starts by preparing and annealing the metal substrate to reduce defects. Hydrocarbon gases, such as methane [61], are then introduced at high temperatures and decompose on the surface to release carbon atoms that form graphene layers. After formation, the graphene can be transferred to other substrates by etching away the metal [62]. The main advantage of CVD is its ability to create continuous graphene films suitable for electronics, sensors, and transparent conductive films [63], although the transfer process can introduce defects or residues that affect quality [64].

2. Antibiotics in Wastewater: A Growing Environmental Concern

The presence of antibiotics in wastewater has emerged as a critical environmental issue, reflecting the unintended consequences of their widespread use and disposal [65]. Antibiotics, which were initially developed as life-saving drugs, have become pervasive in various sectors, including human [66] and veterinary medicine [67], agriculture [68], and aquaculture [69]. These compounds are now frequently detected in wastewater, raising alarms due to their potential impacts on ecosystems and public health [70]. The persistence of antibiotics in the environment, coupled with the inefficiencies of current wastewater treatment technologies, has resulted in their accumulation in natural water bodies, leading to serious ecological and health-related challenges [71]. To address this problem effectively, it is essential to understand the sources and environmental impacts related to antibiotics in wastewater.

2.1. Sources of Antibiotics in Wastewater

The entry of antibiotics into wastewater systems is multifaceted, stemming from various human activities and industrial processes (Figure 3). These sources contribute significantly to the contamination of water resources, making it a pressing environmental concern.

Figure 3. Different sources of antibiotics in wastewater.

2.1.1. Human Medicine

The most significant source of antibiotics in wastewater is human medicine [72]. After administration, a considerable portion of antibiotics consumed by humans are excreted unmetabolized. These excreted antibiotics enter the sewage system through urine and feces and are subsequently conveyed to wastewater treatment plants (WWTPs). Additionally, hospitals and healthcare facilities contribute heavily to this problem. They discharge not only antibiotics excreted by patients but also large quantities of expired or unused medications. In many cases, these drugs are disposed of directly into the sewage system, further increasing the concentration of antibiotics in wastewater [73]. The widespread use of antibiotics in outpatient settings, often without complete courses of treatment, exacerbates this issue by leading to more frequent and varied releases of these compounds into the environment.

2.1.2. Veterinary Medicine and Agriculture

The use of antibiotics in agriculture and animal husbandry is another major contributor to the presence of these drugs in wastewater [74]. In livestock farming, antibiotics are used not only to treat and prevent diseases but also to promote growth. Animals excrete a significant proportion of these antibiotics, which then enter the environment through manure and urine. When manure is used as fertilizer, antibiotics can leach into the soil and subsequently reach groundwater or be carried by runoff into surface water bodies [75]. In aquaculture, antibiotics are used to prevent infections in densely populated fish farms. These drugs are often administered through feed, and excess antibiotics, along with fish excreta, are directly discharged into surrounding water bodies, contributing to the contamination of aquatic environments [76]. The cumulative effect of these practices results in substantial antibiotic loads in the environment, which can eventually make their way into wastewater systems.

2.1.3. Pharmaceutical Manufacturing

The pharmaceutical industry is another significant source of antibiotics in wastewater [77]. During the production of antibiotics, wastewater generated from manufacturing processes can contain high concentrations of these drugs. In regions where environmental regulations are weak or poorly enforced, pharmaceutical plants may discharge their effluents directly into local water bodies or municipal sewage systems without adequate treatment [78]. This practice can lead to localized hotspots of antibiotic contamination, where the concentration of these drugs is much higher than that in typical domestic wastewater [79]. The environmental impact of these discharges is particularly concerning in developing countries, where regulatory oversight may be limited and industrial waste management practices are less stringent [80].

The presence of antibiotics in wastewater has profound implications for environmental health, primarily because of the promotion of antibiotic resistance [81], disruption of natural ecosystems [82], and potential risks to human health [83]. These impacts are complex and multifaceted, reflecting the intricate interplay between chemical pollutants and biological systems.

2.2. *Properties of Graphene-Based Materials for the Treatment of Wastewater with Antibiotics*

The unique properties of GBMs make them highly effective in the treatment of antibiotics in wastewater, addressing one of the most pressing environmental challenges. The extraordinary surface area of graphene, which is among the highest of any material, provides extensive active sites for the adsorption of antibiotic molecules, enabling efficient removal of these contaminants from water [84]. This high adsorption capacity, combined with the chemical stability of graphene [85], ensures that it can effectively capture and retain antibiotics, even in the presence of other competing contaminants. Furthermore, the exceptional electrical conductivity of graphene-based materials enhances their potential use in advanced oxidation processes (AOPs), where graphene can act as a catalyst or support material in the generation of reactive oxygen species (ROS) [86]. These ROS are capable of degrading antibiotics into less harmful compounds, thus reducing their environmental impact. The high thermal conductivity of graphene can also play a role in photothermal processes [87], where it absorbs light and converts it into heat, aiding in the degradation of antibiotic compounds under solar or artificial light. Moreover, the tunable surface chemistry of graphene allows functionalization with specific groups that can target and degrade certain classes of antibiotics, further increasing the specificity and efficiency of the treatment process [88]. These properties collectively position GBMs as powerful tools for developing advanced, sustainable technologies for the effective removal of antibiotics from wastewater, thereby mitigating the risk of environmental contamination and the spread of antibiotic resistance.

3. Research Methods

3.1. Systematic Literature Review

This study conducted a systematic literature review following the Preferred Reporting Items for Systematic Reviews and Meta-Analyses (PRISMA) guidelines [89], which are widely recognized for enhancing the transparency, completeness, and accuracy of systematic reviews and meta-analyses. The PRISMA methodology [90] outlines a structured approach for identifying documents involving sequential steps: identification, screening, and inclusion based on eligibility quality assessment.

3.1.1. Identification

The literature search was carried out via the Scopus database [91], which was selected for its extensive coverage of recent and reputable publications across numerous journals. The search string used was "TITLE (graphene AND wastewater AND antibiotic)", which was restricted to the "Article title" field in Scopus. This search yielded an initial set of 21 documents.

3.1.2. Screening

The 21 identified documents were then subjected to further screening based on predefined criteria: document type (Article), source type (Journal), publication stage (Final), and language (English). The following search string was applied to refine the results:

"TITLE (graphene AND wastewater AND antibiotic) AND (LIMIT-TO (DOCTYPE, "ar")) AND (LIMIT-TO (PUBSTAGE, "final")) AND (LIMIT-TO (SRCTYPE, "j")) AND (LIMIT-TO (LANGUAGE, "English"))".

This screening process, which was conducted on 24 September 2024, narrowed the list to 19 journal articles for further examination.

3.1.3. Eligibility

At the eligibility stage, the focus was on selecting journal articles specifically addressing graphene-based technologies for antibiotic removal from wastewater. The titles, abstracts, and full texts of the 19 articles were reviewed closely. Four articles were excluded because they were outside the scope of the study. Ultimately, 15 articles were deemed eligible for further analysis.

3.1.4. Included Articles and Quality Assessment

A total of 15 articles were retained for the final review. Bibliographic data, including author names, affiliations, article titles, keywords, abstracts, publication years, and journal titles, were extracted from the Scopus database and saved in a .csv file. A quality assessment of these articles was conducted via a checklist (Table 1). All 15 articles met the quality assessment criteria, ensuring their suitability for inclusion in the subsequent analysis and discussion (Scheme 1).

Table 1. Checklist of the quality assessment.

1.	Are the aims and objectives explicitly articulated?
2.	Is the reporting structured in a logical, cohesive, and coherent manner?
3.	Is the proposed technique described in sufficient detail?
4.	Is the research methodology aligned with the study's objectives?
5.	Are the methods for data collection clearly and thoroughly explained?
6.	Do the explanations and conclusions rely appropriately on the data presented?
7.	Does the study make a meaningful contribution to the body of knowledge?
8.	Has the stated aims and objectives been achieved?
9.	Is the research process clearly and comprehensively documented?
10.	Can the study be reproduced based on the information provided?

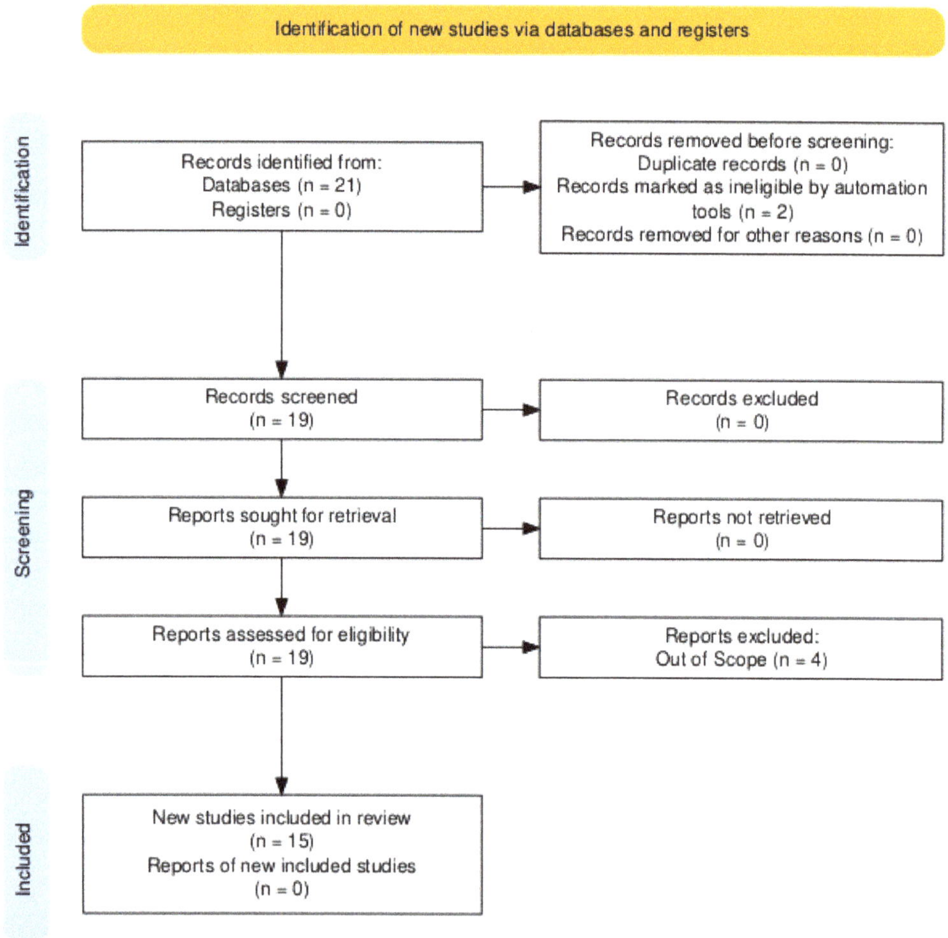

Scheme 1. Three steps of this systematic literature review following PRISMA, 2020 [92].

4. Removal of Antibiotics from Wastewater via Graphene-Based Materials

4.1. Removal of Antibiotics from Wastewater via Photocatalytic Degradation via Graphene-Based Materials

GBMs have demonstrated exceptional efficacy in the photocatalytic degradation of antibiotics from wastewater, leveraging their superior electronic properties and large surface area. These materials facilitate efficient removal processes, significantly reducing antibiotic concentrations and improving wastewater treatment outcomes.

The effectiveness of graphene-based TiO_2 composite photocatalysts for the degradation of the antibiotics sulfamethoxazole (SMX), erythromycin (ERY), and clarithromycin (CLA) under solar radiation in real urban wastewater was evaluated by Karaolia et al. [87]. The investigation involved synthesizing TiO_2–rGO composite photocatalysts via two ex-situ methods, hydrothermal and photocatalytic treatment, with GO and oxide P25 TiO_2 as precursor materials. This study aimed to compare the photocatalytic efficiency of synthesized TiO_2–rGO composites with that of pristine Aeroxide P25 TiO_2 for the removal of these antibiotics from urban wastewater effluents pretreated with a membrane bioreactor and subsequently exposed to simulated solar radiation. The findings revealed that the TiO_2–rGO composite prepared via photocatalytic treatment demonstrated superior photocatalytic degradation of ERY (84 $\pm$ 2%) and CLA (86 $\pm$ 5%), whereas the degradation of SMX (87 $\pm$ 4%) was slightly more effective with Aeroxide P25 TiO_2.

Yazdi et al. [93] investigated a method for synthesizing ternary nanocomposites consisting of TiO_2 nanosheets, rGO, and various amounts of Ag for the photocatalytic treatment of tetracycline (TC) antibiotic wastewater. Plasmonic Ag nanoparticles were incorporated into the TiO_2 nanosheet/rGO nanocomposite through a photo deposition technique (TGA(x) samples). The photocatalytic activity of these nanocomposites was assessed for the degradation of TC at an initial concentration of 30 mg/L, with the TGA (0.076) sample achieving a degradation efficiency of 52.56% after 3 h of visible light exposure. The study revealed that the rGO in the TGA(x) samples significantly facilitated the transfer of photoinduced electrons from the plasmonic Ag nanoparticles to the TiO_2 nanosheets. Additionally, a three-layer artificial neural network model incorporating four input variables, such as irradiation time, catalyst dosage, initial TC concentration, and silver nitrate content, and an output variable, i.e., % degradation, was optimized with 11 hidden neurons. Analysis via Garson's formula revealed that the initial TC concentration was the most influential factor affecting treatment efficiency, accounting for 31% of the relative importance.

In a recent investigation, Yang et al. [94] developed an advanced photocatalytic flow-through system that incorporated a three-dimensional (3D) photoanode along with a Pt/C air-breathing electrode as the cathode (Figure 4). The system's efficacy in degrading ampicillin was evaluated by comparing the photocatalytic performance of a C_3N_4-MoS_2 composite supported on a 3D graphene matrix (C_3N_4-MoS_2/3DG) with that of other 3D materials. The findings revealed that the photoanode achieved a 74.6% removal efficiency for ampicillin within 2 h of the reaction. Subsequent toxicity evaluations, including acute antibacterial potency assessments and ECOSAR model predictions, indicated a reduction in ampicillin toxicity following photocatalytic treatment, likely due to the cleavage of functional groups such as amido and peptide bonds. The superior performance of the flow-through system was attributed to the enhanced adsorption capacity and heterojunction mechanism of the C_3N_4-MoS_2/3DG composite, along with the optimized mass and electron transfer characteristics inherent to the flow-through design. Additionally, the synergistic interaction between the photoanode and cathode was identified as a significant factor contributing to the system's overall efficiency. Consequently, this photoanode-based flow-through cell system is recognized as a promising strategy for the effective removal of organic pollutants in wastewater treatment processes.

In their research, Ivan et al. [22] synthesized hybrid layers composed of Fe oxide, Fe hydroxide, and nitrogen-doped graphene-like platelets through an eco-friendly laser-based method aimed at photocatalytic applications. These composite layers exhibited considerable efficiency in the photodecomposition of antibiotic molecules under visible light irradiation. The efficiency of photodecomposition was systematically analyzed in relation to varying concentrations of base materials, including the Fe oxide nanoparticles and GO platelets used in dispersions subjected to laser irradiation. Although the pure Fe oxide/Fe hydroxide layers strongly absorbed visible light, their photodecomposition efficiency under comparable conditions was negligible. The improved photocatalytic efficiency of the nanohybrid layer, which achieved up to 80% degradation of the initial antibiotic molecules, was attributed to synergistic interactions among the composite materials. These interactions facilitated the effective separation of electron–hole pairs generated on the surfaces of Fe oxide and Fe hydroxide nanoparticles under visible light, with conductive graphene-like platelets playing a critical role. Furthermore, nitrogen-doped graphene-like platelets contributed to the generation of electron–hole pairs under visible light, as demonstrated by the photocatalytic activity observed in the pure, nitrogen-doped graphene-like reference layers. The study also indicated that adsorption processes played a minimal role in the removal of antibiotic molecules from the test solutions, with the observed reduction in antibiotic concentration primarily resulting from photocatalytic decomposition mechanisms.

Lin et al. [95] developed a GO structure-oriented NH_2-MIL-88B(Fe)/GO/sodium alginate (NM88B/GO/SA) aerogel for antibiotic removal (Figure 5). The aerogel was fabricated via GO interface regulation and dual-network crosslinking techniques, resulting in the formation of an NM88B/GO heterostructure via the directional growth of NM88B on the GO surface. Sodium alginate (SA) was incorporated to increase the robustness of the matrix and prevent fragmentation. With a 30 wt% catalyst loading, the composite aerogel demonstrated exceptional photocatalytic performance, achieving over 99% removal of high-concentration (50 ppm) tetracycline hydrochloride (TC-HCl) within 150 min. The aerogel also exhibited notable shape recovery, stability, and reusability, maintaining over 95% degradation efficiency for TC-HCl after five repeated tests. This aerogel has considerable potential as an efficient and reusable photocatalyst for wastewater treatment.

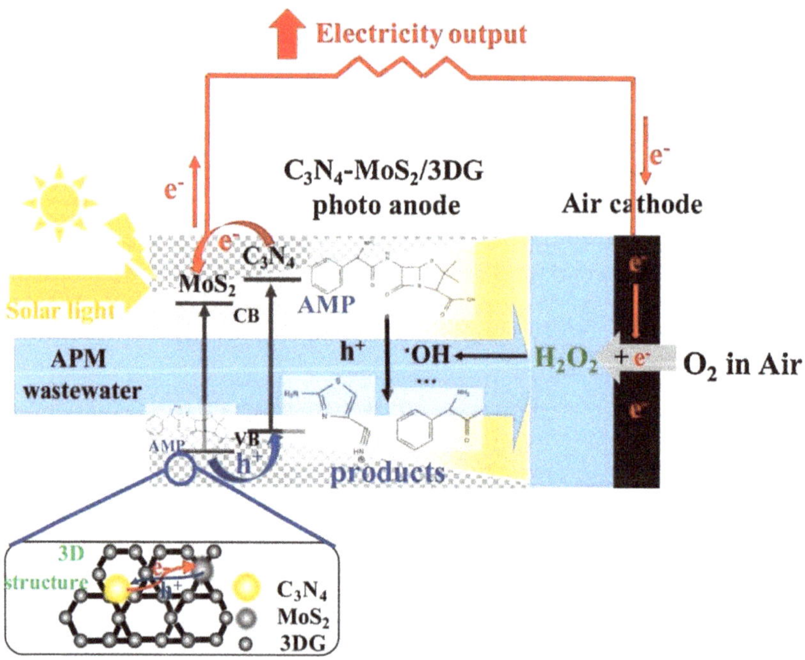

Figure 4. Diagram depicting AMP wastewater treatment via the C_3N_4-MoS_2/3DG flow-through system, enhanced by the addition of trace amounts of H_2O_2 and electricity generation by the air cathode. (Reproduced with permission from [94]).

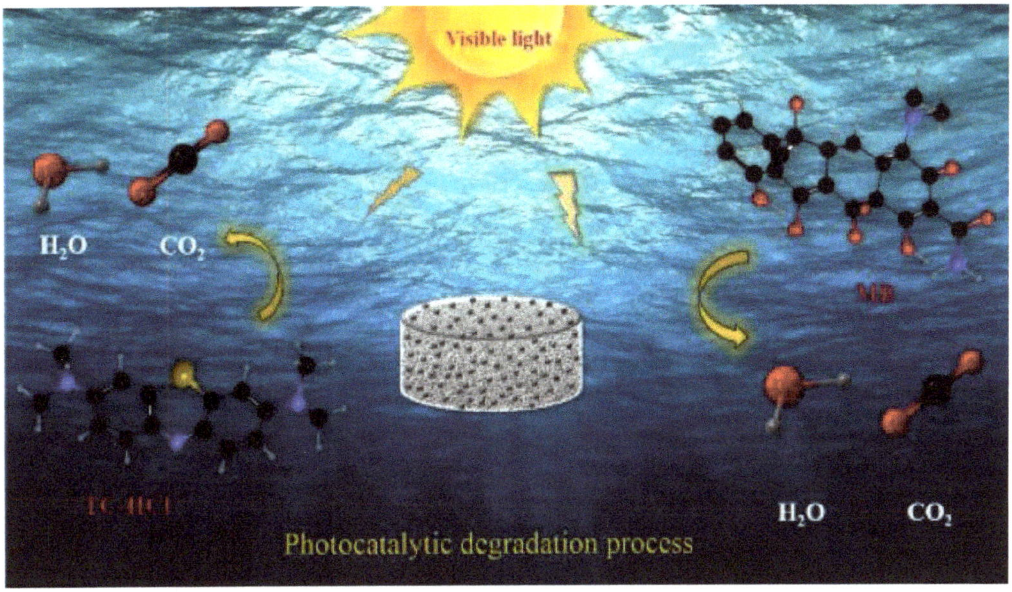

Figure 5. GO-based NM88B/GO/SA aerogels for antibiotic-contaminated wastewater (reproduced with permission from [95]).

4.2. Removal of Antibiotics from Wastewater via Adsorption via Graphene-Based Materials

GBMs have demonstrated exceptional efficacy in the adsorption of antibiotics from wastewater, significantly reducing their concentration. The high surface area and tunable properties of these materials enhance their removal efficiency, making them promising candidates for advanced water treatment technologies.

A one-pot solvothermal synthesis method was employed by Wu et al. [96] to produce RGO-supported ferrite hybrids via the use of GO and Fe^{3+} ions as precursors. The resulting Fe_3O_4 nanoparticles were uniformly dispersed and anchored onto the RGO nanosheets. These nanocomposites exhibited high adsorption efficiency for quantifying three sulfonamides (SAs) in wastewater, which was attributed to the large surface area of the RGO. Several key experimental parameters, including the amount of adsorbent, extraction time, pH, and desorption conditions, were systematically optimized to increase the extraction efficiency. Compared with conventional adsorbents, the RGO–Fe_3O_4 nanocomposites demonstrated superior extraction capabilities. A broad linear detection range of 1 to 200 ng/mL was achieved, with a correlation coefficient above 0.9987, and the detection limits for the three sulfonamides ranged between 0.43 and 0.57 ng/mL. When applied to environmental wastewater samples, the method achieved recoveries ranging from 89.1% to 101.7%, with relative standard deviations below 8.6%, confirming its efficacy and reproducibility across different matrices.

Radmehr et al. [84] synthesized a renewable adsorbent, NiZrAl-layered double hydroxide-graphene oxide-chitosan (NiZrAl-LDH-GO-CS NC), for the effective removal of nalidixic acid (NA) from wastewater. The successful synthesis was confirmed through various characterization techniques. The NiZrAl-LDH-GO-CS NCs exhibited rapid adsorption rates and a maximum adsorption capacity of 277.79 mg/g, surpassing those of other nanobased adsorbents. The adsorption process was analyzed via a central composite design, adaptive network-based fuzzy inference system, and general regression neural network models, with the adaptive network-based fuzzy inference system showing superior performance in predicting NA removal. Optimization through response surface methodology (RSM) achieved over 92% NA removal. Thermodynamic and kinetic assessments revealed chemisorption mechanisms. The study concluded that the LDH-GO-CS adsorbent has significant potential for pharmaceutical wastewater treatment and warrants further research for practical remediation applications.

In their study, Behzadi et al. [97] investigated the use of organic aerogels with high specific surface areas for the removal of TC antibiotics. They synthesized resorcinol formaldehyde (RF) aerogels via a sol-gel process, followed by drying under ambient conditions. To enhance the adsorbent properties, they modified RF aerogels with 1 wt.% graphene and 1 wt.% m-phenylenediamine, producing the RF-G1/PmPDA1 composite. The performance of these modified aerogels was assessed under various conditions, including different pH values (ranging from 2–12), adsorbent doses (4–10 mg), and contact times with antibiotics (3–24 h). Brunauer–Emmett–Teller (BET) analysis indicated that the modification increased the specific surface area of the RF aerogel from 96 to 308 m^2/g. The RF aerogel and RF-G1/PmPDA1 composites achieved TC removal rates of 65.2% and 93.3%, respectively, at an optimal pH of 4.

Kogut et al. [98] evaluated the feasibility of using adsorptive nonwovens as a cost-effective pretreatment strategy in wastewater treatment, highlighting their potential for environmental protection through the efficient adsorption of antibiotics. Research has concentrated on graphene-modified nonwovens (GMNs) and investigated two primary aspects: the adsorption behavior of TC, a common environmental contaminant at concentrations relevant to real-world conditions, and the factors influencing the antibacterial and antifungal properties of these materials. The study revealed that the integration of graphene particles with commercial textile auxiliaries significantly enhanced the antibacterial and antifungal properties of the nonwovens. TC residues were detected at the ng/mL scale via enzyme-linked immunosorbent assay (ELISA), and the adsorption behavior followed Henry and Redlich–Peterson isotherms, demonstrating the effectiveness of the process at low TC concentrations. As a result, well-designed GMNs have emerged as promising candidates for wastewater treatment applications, particularly in sewage treatment facilities. Additionally, statistical analyses of nonwoven and modified nonwoven morphologies, including skewness and kurtosis, provided valuable insights into the parameters influencing fungal growth in these structures. The research concluded that the GMNs exhibited significant antibiotic adsorption capabilities, resulting in a two-fold reduction in the TC concentration during the experiments.

Rajapaksha et al. [99] investigated the impact of pharmaceutical discharge on wastewater contamination, highlighting the significant environmental and public health risks posed by these pollutants. They synthesized GO by oxidizing graphite via a modified Hummers method, which was then used to develop two composite materials (Figure 6): copper oxide-reduced graphene oxide (CuO–rGO) and zinc oxide-reduced graphene oxide (ZnO–rGO). The adsorption capacities of these composites for the antibiotics amoxicillin (AMOX) and TC were evaluated via UV–visible spectroscopy. The CuO–rGO composite exhibited particularly high antibiotic removal efficiencies, achieving capacities of 405 mg/g for AMOX and 552 mg/g for TC at a pH of 7 and a temperature of 333 K. Additionally, this material demonstrated an 80% regeneration efficiency and retained 82% of its adsorption capacity after five reuse cycles. The adsorption behavior of AMOX and TC on GO, CuO–rGO, and ZnO–rGO followed the Langmuir isotherm and pseudo-second-order kinetic models, indicating a chemisorption process. Thermodynamic analyses revealed that these adsorption processes were spontaneous, with exothermic adsorption observed for AMOX and TC on CuO–rGO and ZnO–rGO, whereas adsorption on GO was endothermic. This study emphasized the enhanced efficacy of rGO materials in removing anionic antibiotics from water, suggesting that electrostatic interactions and π–π interactions between antibiotics and adsorbents are key mechanisms. The findings also suggested that, while unmodified GO was effective for the adsorption of cationic contaminants, the rGO composites were more suitable for targeting anionic contaminants in wastewater treatment applications.

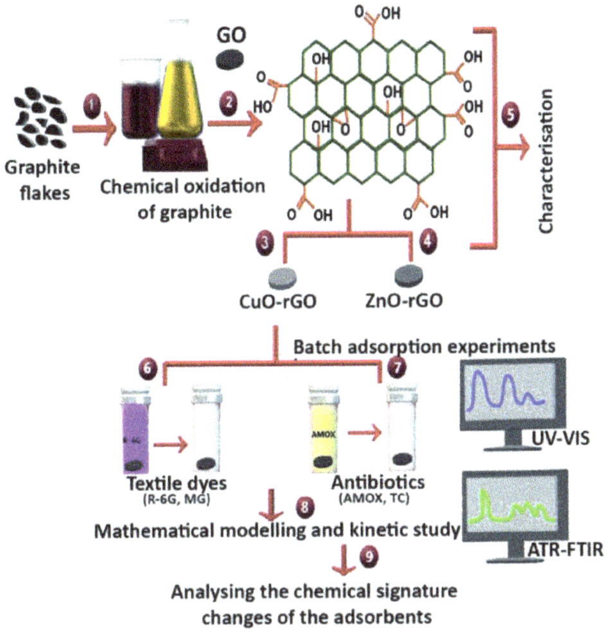

Figure 6. A schematic of the experimental design included the following steps: (1) Synthesizing graphene oxide (GO) from raw graphite flakes via oxidation via a modified Hummers method. (2) Obtaining the resulting GO. (3) Copper oxide-doped reduced graphene oxide (CuO–rGO) was synthesized from synthesized GO and an aqueous $CuSO_4 \cdot 5H_2O$ solution, and (4) zinc oxide-doped reduced graphene oxide (ZnO–rGO) was synthesized from synthesized GO and an aqueous $ZnSO_4 \cdot 7H_2O$ solution through a series of thermal chemical reactions. (5) The chemical and physical properties of GO, CuO–rGO, and ZnO–rGO were characterized via standard microscopic and spectroscopic techniques, including SEM, TEM, ATR-FTIR, and XPS. (6) Batch adsorption experiments were conducted to remove textile dyes (rhodamine 6G (R-6G) and malachite green (MG)) and antibiotics (amoxicillin (AMOX) and tetracycline (TC)) from aqueous solutions via GO, CuO–rGO, and ZnO–rGO adsorbents, followed by analysis via UV–visible spectroscopy. (7) Mathematical modeling and kinetics were applied to study the batch adsorption of textile dyes (R-6G, MG) and antibiotics (AMOX, TC) on GO, CuO–rGO, and ZnO–rGO. (8) Analyzing functional group changes on the GO, CuO–rGO, and ZnO–rGO adsorbents after adsorption of the textile dyes and antibiotics via ATR-FTIR. (Reproduced with permission from [99]).

Taleb et al. [100] addressed the pressing issue of antibiotic removal from contaminated water bodies, which is a critical step toward sustainable development. They synthesized a multifunctional hybrid thin film composed of carboxymethyl cellulose, GO, and polyaniline (CMC/GO/PANI) and conducted a systematic evaluation of its effectiveness in adsorbing oxytetracycline (OTC) from wastewater. The adsorption process was investigated under various conditions, including different reaction times, pH values, concentrations, and temperatures. The findings revealed that the CMC/GO/PANI hybrid thin film exhibited better OTC adsorption than the individual CMC, GO/CMC, and PANI/CMC thin films, which was attributed to the synergistic interaction of its multifunctional components. The adsorption kinetics were best described by the pseudo-second-order model, whereas the Redlich–Peterson isotherm provided an accurate interpretation of the adsorption equilibrium. Thermodynamic analysis indicated that the process was both spontaneous and endothermic. Moreover, the CMC/GO/PANI thin film demonstrated significant reusability, maintaining its adsorption capacity across seven cycles of adsorption and desorption. This study underscores the substantial potential of the CMC/GO/PANI thin film for large-scale wastewater purification, highlighting its durability and enhanced adsorption performance.

AbuZaid et al. [101] investigated the problem of elevated antibiotic levels in wastewater, which exacerbated the emergence of antimicrobial-resistant pathogens and intensified global water scarcity, adversely affecting both environmental systems and human health. Their research focused on the synthesis and assessment of a two-dimensional (2D) lamellar, free-standing GO/Ti_3C_2Tx membrane designed for the effective removal of antibiotics from wastewater (Figure 7). The 50% GO/Ti_3C_2Tx composite membrane significantly increased the water flux, reaching 61.9 L m^{-2} h^{-1}, which markedly exceeded the flux of the pristine GO membrane at 22.8 L m^{-2} h^{-1}. Moreover, the GO/Ti_3C_2Tx composite membranes achieved TC rejection rates consistently above 99%, representing a substantial improvement over the performance of the pristine Ti_3C_2Tx membrane. The membranes with GO to MXene ratios of 50%, 40%, 30%, and 20% were found to have contact angles of 54.5°, 57.4°, 60.6°, and 61.6°, respectively. The study highlighted that the integration of GO and Ti_3C_2Tx enhanced the membrane properties, as indicated by modifications in the interlayer spacing and hydrophilicity. Additionally, compared with both the unmodified GO and Ti_3C_2Tx membranes, the composite membrane exhibited superior antifouling properties.

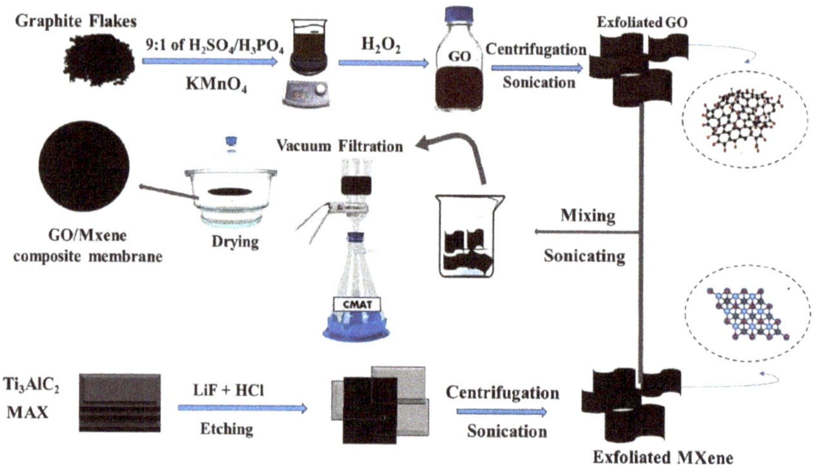

Figure 7. Diagrammatic representation of the synthesis process for free-standing graphene oxide (GO), Ti_3C_2Tx, and GO/Ti_3C_2Tx composite membranes (reproduced with permission from [101]).

Shaker et al. [102] employed a nanocomposite composed of copper nanoparticles immobilized on polyaniline-modified GO (Cu/PANI/GO) to successfully extract TC from environmental water samples. Compared with unmodified GO, the Cu/PANI/GO nanocomposite substantially improved the TC removal efficiency, achieving a remarkable removal rate of 99.6% and an adsorption capacity of 434.78 mg/g. This enhanced performance was attributed to the synergistic effects between the copper nanoparticles and polyaniline components, which led to a TC removal efficiency of approximately 51.5 ± 3.9% greater than that observed with GO alone (Figure 8). The adsorption kinetics and isotherms of TC on the Cu/PANI/GO nanocomposite were accurately modeled via the pseudo-second-order kinetic model and the Langmuir isotherm model, respectively. Additionally, the nanocomposite demonstrated notable durability over five adsorption–desorption cycles, with only a 4.1% reduction in removal efficiency. Researchers have proposed that the strong retention of TC within the Cu/PANI/GO structure results from surface complexation by copper nanoparticles, effective hydrogen bonding with polyaniline and GO, and π–π

interactions with GO. Thus, the Cu/PANI/GO nanocomposite proved to be an effective material for the removal of TCs from wastewater samples.

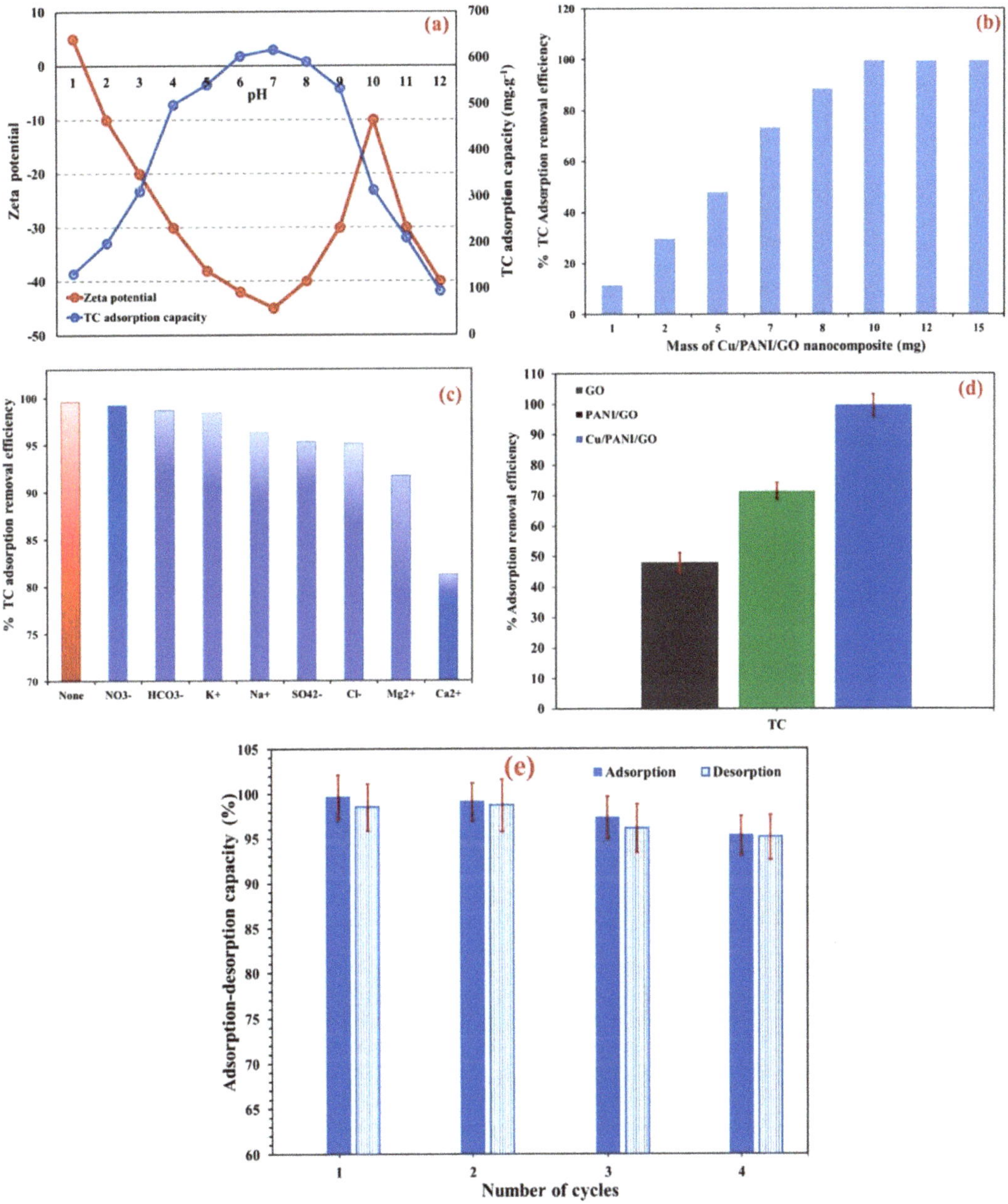

Figure 8. Influence of various factors on the tetracycline adsorption efficiency of the Cu/PANI/GO nanocomposite: (**a**) pH and zeta potential, (**b**) mass dosage (mg), (**c**) presence of interfering ions, (**d**) synergistic impact of Cu nanoparticles and polyaniline on the adsorption efficiency, and (**e**) reusability of the Cu/PANI/GO nanocomposite over four consecutive cycles under optimal conditions (reproduced with permission from [102]).

4.3. Removal of Antibiotics from Wastewater via Other Methods via Graphene-Based Materials

There are alternative methods, in addition to photocatalytic degradation and adsorption, for removing antibiotics from wastewater.

Solís et al. [88] developed magnetic graphene-based catalysts aimed at removing a variety of antibiotics, including SMX, norfloxacin, TC, and flumequine, from water (Figure 9). Their study investigated the catalytic activation of inorganic peroxides, namely, peroxymonosulfate (PMS), peroxydisulfate, and hydrogen peroxide, by altering the ratio of magnetite to graphene from 1:0 to 0:1. The results indicated that graphene played a crucial role in catalytic activation, with the process being most efficient in the presence of PMS. A magnetite concentration of 20% in the solid matrix was adequate for the complete degradation of the antibiotics, and the use of a magnetic field facilitated effective recovery. The performance of the catalyst was further evaluated in simulated urban wastewater, where key factors influencing the process and its stability over multiple reuse cycles were examined. Kinetic scavenging probe tests corroborated the hypothesis of a non-radical mechanism during PMS activation, whereas electron paramagnetic resonance analysis in the presence of D_2O revealed that electron transfer was the predominant reaction mechanism. The magnetic catalyst demonstrated both high catalytic activity and stability, proving to be an effective method for removing antibiotics from water.

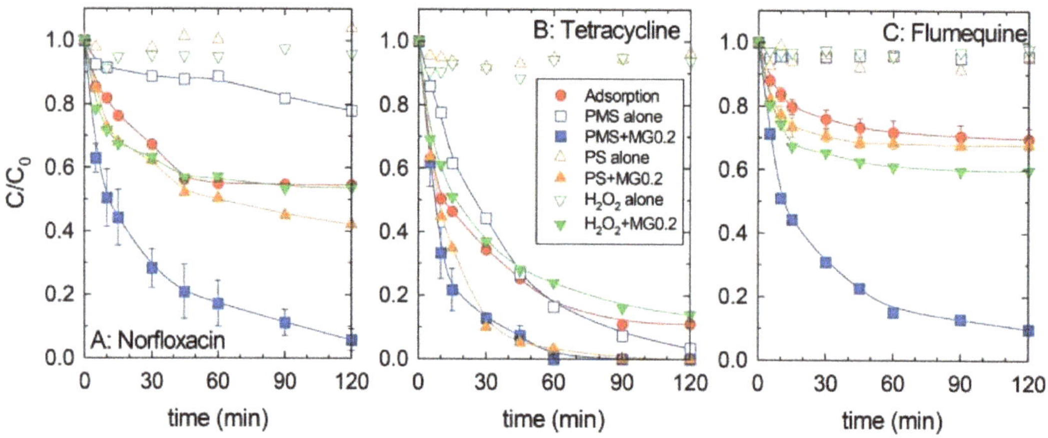

Figure 9. Removal of norfloxacin (**A**), tetracycline (**B**), and flumequine (**C**) via activated inorganic peroxides with magnetic graphene MG0.2 (reproduced with permission from [88]).

Peng et al. [86] developed a flow-through electrochemical system that utilized a series of graphene nanoparticle-loaded PbO_2 reactive electrochemical membrane electrodes (GNP-PbO_2 REMs) mounted on porous titanium substrates with different pore sizes for the treatment of antibiotic-contaminated wastewater. Among these, the GNP-PbO_2 electrodes supported on titanium substrates with a pore size of 150 μm (Ti-150/GNP-PbO_2) exhibited superior electrochemical degradation performance compared with REMs with other pore sizes. This improved performance was attributed to the smaller crystal size, larger electrochemically active surface area, reduced charge-transfer impedance, and higher oxygen evolution potential of Ti-150/GNP-PbO2. Under optimized conditions with an initial pH of 5, the Ti-150/GNP-PbO_2 REM achieved a benzylpenicillin sodium (PNG) removal rate of 99.34%. This performance was attributed to enhanced mass transfer. This study proposed three plausible degradation pathways for PNG within a flow-through electrochemical system and confirmed the stability and safety of the Ti-150/GNP-PbO_2 REM.

A prior discussion on the removal of antibiotics from wastewater via GBMs is summarized in Table 2.

Table 2. Graphene-based materials are used for the removal of antibiotics from wastewater.

Graphene Type	Target	Type of Treatment	Results Achieved	Reference
Reduced Graphene Oxide	sulfamethoxazole (SMX), erythromycin (ERY), and clarithromycin (CLA)	Photocatalytic degradation	Degradation efficiency SMX (87 ± 4%), ERY (84 ± 2%), CLA (86 ± 5%)	Karaolia et al. [87]
Reduced Graphene Oxide	Tetracycline	Photocatalytic degradation	Degradation efficiency 52.56%	Yazdi et al. [93]
Graphene	Ampicillin	Photocatalytic degradation	Removal efficiency 74.6%	Yang et al. [94]
Reduced Graphene Oxide	Chloramphenicol sodium succinate	Photocatalytic degradation	Removal efficiency 80%	Ivan et al. [22]
Graphene Oxide	Tetracycline	Photocatalytic degradation	Removal efficiency > 99%	Lin et al. [95]
Reduced Graphene Oxide	Sulfonamide	Adsorption	Recoveries range from 89.1 and 101.7%	Wu et al. [96]
Graphene Oxide	Nalidixic acid	Adsorption	Adsorption capacity 277.79 mg/g (Removal efficiency 92%)	Radmehr et al. [84]
Graphene	Tetracycline	Adsorption	Removal efficiency 93.3%	Behzadi et al. [97]
Graphene	Tetracycline	Adsorption	NA	Kogut et al. [98]
Graphene Oxide	Amoxicillin and Tetracycline	Adsorption	Adsorption capacity 405 mg/g and 552 mg/g for Amoxicillin and Tetracycline, respectively	Rajapaksha et al. [99]
Graphene Oxide	Oxytetracycline	Adsorption	Adsorption capacity 180.240 mg/g (Removal efficiency > 90%)	Taleb et al. [100]
Graphene Oxide	Tetracycline	Adsorption	Removal efficiency 99.8%	AbuZaid et al. [101]
Graphene Oxide	Tetracycline	Adsorption	Adsorption capacity 434.78 mg/g (Removal efficiency 99.6%)	Shaker et al. [102]
Graphene	Sulfamethoxazole, Norfloxacin, Tetracycline, Flumequine	Catalytic activation of inorganic peroxides	NA	Solís et al. [88]
Graphene	Benzylpenicillin sodium	Electrochemical oxidation	Removal efficiency 99.34%	Peng et al. [86]

Thus, GBMs play a pivotal role in removing antibiotics from wastewater through various mechanisms, including photocatalytic degradation, adsorption, catalytic activation of inorganic peroxides, and electrochemical oxidation. In photocatalysis, graphene enhances the efficiency of semiconductor catalysts by improving charge separation and increasing the generation of ROS, which degrade antibiotics. Its high surface area and functional groups enable strong adsorption of antibiotic molecules, facilitating their removal. Additionally, graphene acts as a catalyst in the activation of inorganic peroxides, generating ROS through Fenton-like reactions that break down antibiotics. In electrochemical oxidation, the excellent conductivity of graphene promotes direct electron transfer and ROS formation, further enhancing antibiotic degradation. These diverse functions make graphene materials highly effective for antibiotic removal, although challenges remain in scaling their use for practical applications.

Antibiotic removal from wastewater is a critical environmental challenge, and two promising approaches, adsorption, and photocatalysis, leverage the unique properties of graphene to address this issue effectively. Adsorption offers a straightforward method for removing antibiotics by capturing contaminants on the graphene surface, benefiting from the high surface area and tunable chemical properties of functionalized graphene. This method is highly effective, especially for a wide range of antibiotics, and allows for simple operation without requiring light sources or complex equipment. However, adsorption has limitations, such as potential saturation of the material, which reduces its effectiveness over time and necessitates regeneration processes that may degrade the material's performance.

Photocatalysis, on the other hand, introduces an additional advantage by degrading antibiotics rather than merely trapping them, breaking them down into less harmful substances using light energy. Graphene combined with metal oxides, like TiO_2 or ZnO, enhances this process due to improved electron transfer, which increases photocatalytic efficiency. Despite this, photocatalysis requires external energy input (light) and specific conditions to function optimally, which can limit its application in large-scale or low-resource settings. Moreover, the potential release of toxic byproducts during degradation is a concern. Thus, while both methods offer significant potential for antibiotic removal, their practical implementation should carefully consider these performance trade-offs.

5. Challenges, Potential Solutions and Mitigation Strategies

The application of advanced graphene-based technologies for antibiotic removal from wastewater faces several challenges. Graphene materials, despite their remarkable properties, often face issues related to scalability [103] and cost-effectiveness [104]. The synthesis and functionalization of GBMs can be complex and expensive, limiting their widespread adoption [105]. Additionally, the removal efficiency of these technologies can be influenced by the presence of other contaminants and the variability in wastewater composition. The effectiveness of graphene-based systems in degrading or adsorbing antibiotics may vary depending on the specific antibiotic and its chemical properties, as well as the operational conditions [106]. Furthermore, potential concerns regarding the stability and longevity of graphene materials in real-world wastewater treatment scenarios need to be addressed to ensure their practical applicability [107].

To address these challenges, research into optimizing the synthesis and functionalization processes of graphene-based materials is essential. The development of cost-effective and scalable methods for producing graphene and its composites could increase their viability for large-scale wastewater treatment applications [108]. Advanced techniques, such as doping [109] or facile functionalization [110], can be employed to tailor graphene materials for improved performance in removing specific antibiotics. Additionally, integrating graphene-based technologies with other advanced treatment methods, such as electrochemical processes [86], could increase their efficiency and broaden their applicability. Collaborative research efforts and innovations in materials science are crucial for overcoming current limitations and maximizing the potential of graphene-based technologies in wastewater treatment.

Effective strategies for overcoming the challenges associated with graphene-based technologies involve a multipronged approach. Regulatory support and funding for research into graphene-based water treatment technologies can accelerate the development of cost-effective solutions [111]. Public and private sector collaborations can facilitate the commercialization of these technologies by addressing scale-up challenges and integrating them into existing treatment infrastructure. Additionally, ongoing research should focus on developing robust methods for evaluating the performance and stability of graphene-based systems in diverse wastewater conditions. Implementing pilot-scale studies and real-world applications will provide valuable insights into the practical challenges and performance of these technologies, guiding future improvements and ensuring their successful deployment in wastewater treatment systems.

GBMs have demonstrated remarkable efficacy in removing antibiotics from wastewater, offering a promising solution to this pressing environmental issue [112]. However, given the current regulatory landscape, specifically the maximum allowable concentration (MAC) values for antibiotics in wastewater, alternative methods that are more cost-effective and can still meet effluent quality standards are often preferred. According to environmental protection agencies, the MAC values for antibiotics in wastewater generally range between 0.01 and 1 μg/L, depending on the antibiotic type and the environmental impact [113].

While alternatives such as activated carbon [114] or biological treatments [115] may meet these regulatory requirements at a lower cost, graphene-based technologies are emerging as viable long-term solutions. Compared with traditional materials, the unique properties of graphene, such as a high surface area, excellent adsorptive capabilities, and the potential for functionalization, provide superior antibiotic removal efficiency [116].

The future direction of wastewater treatment could lean heavily towards graphene-based materials, especially as the costs of production decrease [117] and regulations potentially tighten to require more advanced removal techniques. As more research substantiates the long-term environmental safety and effectiveness of these materials [118], they may become the preferred choice for meeting stringent effluent standards. Furthermore, the multifunctionality of graphene, such as its use in catalytic degradation and its potential for regeneration, makes it a sustainable and scalable solution for antibiotic removal in wastewater [112].

GBMs adsorb antibiotics through various interaction mechanisms, which include π–π stacking [119], hydrogen bonding [120], electrostatic interactions [121], van der Waals forces [120], and hydrophobic interactions [120]. π–π stacking occurs between antibiotics with aromatic rings and the delocalized π–electron system of graphene. Hydrogen bonding forms between functional groups in antibiotics and oxygen-containing groups on graphene oxide, whereas electrostatic interactions depend on the charge of both antibiotics and functionalized graphene surfaces and are influenced by pH. Van der Waals forces contribute to the weaker, nonspecific adsorption and hydrophobic interactions that occur between the hydrophobic portions of antibiotics and the graphene surface. These diverse mechanisms enable GBMs to efficiently adsorb a wide range of antibiotics, making them highly effective in wastewater treatment applications.

GBMs are highly effective at removing antibiotics from wastewater because of their large surface area and versatile adsorption mechanisms. However, despite their efficiency, several limitations arise after their use, particularly concerning environmental and health risks. Key challenges include the regeneration of these materials for reuse, potential secondary contamination, and risks posed by the release of nanomaterials into the environment. Without proper management, these issues could undermine the environmental benefits of using GBMs in wastewater treatment.

To mitigate these risks, several strategies are being explored. Safe disposal methods, such as encapsulating GBMs to prevent antibiotic and nanomaterial leakage, are essential for preventing secondary pollution. Improved regeneration technologies, such as green chemical processes or thermal treatments, can extend the life of GBMs without compromising their adsorption capacity [122]. Research into biodegradable graphene composites also offers a future solution for reducing the long-term risks associated with nanomaterials. These approaches are critical for ensuring that GBMs remain a safe and sustainable option for wastewater treatment.

6. Conclusions

In conclusion, the increasing presence of antibiotics in wastewater presents significant environmental challenges, particularly in fostering the spread of antimicrobial resistance and impacting aquatic ecosystems. The advancements in graphene-based technologies from 2016–2024 have demonstrated substantial potential in addressing these issues, particularly through the utilization of GO, rGO, and various graphene composites. These materials have exhibited remarkable adsorption capacities and photocatalytic degradation efficiencies, particularly when functionalized or integrated with other substances such as metal oxides and polymers. Despite the challenges related to scalability, cost, and environmental impact, the progress in this field underscores the promising role of GBMs' ineffective antibiotic removal from wastewater. Future efforts should focus on optimizing these technologies for large-scale applications, ensuring their sustainability, and integrating them into broader environmental management practices. These advancements constitute critical steps toward safer and more sustainable approaches to managing wastewater contamination.

Funding: This research received no external funding.

Conflicts of Interest: The authors declare no conflict of interest.

References

1. Samal, K.; Mahapatra, S.; Hibzur Ali, M. Pharmaceutical Wastewater as Emerging Contaminants (EC): Treatment Technologies, Impact on Environment and Human Health. *Energy Nexus* **2022**, *6*, 100076. [CrossRef]
2. Sambaza, S.S.; Naicker, N. Contribution of Wastewater to Antimicrobial Resistance: A Review Article. *J. Glob. Antimicrob. Resist.* **2023**, *34*, 23–29. [CrossRef] [PubMed]
3. Kayode-Afolayan, S.D.; Ahuekwe, E.F.; Nwinyi, O.C. Impacts of Pharmaceutical Effluents on Aquatic Ecosystems. *Sci. Afr.* **2022**, *17*, e01288. [CrossRef]
4. Kumar, S.B.; Arnipalli, S.R.; Ziouzenkova, O. Antibiotics in Food Chain: The Consequences for Antibiotic Resistance. *Antibiotics* **2020**, *9*, 688. [CrossRef]
5. Zinicovscaia, I. Conventional Methods of Wastewater Treatment. In *Cyanobacteria for Bioremediation of Wastewaters*; Zinicovscaia, I., Cepoi, L., Eds.; Springer International Publishing: Cham, Switzerland, 2016; pp. 17–25. ISBN 978-3-319-26751-7.
6. Sengupta, J.; Hussain, C.M. Sensitive and Selective Detection of Heavy Metal Ions and Organic Pollutants with Graphene-Integrated Sensing Platforms. *Nanoscale* **2024**, *16*, 14195–14212. [CrossRef]
7. Avornyo, A.; Chrysikopoulos, C.V. Applications of Graphene Oxide (GO) in Oily Wastewater Treatment: Recent Developments, Challenges, and Opportunities. *J. Environ. Manag.* **2024**, *353*, 120178. [CrossRef]
8. Wang, X.; Yin, R.; Zeng, L.; Zhu, M. A Review of Graphene-Based Nanomaterials for Removal of Antibiotics from Aqueous Environments. *Environ. Pollut.* **2019**, *253*, 100–110. [CrossRef]
9. Page, M.J.; McKenzie, J.E.; Bossuyt, P.M.; Boutron, I.; Hoffmann, T.C.; Mulrow, C.D.; Shamseer, L.; Tetzlaff, J.M.; Akl, E.A.; Brennan, S.E.; et al. The PRISMA 2020 Statement: An Updated Guideline for Reporting Systematic Reviews. *BMJ* **2021**, *372*, n71. [CrossRef]
10. Novoselov, K.S.; Geim, A.K.; Morozov, S.V.; Jiang, D.; Zhang, Y.; Dubonos, S.V.; Grigorieva, I.V.; Firsov, A.A. Electric Field Effect in Atomically Thin Carbon Films. *Science* **2004**, *306*, 666–669. [CrossRef]
11. Tiwari, S.K.; Sahoo, S.; Wang, N.; Huczko, A. Graphene Research and Their Outputs: Status and Prospect. *J. Sci. Adv. Mater. Devices* **2020**, *5*, 10–29. [CrossRef]
12. Olabi, A.G.; Abdelkareem, M.A.; Wilberforce, T.; Sayed, E.T. Application of Graphene in Energy Storage Device—A Review. *Renew. Sustain. Energy Rev.* **2021**, *135*, 110026. [CrossRef]
13. Aykaç, A.; Gergeroglu, H.; Beşli, B.; Akkaş, E.Ö.; Yavaş, A.; Güler, S.; Güneş, F.; Erol, M. An Overview on Recent Progress of Metal Oxide/Graphene/CNTs-Based Nanobiosensors. *Nanoscale Res. Lett.* **2021**, *16*, 65. [CrossRef]
14. Tarhini, A.; Tehrani-Bagha, A.R. Advances in Preparation Methods and Conductivity Properties of Graphene-Based Polymer Composites. *Appl. Compos. Mater.* **2023**, *30*, 1737–1762. [CrossRef]
15. Mazilova, T.I.; Sadanov, E.V.; Mikhailovskij, I.M. Tensile Strength of Graphene Nanoribbons: An Experimental Approach. *Mater. Lett.* **2019**, *242*, 17–19. [CrossRef]
16. Liu, D.; Fu, H.; Yang, T.; Wang, W.; Zhao, J.; Wu, K.; Wu, C.; Yong, Z.; Zhang, Y. A Modified Spin-Casting Approach for Scalable Preparation of Ultra-Thick Reduced Graphene Oxide Films with High Thermal Conductivity. *Mater. Res. Express* **2022**, *9*, 036405. [CrossRef]
17. Sun, B.; Pang, J.; Cheng, Q.; Zhang, S.; Li, Y.; Zhang, C.; Sun, D.; Ibarlucea, B.; Li, Y.; Chen, D.; et al. Synthesis of Wafer-Scale Graphene with Chemical Vapour Deposition for Electronic Device Applications. *Adv. Mater. Technol.* **2021**, *6*, 2000744. [CrossRef]
18. Kumar, A.; Sharma, K.; Dixit, A.R. Tensile, Flexural and Interlaminar Shear Strength of Carbon Fibre Reinforced Epoxy Composites Modified by Graphene. *Polym. Bull.* **2023**, *80*, 7469–7490. [CrossRef]
19. Xu, J.; Wang, L. Chapter 1—Carbon Nanomaterials. In *Nano-Inspired Biosensors for Protein Assay with Clinical Applications*; Li, G., Ed.; Elsevier: Amsterdam, The Netherlands, 2019; pp. 3–38. ISBN 978-0-12-815053-5.
20. Belyaeva, L.A.; Schneider, G.F. Wettability of Graphene. *Surf. Sci. Rep.* **2020**, *75*, 100482. [CrossRef]
21. Liu, Z.; Navik, R.; Tan, H.; Xiang, Q.; Wahyudiono; Goto, M.; Ibarra, R.M.; Zhao, Y. Graphene-Based Materials Prepared by Supercritical Fluid Technology and Its Application in Energy Storage. *J. Supercrit. Fluids* **2022**, *188*, 105672. [CrossRef]

22. Ivan, R.; Popescu, C.; Antohe, V.A.; Antohe, S.; Negrila, C.; Logofatu, C.; del Pino, A.P.; György, E. Iron Oxide/Hydroxide–Nitrogen Doped Graphene-like Visible-Light Active Photocatalytic Layers for Antibiotics Removal from Wastewater. *Sci. Rep.* **2023**, *13*, 2740. [CrossRef]
23. Utkan, G.; Yumusak, G.; Tunali, B.C.; Ozturk, T.; Turk, M. Production of Reduced Graphene Oxide by Using Three Different Microorganisms and Investigation of Their Cell Interactions. *ACS Omega* **2023**, *8*, 31188–31200. [CrossRef] [PubMed]
24. Gul, W.; Akbar Shah, S.R.; Khan, A.; Ahmad, N.; Ahmed, S.; Ain, N.; Mehmood, A.; Salah, B.; Ullah, S.S.; Khan, R. Synthesis of Graphene Oxide (GO) and Reduced Graphene Oxide (rGO) and Their Application as Nano-Fillers to Improve the Physical and Mechanical Properties of Medium Density Fiberboard. *Front. Mater.* **2023**, *10*, 1206918. [CrossRef]
25. Morimoto, N.; Kubo, T.; Nishina, Y. Tailoring the Oxygen Content of Graphite and Reduced Graphene Oxide for Specific Applications. *Sci. Rep.* **2016**, *6*, 21715. [CrossRef]
26. Ashan, M.; Alharbi, F.F.; Aman, S.; Al-Sehemi, A.G.; Ahmad, K.; Aslam, M.; Saleem, M.I. Elevating Capacitive Features of Hydrothermally Produced CeVS3 Utilizing rGO for Supercapacitor Applications. *Diam. Relat. Mater.* **2024**, *147*, 111319. [CrossRef]
27. Tan, B.; Han, S.; Luo, W.; Chao, Z.; Fan, J.; Wang, M. Synthesis of RGO-Supported Layered MoS2 with Enhanced Electrochemical Performance for Aluminum Ion Batteries. *J. Alloys Compd.* **2020**, *841*, 155732. [CrossRef]
28. Habte, A.T.; Ayele, D.W. Synthesis and Characterization of Reduced Graphene Oxide (rGO) Started from Graphene Oxide (GO) Using the Tour Method with Different Parameters. *Adv. Mater. Sci. Eng.* **2019**, *2019*, 5058163. [CrossRef]
29. Yan, Y.; Chen, J.; Li, N.; Tian, J.; Li, K.; Jiang, J.; Liu, J.; Tian, Q.; Chen, P. Systematic Bandgap Engineering of Graphene Quantum Dots and Applications for Photocatalytic Water Splitting and CO2 Reduction. *ACS Nano* **2018**, *12*, 3523–3532. [CrossRef]
30. Rai, D.; Jaiswal, Y.; Sinha, S. Graphene Quantum Dots Synthesis Using Waste Unburnt Carbon: Implications for Optoelectronics. *Appl. Surf. Sci.* **2024**, *653*, 159386. [CrossRef]
31. Rasheed, P.A.; Ankitha, M.; Pillai, V.K.; Alwarappan, S. Graphene Quantum Dots for Biosensing and Bioimaging. *RSC Adv.* **2024**, *14*, 16001–16023. [CrossRef]
32. Chen, Y.-L.; Kurniawan, D.; Tsai, M.-D.; Chang, J.-W.; Chang, Y.-N.; Yang, S.-C.; Chiang, W.-H.; Kung, C.-W. Two-Dimensional Metal–Organic Framework for Post-Synthetic Immobilization of Graphene Quantum Dots for Photoluminescent Sensing. *Commun. Chem.* **2024**, *7*, 108. [CrossRef]
33. El-Mahalawy, A.M.; Abbas, W.; Mostafa, O.; Zidan, N.A.; Issa, H.H.; Fedawy, M.; Wassel, A.R. Integrative Role of PEDOT: PSS in Adjusting the Photoresponse Efficiency of Novel Reduced Graphene Quantum Dots/Silicon Heterojunction for Optoelectronics and Solar Energy Conversion Applications. *Surf. Interfaces* **2024**, *46*, 103946. [CrossRef]
34. Cui, Y.; Liu, L.; Shi, M.; Wang, Y.; Meng, X.; Chen, Y.; Huang, Q.; Liu, C. A Review of Advances in Graphene Quantum Dots: From Preparation and Modification Methods to Application. *C* **2024**, *10*, 7. [CrossRef]
35. Wang, X.-Y.; Narita, A.; Müllen, K. Precision Synthesis versus Bulk-Scale Fabrication of Graphenes. *Nat. Rev. Chem.* **2017**, *2*, 1–10. [CrossRef]
36. Sohrabi, B.; Poorsargol, M.; Ingram, S.; Roudsari, G. Surfactants-Emerging Amphiphiles in Liquid Phase Exfoliation Method for Dispersing Carbon Nanotubes and Graphene: Experimental and Molecular Dynamics Simulation Studies. *J. Mol. Liq.* **2024**, *409*, 125493. [CrossRef]
37. Gharib, D.H.; Gietman, S.; Malherbe, F.; Moulton, S.E. High Yield, Solid Exfoliation and Liquid Dispersion of Graphite Driven by a Donor-Acceptor Interaction. *Carbon* **2017**, *123*, 695–707. [CrossRef]
38. Sandhya, M.; Ramasamy, D.; Sudhakar, K.; Kadirgama, K.; Harun, W.S.W. Ultrasonication an Intensifying Tool for Preparation of Stable Nanofluids and Study the Time Influence on Distinct Properties of Graphene Nanofluids—A Systematic Overview. *Ultrason. Sonochem.* **2021**, *73*, 105479. [CrossRef]
39. Biranje, P.M.; Patwardhan, A.W.; Joshi, J.B.; Dasgupta, K. Exfoliated Graphene and Its Derivatives from Liquid Phase and Their Role in Performance Enhancement of Epoxy Matrix Composite. *Compos. Part A Appl. Sci. Manuf.* **2022**, *156*, 106886. [CrossRef]
40. Tyurnina, A.V.; Tzanakis, I.; Morton, J.; Mi, J.; Porfyrakis, K.; Maciejewska, B.M.; Grobert, N.; Eskin, D.G. Ultrasonic Exfoliation of Graphene in Water: A Key Parameter Study. *Carbon* **2020**, *168*, 737–747. [CrossRef]
41. Xu, Y.; Cao, H.; Xue, Y.; Li, B.; Cai, W. Liquid-Phase Exfoliation of Graphene: An Overview on Exfoliation Media, Techniques, and Challenges. *Nanomaterials* **2018**, *8*, 942. [CrossRef]
42. Chen, X.; Qu, Z.; Liu, Z.; Ren, G. Mechanism of Oxidization of Graphite to Graphene Oxide by the Hummers Method. *ACS Omega* **2022**, *7*, 23503–23510. [CrossRef]
43. Li, C.; Chen, X.; Shen, L.; Bao, N. Revisiting the Oxidation of Graphite: Reaction Mechanism, Chemical Stability, and Structure Self-Regulation. *ACS Omega* **2020**, *5*, 3397–3404. [CrossRef] [PubMed]
44. Guex, L.G.; Sacchi, B.; Peuvot, K.F.; Andersson, R.L.; Pourrahimi, A.M.; Ström, V.; Farris, S.; Olsson, R.T. Experimental Review: Chemical Reduction of Graphene Oxide (GO) to Reduced Graphene Oxide (rGO) by Aqueous Chemistry. *Nanoscale* **2017**, *9*, 9562–9571. [CrossRef] [PubMed]
45. Hidayat, R.; Wahyuningsih, S.; Ramelan, A.H. Simple Synthesis of rGO (Reduced Graphene Oxide) by Thermal Reduction of GO (Graphene Oxide). *IOP Conf. Ser. Mater. Sci. Eng.* **2020**, *858*, 012009. [CrossRef]
46. Zhou, A.; Bai, J.; Hong, W.; Bai, H. Electrochemically Reduced Graphene Oxide: Preparation, Composites, and Applications. *Carbon* **2022**, *191*, 301–332. [CrossRef]

47. Liu, W.; Speranza, G. Tuning the Oxygen Content of Reduced Graphene Oxide and Effects on Its Properties. *ACS Omega* **2021**, *6*, 6195–6205. [CrossRef]
48. Ahmad, R.T.M.; Hong, S.-H.; Shen, T.-Z.; Song, J.-K. Water-Assisted Stable Dispersal of Graphene Oxide in Non-Dispersible Solvents and Skin Formation on the GO Dispersion. *Carbon* **2016**, *98*, 188–194. [CrossRef]
49. Mohan, V.B.; Lau, K.; Hui, D.; Bhattacharyya, D. Graphene-Based Materials and Their Composites: A Review on Production, Applications and Product Limitations. *Compos. Part B Eng.* **2018**, *142*, 200–220. [CrossRef]
50. Phiri, J.; Johansson, L.-S.; Gane, P.; Maloney, T. A Comparative Study of Mechanical, Thermal and Electrical Properties of Graphene-, Graphene Oxide- and Reduced Graphene Oxide-Doped Microfibrillated Cellulose Nanocomposites. *Compos. Part B Eng.* **2018**, *147*, 104–113. [CrossRef]
51. Yi, M.; Shen, Z. A Review on Mechanical Exfoliation for the Scalable Production of Graphene. *J. Mater. Chem. A* **2015**, *3*, 11700–11715. [CrossRef]
52. Cao, X.; Jiang, C.; Tan, D.; Li, Q.; Bi, S.; Song, J. Recent Mechanical Processing Techniques of Two-Dimensional Layered Materials: A Review. *J. Sci. Adv. Mater. Devices* **2021**, *6*, 135–152. [CrossRef]
53. Madurani, K.A.; Suprapto, S.; Machrita, N.I.; Bahar, S.L.; Illiya, W.; Kurniawan, F. Progress in Graphene Synthesis and Its Application: History, Challenge and the Future Outlook for Research and Industry. *ECS J. Solid. State Sci. Technol.* **2020**, *9*, 093013. [CrossRef]
54. Norimatsu, W. A Review on Carrier Mobilities of Epitaxial Graphene on Silicon Carbide. *Materials* **2023**, *16*, 7668. [CrossRef]
55. Han, D.; Wang, X.; Zhao, Y.; Chen, Y.; Tang, M.; Zhao, Z. High-Quality Graphene Synthesis on Amorphous SiC through a Rapid Thermal Treatment. *Carbon* **2017**, *124*, 105–110. [CrossRef]
56. Huang, H.; Chen, S.; Wee, A.T.S.; Chen, W. 1—Epitaxial Growth of Graphene on Silicon Carbide (SiC). In *Graphene*; Skákalová, V., Kaiser, A.B., Eds.; Woodhead Publishing: Cambridge, UK, 2014; pp. 3–26. ISBN 978-0-85709-508-4.
57. Zhao, J.; Ji, P.; Li, Y.; Li, R.; Zhang, K.; Tian, H.; Yu, K.; Bian, B.; Hao, L.; Xiao, X.; et al. Ultrahigh-Mobility Semiconducting Epitaxial Graphene on Silicon Carbide. *Nature* **2024**, *625*, 60–65. [CrossRef] [PubMed]
58. Endoh, N.; Akiyama, S.; Tashima, K.; Suwa, K.; Kamogawa, T.; Kohama, R.; Funakubo, K.; Konishi, S.; Mogi, H.; Kawahara, M.; et al. High-Quality Few-Layer Graphene on Single-Crystalline SiC Thin Film Grown on Affordable Wafer for Device Applications. *Nanomaterials* **2021**, *11*, 392. [CrossRef]
59. Pedrazzetti, L.; Gibertini, E.; Bizzoni, F.; Russo, V.; Lucotti, A.; Nobili, L.; Magagnin, L. Graphene Growth on Electroformed Copper Substrates by Atmospheric Pressure CVD. *Materials* **2022**, *15*, 1572. [CrossRef]
60. Lee, D.Y.; Nam, J.; Lee, G.Y.; Lee, I.; Jang, A.-R.; Kim, K.S. Conveyor CVD to High-Quality and Productivity of Large-Area Graphene and Its Potentiality. *Nano Converg.* **2024**, *11*, 32. [CrossRef]
61. Saeed, M.; Alshammari, Y.; Majeed, S.A.; Al-Nasrallah, E. Chemical Vapour Deposition of Graphene—Synthesis, Characterization, and Applications: A Review. *Molecules* **2020**, *25*, 3856. [CrossRef]
62. Dhiman, G.; Kumar, S.; Kumar, R.; Brajpuriya, R. An Improved CVD Design for Graphene Growth and Transfer Improvements. *J. Electron. Mater.* **2024**, *53*, 5916–5925. [CrossRef]
63. Yan, Z.; Wang, S.; Chen, X.; Poduval, G.K.; Stride, J.A. Optimizing the Seeded CVD-Growth of Uniform Graphene Films on Silicon. *J. Mater. Sci.* **2023**, *58*, 9434–9445. [CrossRef]
64. Ullah, S.; Yang, X.; Ta, H.Q.; Hasan, M.; Bachmatiuk, A.; Tokarska, K.; Trzebicka, B.; Fu, L.; Rummeli, M.H. Graphene Transfer Methods: A Review. *Nano Res.* **2021**, *14*, 3756–3772. [CrossRef]
65. Kraemer, S.A.; Ramachandran, A.; Perron, G.G. Antibiotic Pollution in the Environment: From Microbial Ecology to Public Policy. *Microorganisms* **2019**, *7*, 180. [CrossRef] [PubMed]
66. Cook, M.A.; Wright, G.D. The Past, Present, and Future of Antibiotics. *Sci. Transl. Med.* **2022**, *14*, eabo7793. [CrossRef]
67. Caneschi, A.; Bardhi, A.; Barbarossa, A.; Zaghini, A. The Use of Antibiotics and Antimicrobial Resistance in Veterinary Medicine, a Complex Phenomenon: A Narrative Review. *Antibiotics* **2023**, *12*, 487. [CrossRef] [PubMed]
68. Batuman, O.; Britt-Ugartemendia, K.; Kunwar, S.; Yilmaz, S.; Fessler, L.; Redondo, A.; Chumachenko, K.; Chakravarty, S.; Wade, T. The Use and Impact of Antibiotics in Plant Agriculture: A Review. *Phytopathology* **2024**, *114*, 885–909. [CrossRef]
69. Tian, L.; Fang, G.; Li, G.; Li, L.; Zhang, T.; Mao, Y. Metagenomic Approach Revealed the Mobility and Co-Occurrence of Antibiotic Resistomes between Non-Intensive Aquaculture Environment and Human. *Microbiome* **2024**, *12*, 107. [CrossRef]
70. Yang, Y.; Ji, Y.; Gao, Y.; Lin, Z.; Lin, Y.; Lu, Y.; Zhang, L. Antibiotics and Antimycotics in Waste Water Treatment Plants: Concentrations, Removal Efficiency, Spatial and Temporal Variations, Prediction, and Ecological Risk Assessment. *Environ. Res.* **2022**, *215*, 114135. [CrossRef]
71. de Ilurdoz, M.S.; Sadhwani, J.J.; Reboso, J.V. Antibiotic Removal Processes from Water & Wastewater for the Protection of the Aquatic Environment—A Review. *J. Water Process Eng.* **2022**, *45*, 102474. [CrossRef]
72. Polianciuc, S.I.; Gurzău, A.E.; Kiss, B.; Ștefan, M.G.; Loghin, F. Antibiotics in the Environment: Causes and Consequences. *Med. Pharm. Rep.* **2020**, *93*, 231–240. [CrossRef]
73. Samrot, A.V.; Wilson, S.; Sanjay Preeth, R.S.; Prakash, P.; Sathiyasree, M.; Saigeetha, S.; Shobana, N.; Pachiyappan, S.; Rajesh, V.V. Sources of Antibiotic Contamination in Wastewater and Approaches to Their Removal—An Overview. *Sustainability* **2023**, *15*, 12639. [CrossRef]

74. Robles-Jimenez, L.E.; Aranda-Aguirre, E.; Castelan-Ortega, O.A.; Shettino-Bermudez, B.S.; Ortiz-Salinas, R.; Miranda, M.; Li, X.; Angeles-Hernandez, J.C.; Vargas-Bello-Pérez, E.; Gonzalez-Ronquillo, M. Worldwide Traceability of Antibiotic Residues from Livestock in Wastewater and Soil: A Systematic Review. *Animals* **2022**, *12*, 60. [CrossRef] [PubMed]
75. Gros, M.; Catalán, N.; Mas-Pla, J.; Čelić, M.; Petrović, M.; Farré, M.J. Groundwater Antibiotic Pollution and Its Relationship with Dissolved Organic Matter: Identification and Environmental Implications. *Environ. Pollut.* **2021**, *289*, 117927. [CrossRef]
76. Suyamud, B.; Chen, Y.; Quyen, D.T.T.; Dong, Z.; Zhao, C.; Hu, J. Antimicrobial Resistance in Aquaculture: Occurrence and Strategies in Southeast Asia. *Sci. Total Environ.* **2024**, *907*, 167942. [CrossRef] [PubMed]
77. Rayan, R.A. Pharmaceutical Effluent Evokes Superbugs in the Environment: A Call to Action. *Biosaf. Health* **2023**, *5*, 363–371. [CrossRef]
78. Thai, P.K.; Ky, L.X.; Binh, V.N.; Nhung, P.H.; Nhan, P.T.; Hieu, N.Q.; Dang, N.T.T.; Tam, N.K.B.; Anh, N.T.K. Occurrence of Antibiotic Residues and Antibiotic-Resistant Bacteria in Effluents of Pharmaceutical Manufacturers and Other Sources around Hanoi, Vietnam. *Sci. Total Environ.* **2018**, *645*, 393–400. [CrossRef] [PubMed]
79. Singh, B.J.; Chakraborty, A.; Sehgal, R. A Systematic Review of Industrial Wastewater Management: Evaluating Challenges and Enablers. *J. Environ. Manag.* **2023**, *348*, 119230. [CrossRef]
80. Singh, S.; Sharma, P.; Pal, N.; Sarma, D.K.; Kumar, M. Antibiotic Disposal Challenges in India: Investigating Causes and Effects. *Environ. Monit. Assess.* **2024**, *196*, 325. [CrossRef]
81. Blair, J.M.A.; Webber, M.A.; Baylay, A.J.; Ogbolu, D.O.; Piddock, L.J.V. Molecular Mechanisms of Antibiotic Resistance. *Nat. Rev. Microbiol.* **2015**, *13*, 42–51. [CrossRef] [PubMed]
82. Grenni, P.; Ancona, V.; Barra Caracciolo, A. Ecological Effects of Antibiotics on Natural Ecosystems: A Review. *Microchem. J.* **2018**, *136*, 25–39. [CrossRef]
83. Ben, Y.; Fu, C.; Hu, M.; Liu, L.; Wong, M.H.; Zheng, C. Human Health Risk Assessment of Antibiotic Resistance Associated with Antibiotic Residues in the Environment: A Review. *Environ. Res.* **2019**, *169*, 483–493. [CrossRef]
84. Radmehr, S.; Hosseini Sabzevari, M.; Ghaedi, M.; Ahmadi Azqhandi, M.H.; Marahel, F. Adsorption of Nalidixic Acid Antibiotic Using a Renewable Adsorbent Based on Graphene Oxide from Simulated Wastewater. *J. Environ. Chem. Eng.* **2021**, *9*, 105975. [CrossRef]
85. Kumar, R.; Samykano, M.; Ngui, W.K.; Pandey, A.K.; Kalidasan, B.; Kadirgama, K.; Tyagi, V.V. Investigation of Thermal Performance and Chemical Stability of Graphene Enhanced Phase Change Material for Thermal Energy Storage. *Phys. Chem. Earth Parts A/B/C* **2022**, *128*, 103250. [CrossRef]
86. Peng, Y.; Yan, Y.; Ma, X.; Jiang, B.; Chen, R.; Feng, H.; Xia, Y. Efficient Electrochemical Oxidation of Antibiotic Wastewater Using a Graphene-Loaded PbO2 Membrane Anode: Mechanisms and Applications. *Environ. Res.* **2024**, *259*, 119517. [CrossRef]
87. Karaolia, P.; Michael-Kordatou, I.; Hapeshi, E.; Drosou, C.; Bertakis, Y.; Christofilos, D.; Armatas, G.S.; Sygellou, L.; Schwartz, T.; Xekoukoulotakis, N.P.; et al. Removal of Antibiotics, Antibiotic-Resistant Bacteria and Their Associated Genes by Graphene-Based TiO_2 Composite Photocatalysts under Solar Radiation in Urban Wastewaters. *Appl. Catal. B Environ.* **2018**, *224*, 810–824. [CrossRef]
88. Solís, R.R.; Dinc, Ö.; Fang, G.; Nadagouda, M.N.; Dionysiou, D.D. Activation of Inorganic Peroxides with Magnetic Graphene for the Removal of Antibiotics from Wastewater. *Environ. Sci. Nano* **2021**, *8*, 960–977. [CrossRef]
89. Dickson, K.; Yeung, C.A. PRISMA 2020 Updated Guideline. *Br Dent J* **2022**, *232*, 760–761. [CrossRef]
90. López-Sánchez, J.A.; Patiño-Vanegas, J.C.; Valencia-Arias, A.; Valencia, J. Use and Adoption of ICTs Oriented to University Student Learning: Systematic Review Using PRISMA Methodology. *Cogent Educ.* **2023**, *10*, 2288490. [CrossRef]
91. Scopus | Abstract and Citation Database | Elsevier. Available online: https://www.elsevier.com/en-in/products/scopus (accessed on 24 September 2024).
92. Haddaway, N.R.; Page, M.J.; Pritchard, C.C.; McGuinness, L.A. PRISMA2020: An R Package and Shiny App for Producing PRISMA 2020-Compliant Flow Diagrams, with Interactivity for Optimized Digital Transparency and Open Synthesis. *Campbell Syst. Rev.* **2022**, *18*, e1230. [CrossRef] [PubMed]
93. Tabatabai-Yazdi, F.-S.; Ebrahimian Pirbazari, A.; Esmaeili Khalilsaraei, F.; Asasian Kolur, N.; Gilani, N. Photocatalytic Treatment of Tetracycline Antibiotic Wastewater by Silver/TiO2 Nanosheets/Reduced Graphene Oxide and Artificial Neural Network Modelling. *Water Environ. Res.* **2020**, *92*, 662–676. [CrossRef]
94. Yang, W.; Wang, Y. Enhanced Electron and Mass Transfer Flow-through Cell with C3N4-MoS2 Supported on Three-Dimensional Graphene Photoanode for the Removal of Antibiotic and Antibacterial Potencies in Ampicillin Wastewater. *Appl. Catal. B Environ.* **2021**, *282*, 119574. [CrossRef]
95. Lin, X.; Zhao, J.; Zhang, Y.; Li, Y.; Liao, Y.; Zhang, H. Graphene Oxide Structure-Oriented NM88B/GO/SA Aerogel for Highly Efficient Degradation of Dye and Antibiotic Wastewater. *J. Polym. Environ.* **2024**, *32*, 2091–2104. [CrossRef]
96. Wu, J.; Zhao, H.; Chen, R.; Pham-Huy, C.; Hui, X.; He, H. Adsorptive Removal of Trace Sulfonamide Antibiotics by Water-Dispersible Magnetic Reduced Graphene Oxide-Ferrite Hybrids from Wastewater. *J. Chromatogr. B* **2016**, *1029–1030*, 106–112. [CrossRef]
97. Behzadi, A.; Yazdanbakhsh, A. Synthesis and Characterization of Modified Resorcinol Formaldehyde Aerogel by Graphene/m-Phenylenediamine as a Novel Adsorbent to Remove Tetracycline Antibiotics from Wastewater. *J. Water Environ. Nanotechnol.* **2022**, *7*, 44–54. [CrossRef]

98. Kogut, I.; Armbruster, F.; Polak, D.; Kaur, S.; Hussy, S.; Thiem, T.; Gerhardts, A.; Szwast, M. Antibacterial, Antifungal, and Antibiotic Adsorption Properties of Graphene-Modified Nonwoven Materials for Application in Wastewater Treatment Plants. *Processes* **2022**, *10*, 2051. [CrossRef]
99. Rajapaksha, P.; Orrell-Trigg, R.; Truong, Y.B.; Cozzolino, D.; Truong, V.K.; Chapman, J. Wastewater Depollution of Textile Dyes and Antibiotics Using Unmodified and Copper Oxide/Zinc Oxide Nanofunctionalised Graphene Oxide Materials. *Environ. Sci. Adv.* **2022**, *1*, 456–469. [CrossRef]
100. Taleb, M.A.; Kumar, R.; Barakat, M.A. Multifunctional Carboxymethyl Cellulose/Graphene Oxide/Polyaniline Hybrid Thin Film for Adsorptive Removal of Cu(II) and Oxytetracycline Antibiotic from Wastewater. *Int. J. Biol. Macromol.* **2023**, *253*, 126699. [CrossRef] [PubMed]
101. AbuZaid, M.; Pandey, R.P.; Hasan, S.W. Efficient Antibiotic Remediation from Wastewater Using a Lamellar Free-Standing 2D Graphene Oxide/Ti3C2Tx Hybrid Membrane. *Surf. Interfaces* **2024**, *49*, 104431. [CrossRef]
102. Shaker, M.A.; Alshitari, W.H.; Basha, M.T.; Aly, N.A.; Asim, M.; Albishri, H.M.; Bhawani, S.A.; Yakout, A.A. Synergetic Impact of Copper Nanoparticles and Polyaniline Reinforced Graphene Oxide Nanocomposite on the Sequestration of Tetracycline Antibiotic from Milk and Wastewaters Samples. *Mater. Today Commun.* **2024**, *38*, 107869. [CrossRef]
103. Safian, M.T.; Umar, K.; Mohamad Ibrahim, M.N. Synthesis and Scalability of Graphene and Its Derivatives: A Journey towards Sustainable and Commercial Material. *J. Clean. Prod.* **2021**, *318*, 128603. [CrossRef]
104. Lv, H.; Du, M.; Li, Z.; Xiao, L.; Zhou, S. Cost Optimization of Graphene Oxide-Modified Ultra-High-Performance Concrete Based on Machine Learning Methods. *Inorganics* **2024**, *12*, 181. [CrossRef]
105. Xiao, Y.; Pang, Y.X.; Yan, Y.; Qian, P.; Zhao, H.; Manickam, S.; Wu, T.; Pang, C.H. Synthesis and Functionalization of Graphene Materials for Biomedical Applications: Recent Advances, Challenges, and Perspectives. *Adv. Sci.* **2023**, *10*, 2205292. [CrossRef] [PubMed]
106. Mohammed, H.; Kumar, A.; Bekyarova, E.; Al-Hadeethi, Y.; Zhang, X.; Chen, M.; Ansari, M.S.; Cochis, A.; Rimondini, L. Antimicrobial Mechanisms and Effectiveness of Graphene and Graphene-Functionalized Biomaterials. A Scope Review. *Front. Bioeng. Biotechnol.* **2020**, *8*, 465–486. [CrossRef] [PubMed]
107. Costinas, C.; Salagean, C.A.; Cotet, L.C.; Baia, M.; Todea, M.; Magyari, K.; Baia, L. Insights into the Stability of Graphene Oxide Aqueous Dispersions. *Nanomaterials* **2022**, *12*, 4489. [CrossRef]
108. Chen, S.; Wang, Q.; Liu, C.; Zhang, J.; Wang, L.; Luo, J.-L.; Fu, X.-Z. Roll-to-Roll Scale Fabrication of High-Performance Graphene-Assembled Film Cathode Current Collectors for Lithium-Ion Batteries. *ACS Sustain. Chem. Eng.* **2023**, *11*, 13483–13491. [CrossRef]
109. Su, L.; Ning, Y.; Ma, Z.; Zhang, Y.; Liu, C.; Zhang, Y.; Miao, L.; Zhou, J.; Wu, B.; Qian, J. Polypyrrole-Reinforced N,S-Doping Graphene Foam for Efficient Solar Purification of Wastewater. *Sol. RRL* **2021**, *5*, 2100210. [CrossRef]
110. Al-Qadri, A.A.Q.; Drmosh, Q.A.; Onaizi, S.A. Enhancement of Bisphenol a Removal from Wastewater via the Covalent Functionalization of Graphene Oxide with Short Amine Molecules. *Case Stud. Chem. Environ. Eng.* **2022**, *6*, 100233. [CrossRef]
111. Pan, X.; Ji, J.; Zhang, N.; Xing, M. Research Progress of Graphene-Based Nanomaterials for the Environmental Remediation. *Chin. Chem. Lett.* **2020**, *31*, 1462–1473. [CrossRef]
112. Rana, K.; Kaur, H.; Singh, N.; Sithole, T.; Siwal, S.S. Graphene-Based Materials: Unravelling Its Impact in Wastewater Treatment for Sustainable Environments. *Next Mater.* **2024**, *3*, 100107. [CrossRef]
113. Ågerstrand, M.; Josefsson, H.; Wernersson, A.-S.; Larsson, D.G.J. Opportunities to Tackle Antibiotic Resistance Development in the Aquatic Environment through the Water Framework Directive. *Ambio* **2023**, *52*, 941–951. [CrossRef]
114. Wang, J.; Wang, R.; Ma, J.; Sun, Y. Study on the Application of Shell-Activated Carbon for the Adsorption of Dyes and Antibiotics. *Water* **2022**, *14*, 3752. [CrossRef]
115. Lu, Z.; Liu, G.; Xie, H.; Zhai, Y.; Li, X. Advances and Solutions in Biological Treatment for Antibiotic Wastewater with Resistance Genes: A Review. *J. Environ. Manag.* **2024**, *368*, 122115. [CrossRef]
116. Kong, Q.; Shi, X.; Ma, W.; Zhang, F.; Yu, T.; Zhao, F.; Zhao, D.; Wei, C. Strategies to Improve the Adsorption Properties of Graphene-Based Adsorbent towards Heavy Metal Ions and Their Compound Pollutants: A Review. *J. Hazard. Mater.* **2021**, *415*, 125690. [CrossRef] [PubMed]
117. What Factors Impact Graphene Cost? (Updated 2024) | INN. Available online: https://investingnews.com/daily/tech-investing/nanoscience-investing/graphene-investing/graphene-cost/ (accessed on 28 September 2024).
118. Ding, X.; Pu, Y.; Tang, M.; Zhang, T. Environmental and Health Effects of Graphene-Family Nanomaterials: Potential Release Pathways, Transformation, Environmental Fate and Health Risks. *Nano Today* **2022**, *42*, 101379. [CrossRef]
119. Tang, R.; Gong, D.; Deng, Y.; Xiong, S.; Zheng, J.; Li, L.; Zhou, Z.; Su, L.; Zhao, J. π-π Stacking Derived from Graphene-like Biochar/g-C3N4 with Tunable Band Structure for Photocatalytic Antibiotics Degradation via Peroxymonosulfate Activation. *J. Hazard. Mater.* **2022**, *423*, 126944. [CrossRef] [PubMed]
120. Kern, M.; Škulj, S.; Rožman, M. Adsorption of a Wide Variety of Antibiotics on Graphene-Based Nanomaterials: A Modelling Study. *Chemosphere* **2022**, *296*, 134010. [CrossRef]

121. Yasmin, S.; Azam, M.G.; Hossain, M.S.; Akhtar, U.S.; Kabir, M.H. Efficient Removal of Ciprofloxacin from Aqueous Solution Using Zn–C Battery Derived Graphene Oxide Enhanced by Hydrogen Bonding, Electrostatic and π-π Interaction. *Heliyon* **2024**, *10*, e33317. [CrossRef]
122. Liu, Y.; Biswas, B.; Hassan, M.; Naidu, R. Green Adsorbents for Environmental Remediation: Synthesis Methods, Ecotoxicity, and Reusability Prospects. *Processes* **2024**, *12*, 1195. [CrossRef]

C Journal of Carbon Research

MDPI

Review

Research Progress in Graphene-Based Adsorbents for Wastewater Treatment: Preparation, Adsorption Properties and Mechanisms for Inorganic and Organic Pollutants

Guangqian Li [1], Ruiling Du [1], Zhanfang Cao [1], Changxin Li [2], Jianrong Xue [3], Xin Ma [1,*] and Shuai Wang [1,*]

[1] Hunan Provincial Key Laboratory of Efficient and Clean Utilization of Manganese Resources, College of Chemistry and Chemical Engineering, Central South University, Changsha 410083, China; 15567133352@163.com (G.L.); 13856761570@163.com (R.D.); zfcao1980@csu.edu.cn (Z.C.)

[2] College of Safety Science and Engineering, Nanjing Tech University, Nanjing 211816, China; lichangxin20160706@163.com

[3] College of Chemistry and Chemical Engineering, Hunan University of Science and Technology, Xiangtan 411201, China; jianrongxue@hnust.edu.cn

* Correspondence: maxin2013@csu.edu.cn (X.M.); wangshuai@csu.edu.cn (S.W.); Tel.: +86-731-88879616 (S.W. & X.M.)

Abstract: Graphene-based adsorbents show great potential for application in the field of environmental pollution treatment due to their unique two-dimensional structure, high specific surface area, and tunable surface chemistry. This paper reviews the research on the application of graphene and its derivatives as novel adsorbents in the field of wastewater treatment in recent years, focusing on the preparation and functionalization of graphene-based adsorbents, as well as their adsorption performance and mechanism of action in the removal of inorganic and organic pollutants, and provides an outlook on the future directions of the research on graphene-based adsorbents. The analysis in this paper focuses on the functionalization of graphene-based adsorbents by introducing magnetic particles, hybridization with other materials, and grafting with polymers. The modified graphene-based adsorbents showed significant adsorption and removal of pollutants and were easy to recycle and regenerate. The adsorption of pollutants on graphene-based adsorbents is mainly carried out through π–π interaction, hydrogen bonding, and electrostatic interaction, which is related to the structure of the pollutants. Future research directions on graphene-based adsorbents should focus on in-depth adsorption mechanism studies and the development of cost-effective graphene-based adsorbents for wastewater treatment.

Keywords: graphene; adsorbent; ion exchange; adsorption; wastewater treatment

1. Introduction

With the rapid development of industry and the fast increase in population, the problem of increasing pollutants in water bodies is becoming more and more serious. The pollutants in water can be divided into inorganic pollutants and organic pollutants, among which inorganic pollutants mainly include heavy metals, rare metals, and non-metallic inorganic compounds such as fluorides and phosphates, and so on, while organic pollutants mainly include organic dyes, antibiotics and other organic pharmaceuticals, aromatic compounds, etc. These pollutants not only cause certain damage to our environment but also have many adverse effects on human health [1–4]. The issue of pollution of the water environment has slowly come to affect people's daily lives in all aspects. Therefore, addressing the issue of contaminated water and making it cleaner and safer water is a current focus and concern for people in general [5]. When the concentration of pollutants in the human body is higher than the maximum limit that the human body can withstand, these toxic pollutants will have negative impacts on health, with the more serious impacts even leading to cancer [6–8]. As such, the phenomenon of water pollution is a challenge that needs to be urgently resolved in order to prevent the environment from becoming more hazardous to humans. Thus, the establishment of cost-effective strategies to remove pollutants from wastewater is a global requirement. Today, chemical precipitation [9], ion exchange and adsorption [10], solvent extraction [11], membrane separation [12], electrocatalysts [13], etc., are used for the separation of pollutants in wastewater, among which ion exchange and adsorption are widely used for their advantages relating to easy operation, high efficiency, environmental friendliness, low cost, and the good reusability of adsorbents [14–17]. Carbon materials are popular commodity adsorption materials, such as activated carbon, which has been widely used in wastewater treatment [18–21]. However, the adsorption capacity of activated carbon is limited by its surface characteristics and pore structure, and issues may arise, such as blockages and difficulties in regeneration during its use. In comparison, graphene-based adsorbents demonstrate a wider application potential in the field of wastewater treatment due to their unique structure and excellent chemical stability.

Graphene can be viewed as the basic unit of carbon materials, which can be enclosed, curled, and stacked to form other carbon nanomaterials in other dimensions, such as zero-dimensional fullerenes, one-dimensional carbon nanotubes, and three-dimensional (3D) graphite, as shown in Figure 1a. One-dimensional carbon-based nanomaterials, such as carbon nanotubes (CNTs) and carbon nanofibers (CNFs), have attracted much attention due to their nanoscale dimensions and homogeneous microporous structures with large surface areas. In particular, CNFs are a class of lightweight carbon nanomaterials with excellent specific surface area properties [22]. Until recently, however, CNFs have been mainly associated with catalysis [23], with very few reports of their use as adsorbents. Two-dimensional graphene crystals are formed from closely packed carbon atoms, with a thickness of one carbon atom for each layer of the crystal. The surface is not completely flat, with local microscopic wrinkles and corrugations. However, the hydrophobicity of graphene sometimes hinders its application in water and wastewater treatment. Figure 1b–e represents the molecular structures of graphite, graphene, GO, and rGO, respectively. Among them, graphene oxide (GO) is an important derivative of graphene with a high theoretical specific surface area and a structure similar to that of graphene [24,25], and the structural layers of GO are rich in oxygen-containing functional groups, such as C=O, C–OH, –COOH, and C–O–C [26,27]. Among them, the hydroxyl and epoxy groups are mainly distributed inside the lamellar structure of graphite oxide, while carbonyl and carboxyl groups are mainly distributed at the edges [28–30], as shown in Figure 1d. These functional groups provide good adhesion sites for the adsorption of pollutants in water and have broad application prospects in the field of wastewater treatment [31]. GO materials can also be used as carriers and combined with other materials to form composites, thus improving the performance of the materials themselves. GO-based composites are novel materials that can further improve the adsorption properties and selectivity of graphene through functionalized modification of GO to form new active groups and topologies. The functionalized modification of graphene serves mainly to retain the good physicochemical properties of GO, while effectively compensating for its application limitations. In conclusion, GO and its composites are typical carbon-based adsorbent materials that can effectively remove toxic pollutants from water [32–34].

The aim of this paper is to review recent research advances in the use of graphene and its derivatives for the removal of heavy metals and organic pollutants from water and wastewater. The discussion will focus on the preparation methods and functionalization strategies for graphene-based adsorbents, as well as their performance and mechanism of action in practical applications. Through a comprehensive analysis of the existing literature, this paper will reveal the prospects for the application of graphene in the field of environmental adsorption, and it will point out the problems in the current research and future development directions.

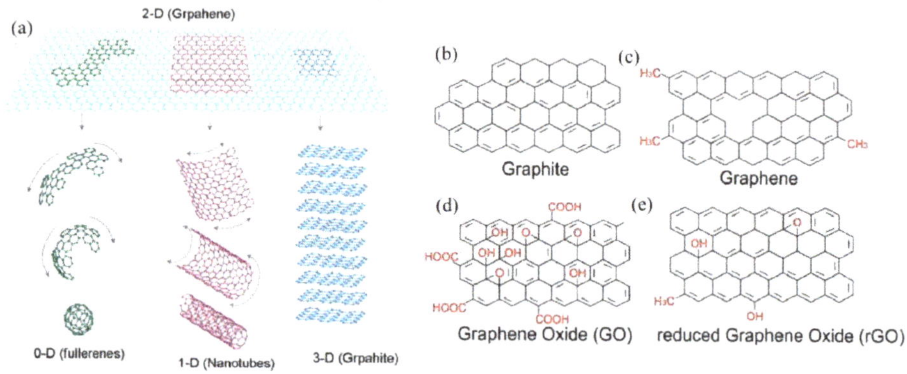

Figure 1. (**a**) Carbon allotropes: graphene to fullerene, nanotubes and graphite. Reprinted with permission from reference [35]. Copyright 2007, Springer Nature. The molecular structure of (**b**) graphite, (**c**) graphene, (**d**) GO, and (**e**) rGO [36]. Copyright Taylor and Francis, 2022.

2. Preparation of Graphene and Its Derivatives

2.1. Preparation of Graphene and GO

2.1.1. Graphene

Graphene has an sp^2 carbon lattice with an extremely high specific surface area, excellent electrical and thermal conductivity, and superior mechanical strength [37]. These properties enable graphene to show great potential for applications in a wide range of fields, including electronic devices, energy storage, composite materials, and environmental governance [38]. Graphene is often referred to as the material of the future due to its chemical stability and many other key advantages [39].

Figure 2 illustrates the two main methods of graphene synthesis: bottom-up and top-down methods [40]. Bottom-up methods involve the continuous growth of graphene through methods that break the chemical bonds of carbon-containing compounds and deposit carbon atoms on a substrate. Bottom-up methods include techniques such as chemical vapor deposition (CVD) [41] and epitaxial growth [42], which build graphene structures at the atomic or molecular level. In contrast, top-down methods are used to obtain graphene from bulk graphite by breaking the van der Waals forces between graphite layers with the help of external forces. Top-down methods include the reduction of graphene oxide via the Hummers method and the modified Hummers method to obtain reduced graphene oxide (rGO) [43]. There are also techniques such as ultrasonic exfoliation [44], ball-milling exfoliation [45], and electrochemical exfoliation [46]. In addition, there are techniques using graphene composites combined with nanoparticles (NPs), such as graphene encapsulation, graphene anchoring, and hybrid hybridization, through which sandwich structures and laminated hybrid NPs are prepared. For the analysis of their physicochemical properties, a variety of microscopic (e.g., scanning electron microscopy, transmission electron microscopy, atomic-force microscopy) and spectroscopic (e.g., X-ray photoelectron spectroscopy, X-ray absorption spectroscopy, Raman spectroscopy, Fourier transform infrared spectroscopy) techniques can be used [47]. Overall, the top-down preparation of graphene is inexpensive, simple, energy-efficient, and easy to scale up.

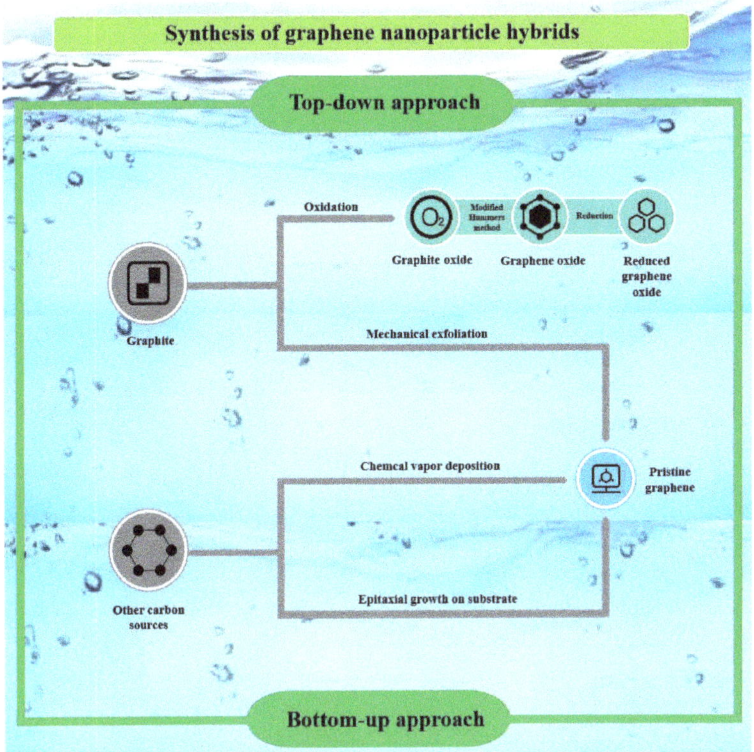

Figure 2. Synthesis of graphene nanoparticle hybrids.

2.1.2. Graphene Oxide

GO has a lower production cost, larger-scale production, and easier processing compared to graphene. It is used as a precursor for the preparation of rGO.

In the conventional Hummers method of preparing GO, it is necessary to reduce residual potassium permanganate with hydrogen peroxide, which is both risky and difficult to control. The use of strong oxidizing agents poses a danger, so a great deal of research has been conducted to find safe and straightforward methods for reducing the risk while preparing high-quality GO. Among them, the Hummers method is the focus of related research. Marcano et al. [48] produced GO using H_2SO_4, H_3PO_4, and $KMnO_4$, while Peng et al. [49] used K_2FeO_4 instead of $KMnO_4$ to reduce the risk caused by high valent manganese ions. Ye et al. [50] designed a microfluidic oxidation method to solve the problems of a long reaction time and poor safety. Zhu et al. [51], meanwhile, prepared acid-less GO by recycling waste acid to reduce environmental pollution. However, the process of adding hydrogen peroxide was still required. Zhou et al. [52] prepared GO without hydrogen peroxide, thus avoiding the high risk of a sharp exothermic reduction reaction due to the quenching reaction with the addition of hydrogen peroxide, whereby they induced the spontaneous reduction of high valent manganese ions by simply adjusting the reaction temperature/time and the amount of concentrated sulfuric acid. It was noteworthy that they achieved a C/O ratio of 1.986, which is lower than that of commercial GO. To promote environmental sustainability, many researchers have explored new production methods for synthesizing GO that offer economic and environmental benefits. Recent researchers have begun to focus on synthesizing GO using alternative, low-cost, carbon-rich materials, including coconut shells, bagasse, tea leaves, pine leaves, and waste tires, rather than graphite. These new preparation methods offer economic and environmental benefits [53,54]. Many applications can be realized by using waste carbon and converting it into GO [55,56].

2.2. Preparation of GO Derivatives

2.2.1. GO Inorganic Composites

GO inorganic composites play a crucial role in wastewater treatment. These composites endow graphene with unique properties, which not only improve the treatment efficiency and effectiveness but also expand the range of applications.

What is widely studied currently is the modification of GO onto magnetic nanoparticles to separate GO from water using gravity or applied magnetic force [57]. Figure 3a shows an efficient procedure for the preparation of magnetically reduced GO composites (MN–CCG) though ultrasonication of a suspension of iron(II) sulfate ($FeSO_4$) with GO in ethanol [58]. The method starts by mixing a GO powder suspension with Fe (II) sulfate, and then ethanol is added for stirring. After sonication, the color of the mixture changes from the original dispersed system to a black precipitate, which indicates the formation of MN–CCG. The MN–CCG composites show scalability and industrial compatibility for the remediation of pollutants.

A number of studies have been conducted on the application of Fe_3O_4/GO and MnO_2/GO in the removal of organic pollutants. Wang et al. [59] prepared Fe_3O_4/GO magnetic composites by sufficiently mixing GO, $FeCl_2$, and $FeCl_3$, adding ammonia drop by drop to adjust the pH to 9 with N_2 being passed through the mixture, and then aging the mixture for 3 h. Chong et al. [60] first synthesized Fe_3O_4/GO though a similar classical co-precipitation method and then proposed preparing Fe^0/Fe_3O_4/GO composites by using a one-step reduction method and investigated the removal properties of typical dyes. As shown in Figure 3b, Fe^0/Fe_3O_4/GO composites were prepared though this method, which has a cumbersome process and needs the toxic chemical hydrazine hydrate in the preparation of Fe^0/Fe_3O_4/GO, as is the case for the one-step reduction method, and at the same time, the new approach maintained the good properties of Fe^0/Fe_3O_4/GO. Das et al. [61] successfully synthesized GO iron nanohybrid (GFeN) materials by modifying iron/iron oxide (Fe/Fe_xO_y) nanoparticles on the surface of GO using a sol–gel process, and they studied these materials' efficiency and performance in removing arsenic from water. Mosleh et al. [62] synthesized iron-containing GO-doped lanthanum titanate ($LaTiO_3$) nanoparticles (FeGO–TiLa) for the effective removal of harmful lead ions from water samples. By utilizing ultrasonic techniques, La(III) ions were coordinated with oxygen-containing ligands to achieve the surface modification of GO through soft–soft interactions. The synthesis of magnetic graphene-oxide-doped titanium lanthanide oxide nanoparticles is shown in Figure 3c. Pan et al. [63] used a mixture of $MnCl_2{\cdot}4H_2O$, $KMnO_4$, and GO to obtain α–MnO_2/GO composites (α–GOM_2) thought an in situ redox reaction. The specific surface area of α-GOM_2 was 2001 $m^2{\cdot}g^{-1}$, which was higher than that of GO (1711 $m^2{\cdot}g^{-1}$).

GO composites with non-magnetic inorganic materials such as SiO_2, ZnO, and molecular sieves can adsorb pollutants in wastewater very well. Bagbi et al. [64] synthesized and characterized mesoporous silica nanoparticle-decorated GO nanosheets ($MSiO_2$–GO) for the removal of Pb(II) from water. Magnesium dioxide nanoparticles were prepared from ethyl orthosilicate (TEOS) though an ultrasonic method, and GO sheets were prepared using a modified Hummers method. Singh et al. [65] prepared a composite material consisting of zinc oxide (ZnO) nanoparticles and GO through a solvothermal method, and they showed that the material can be used as an adsorbent for the removal of hexavalent chromium ions from an aqueous solution. Li et al. [66] employed the mesoporous material SBA-15 covalently linked with GO to prepare GO–SBA–15. GO–SBA–15 showed good adsorption selectivity for Pb(II) in the presence of Li(I), Na(I), K(I), Ca(II), Mg(II), Cd(II), Cr(III), Co(II), Hg(II), As(III), Mn(II), Ni(II), and Zn(II) ions together.

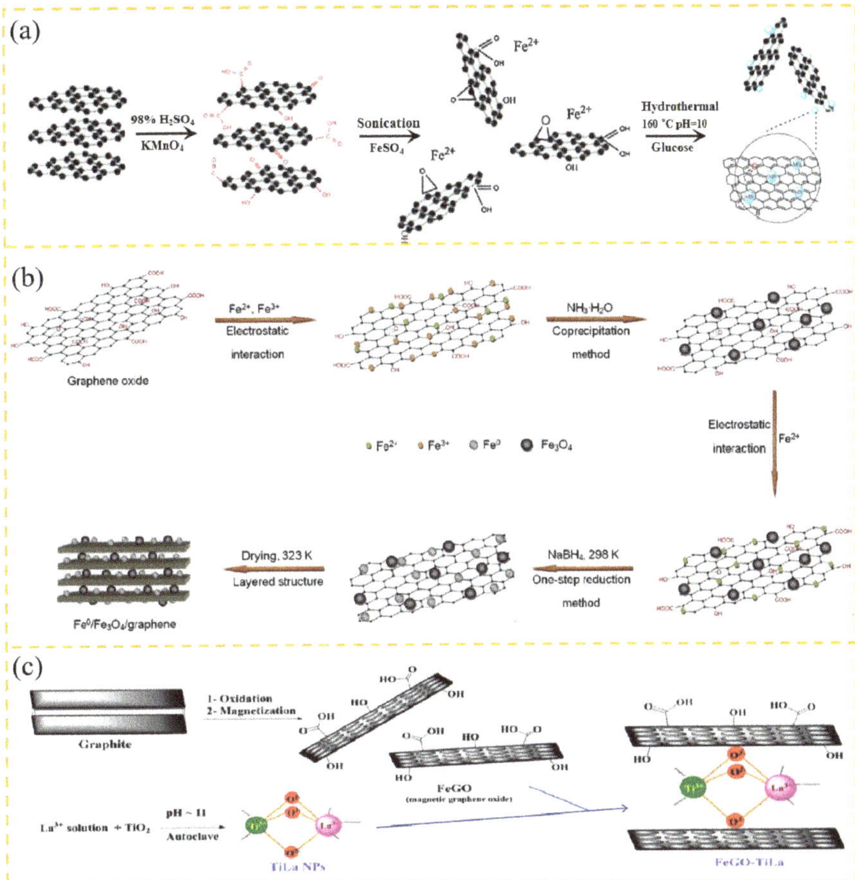

Figure 3. (**a**) Experimental method for synthesizing magnetic GO though the hydrothermal method. Reprinted with permission from reference [58]. Copyright 2011, Elsevier. (**b**) Preparation of $Fe^0/Fe_3O_4/GO$. Reprinted with permission from reference [60]. Copyright 2016, Elsevier. (**c**) Magnetic graphene oxide dopes titanium lanthanum oxide nanoparticle synthesis. Reprinted with permission from reference [62]. Copyright 2022, Elsevier.

2.2.2. GO Polymer Composites

GO, when composited with other polymeric materials, can significantly increase the adsorption capacity for specific pollutants. Such composites typically exhibit higher adsorption capacities and faster adsorption rates. Though the selection of polymer materials with different functions, GO composites can be functionally customized to remove different types of pollutants from wastewater in a targeted manner.

There are a large number of –OH and $–NH_2$ in chitosan (CS), and these functional groups can combine with oxygen-containing groups on GO through hydrogen bonding or other interactions to improve the stability and functionality of the composites. Li et al. [67] successfully developed GO/CS composites (GCs) though the dropwise addition of GO hydrosol onto the surface of CS. The prepared adsorbent was used for simulated static adsorption of copper, lead, and cadmium ions in 50 $mg \cdot L^{-1}$ simulated wastewater. Sherlala et al. [68] prepared CS magnetic GO nanocomposites (CMGO) by using a chemical coprecipitation method, and they studied their adsorption properties for removing arsenic from water.

However, the CS/GO composites showed a low adsorption capacity and poor selectivity. Therefore, it seems necessary to graft specific functional groups on CS/GO composites (CGs) to enhance their adsorption performance for pollutants in wastewater. Xia et al. [69] used GO, CS, and phytic acid as raw materials to directly prepare a novel CS/GO composite containing phytic acid (PCG) though a hydrothermal reaction for the removal of uranium (VI) from an aqueous solution where the phytic-acid-containing CS/GO composite was prepared as shown in Figure 4a. The FTIR and XPS survey spectra of GO, CG, and PCG are shown in Figure 4b,c, demonstrating the successful synthesis of PCG. The results showed that the PCG produced via chemical bonding of the hydroxyl group in the phytic acid molecule with the amino group in CS had a porous structure with a lamellar surface, good thermal stability, and chelating ability for uranium (VI). Zhou et al. [6] prepared a novel composite polyamide oxime/GO/CS composite (PAO/GO/CS) though crosslinking and in situ polymerization, and the synthesis route is shown in Figure 4d. The FTIR and XRD analysis of PAO/GO/CS and its precursor composites before adsorption of U(VI) are shown in Figure 4e,f. This indicates that the amidoxime groups were successfully grafted onto the surface of GO/CS. The resulting defects in uranium (VI) adsorption could be compensated by grafting amidoxime on the surface of GO/CS. According to the characterization and density-functional theory (DFT), uranium (VI) was spontaneously adsorbed onto the oxygen-containing groups in PAO/GO/CS mainly though chemisorption. Zhang et al. [70] prepared a GO suspension by using the improved Hummers method combined with ultrasonic dispersion technology, and then they successfully synthesized a new material, GEC, through covalent binding and electrostatic interaction with CS and EDTA-2Na.

GO can also be composited with other polymeric materials such as sodium alginate, resins, polyethyleneimine, and polyamide dendritic polymers. Abd-Elhamid et al. [71] studied a new strategy for modifying GO with sodium alginate (SA) using ethyl orthosilicate (TEOS) as a binding agent. The GO–sodium alginate composites contained a high number of carboxylate (–COO–Na^+) groups, which can interact with metal ions quickly and efficiently. The preparation of GO–SA is shown in Figure 5a. Ma et al. [72] utilized Amberlite IRC748 resin to modify GO to prepare a multifunctional composite adsorbent (IRC748–GO), and the mode of IRC748–GO synthesis is shown in Figure 5b. Arshad et al. [73], meanwhile, synthesized a composite of polyethyleneimine and GO, which showed an excellent adsorption capacity and stability for heavy metal ions such as Pb(II), Hg(II), and Cd(II) in an aqueous solution. Peer et al. [74] prepared a magnetic GO–polyamide dendrimer (mGO–PAMAM) composite by using a "grafting" strategy and investigated its adsorption behavior on Cu(II), Pb(II), and Cd(II). Moreover, Ma et al. [75] synthesized graphene-loaded Fe_3O_4 nanoparticles as carriers and ethylenediamine and methyl acrylate as functional monomers by using stepwise growth chemistry to synthesize polyamide amide dendritic polymer-functionalized magnetic GO with different generations of polyamide amide dendritic polymers (MGO–PAMAM). The synthesis process of MGO–PAMAM is shown in Figure 5c.

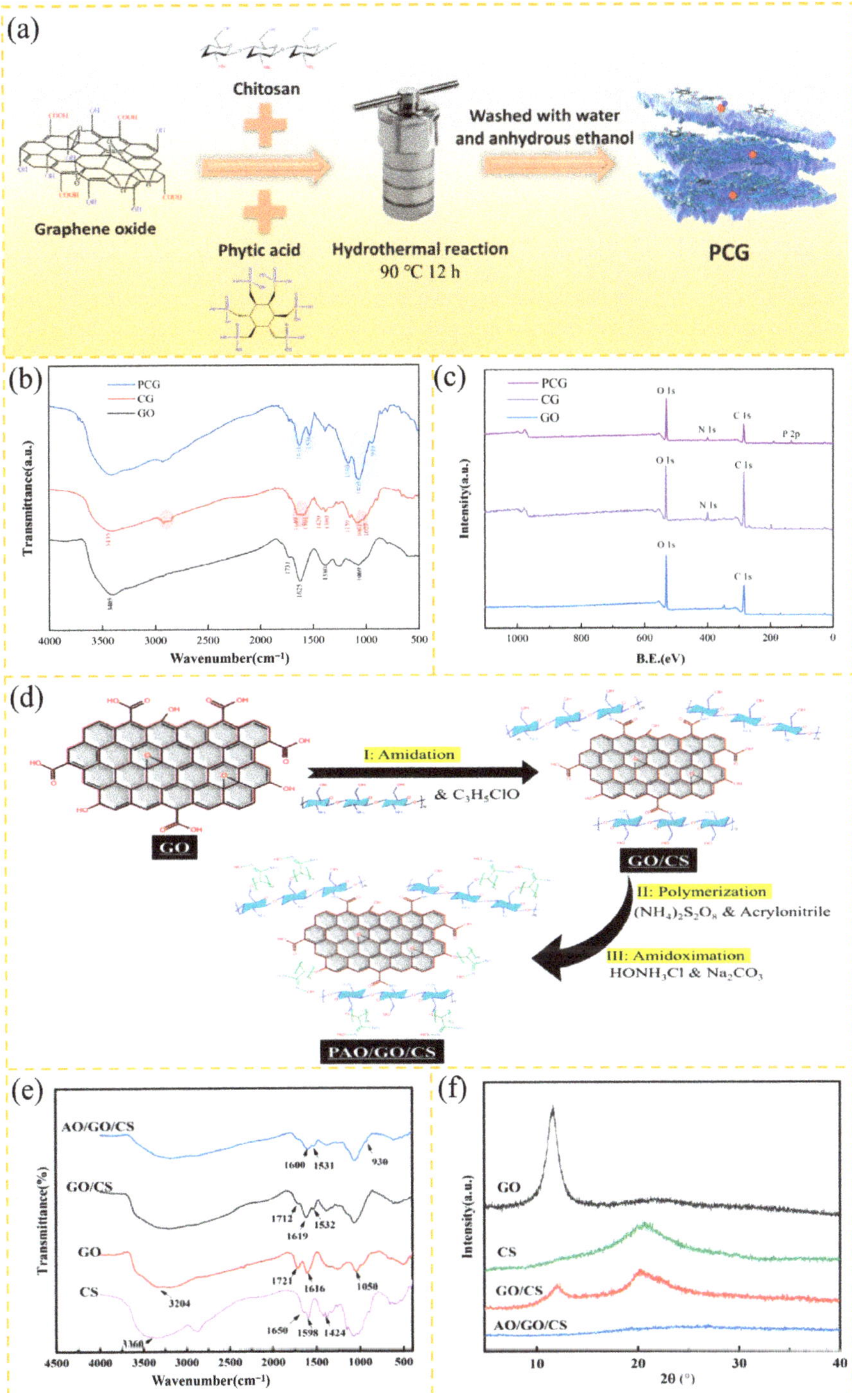

Figure 4. (**a**) Flowchart for the preparation of CS/GO composites containing phytic acid. (**b**) FTIR spectra of GO, CG, and PCG. (**c**) XPS survey spectra of GO, CG, and PCG. Reprinted with permission

from reference [69]. Copyright 2022, Elsevier. (**d**) Roadmap for synthesizing PAO/GO/CS composites. (**e**) FTIR spectra and (**f**) XRD spectra of PAO/GO/CS and its precursor composites. Reprinted with permission from reference [6]. Copyright 2021, Elsevier.

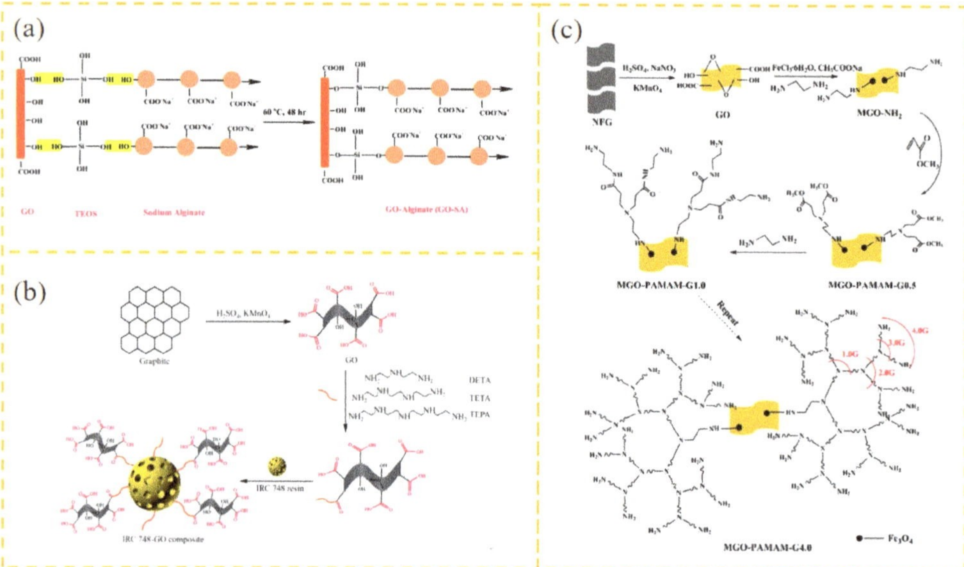

Figure 5. (**a**) Preparation of GO–SA adsorbent [71]. Copyright 2016, Springer Nature. (**b**) Synthesis of IRC748–GO composite. Reprinted with permission from reference [72]. Copyright 2024, Springer Nature. (**c**) Synthesis process of MGO-PAMAM. Reprinted with permission from reference [75]. Copyright 2024, Elsevier.

2.3. *Preparation of 3D Graphene-Based Adsorbents*

Due to π–π interaction, GO is easily stacked and aggregated, which greatly reduces the adsorption capacity and limits its further application, and the need for centrifugal and magnetic separation techniques complicates the material recovery process. Instead, assembling graphene nanosheets into 3D monolithic porous macrostructures prevents agglomeration and promotes mass transfer. This bulky structure also facilitates their collection and manipulation in water, which may open up a new avenue for graphene materials, one that works toward developing green and sustainable catalysis. 3D GO materials not only have a high specific surface area and porous structure [76], which enhance the adsorption capacity, but also can be effectively reused through a simple recycling process without any loss of material due to their 3D structure, which further expands the scope of GO applications. The preparation method for 3D GO and its composites and their application in the field of water treatment has come to form a hot spot of current research [77]. A large number of literature studies have shown that 3D graphene with such structures can effectively adsorb organic solvents [78], dyes [79], heavy metals [80], and antibiotics [81] in water.

2.3.1. Graphene Foam

Graphene foam is a material with a stable honeycomb structure made from tubular graphene through specific physical and chemical means. This material has a large specific surface area, light weight, and has high mechanical strength. As an advanced 3D-structured carbon material, graphene foam has a variety of remarkable structural properties and a wide range of preparation methods, which give it great potential for application in many fields.

Foamed graphene via freeze-drying, chemical vapor deposition, hydrothermal methods and so on. Jayanthi et al. [82] employed a freeze-drying technique to prepare porous 3D graphene foam. The material has a macroporous structure that allows the diffusion of pollutants within its pores, which is effective in the adsorption of Rhodamine B, malachite green, and acridine yellow and avoids the problem of decreasing adsorption capacity due to the stacking of single layers of graphene. Lei et al. [83] obtained a graphene oxide foam (GOF) by using chemical vapor deposition. This material has a high surface area (578.4 $m^2 \cdot g^{-1}$), is rich in oxygen-containing functional groups, and its adsorption capacities of Zn(II), Cd(II), Pb(II), and Fe(III) were higher than those of activated carbon, CNs, carbon foam, and GO adsorbents.

2.3.2. Graphene Sponge

Graphene sponge (GS) is a material constructed from two-dimensional graphene nanosheets built on top of each other, which can effectively impede agglomeration between graphene sheet layers and has excellent properties such as a large specific surface area, high mechanical strength, good electrical conductivity, large adsorption capacity, and good thermal stability.

Graphene sponge preparation methods include the hydrothermal method, freeze-drying, chemical vapor deposition, physical mixing, and so on. Wu et al. [84] found that GO can be hydrothermally reduced by glucose to form a graphene sponge for the adsorption of various organic solvents. During hydrothermal treatment, graphene sheets were reduced by glucose and formed a 3D porous structure. Bagoole et al. [85] used a modified Hummers method to synthesize GO solutions, followed by high-pressure hydrothermal treatment, freeze-drying, and chemical vapor-phase functionalization to prepare 3D graphene sponges. The surface-functionalized materials proved to have an excellent performance in terms of high-efficiency adsorption. These findings provide a valuable reference for the future development of novel separation materials. Yu et al. [81], meanwhile, prepared a magnetic GO sponge (MGOS) for the adsorption of tetracycline by freeze-drying dispersion of Fe_3O_4 nanoparticles and GO. The small size and negative surface charge of Fe_3O_4, the high degree of oxidation of graphene, and freeze-drying were responsible for the high adsorption capacity of MGOS. The adsorption behavior of tetracycline on MGOS suggests that MGOS holds promise for the remediation of tetracycline-based pollutants. Maimaiti et al. [86] designed a magnetic Fe_3O_4/TiO_2/graphene sponge (MFTGS) with a good regeneration ability in efforts to remove dyes. Fe_3O_4 was mixed with GO, and then tetrabutyl titanate was hydrolyzed to deposit titanium dioxide on the Fe_3O_4/GO mixture. The results showed that an MFTGS has the potential to remove dye contaminants from aqueous environments and can be regenerated through simple washing conditions. To further reduce the cost, Sun et al. [87] developed a flexible one-pot method to synthesize a novel amino-functionalized graphene sponge sorbent (AFGS) by taking advantage of the low-cost and super-hydrophilicity of the commercially available melamine sponge (MF), in which the graphene nanosheets were assembled into a 3D network on the backbone of the melamine sponge with the assistance of ethylenediamine. The formation process and morphology of AFGS, which is different from graphene sponge, are shown in Figure 6. Figure 6a shows that the MF backbone formed a 3D porous structure. Under the conventional dip-coating method (Route 1 and Figure 6b,c), the surface of the sponge skeleton was encapsulated by graphene fragments. The SEM image of graphene sponge at high magnification (Figure 6c) is shown. In contrast, the AFGS obtained by route 2 showed a leafy graphene structure grown on spongy branches (Figure 6d). The size of leafy graphene sheets is about tens to hundreds of micrometers (Figure 6e).

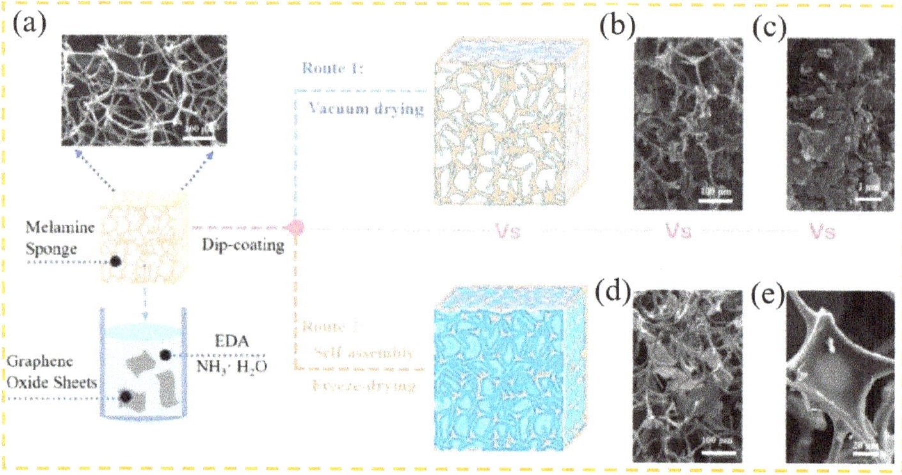

Figure 6. (**a**) Synthetic route and SEM images of melamine sponge. (**b**,**c**) SEM images of graphene sponge at low magnification and high magnification. (**d**,**e**) SEM images of AFGS at low magnification and high magnification. Reprinted with permission from reference [87]. Copyright 2021, Elsevier.

2.3.3. Graphene Gel

Graphene is highly hydrophobic and has a low adsorption capacity. In order to maintain a stable dispersion of graphene-based adsorbent materials, various methods have been employed to synthesize 3D graphene-based gels, such as hydrothermal reduction [88], chemical reduction [89], and chemical vapor deposition [90].

Hydrogels have attracted much attention due to their simple synthesis, ease of handling, low cost, and high adsorption efficiency [91]. Hydrogels can absorb large amounts of liquids and low-molecular-weight compounds without losing their original shape [92]. Due to their unique functionality, reusability, and high adsorption capacity, hydrogels are considered ideal adsorbents for future adsorption processes [93]. Developing hydrogel-based adsorbents for the adsorption of various aquatic pollutants has now become a focus of research. Taye et al. [94] prepared a hydrogel named water hyacinth leaf protein concentrate/GO (WHLPC/GO) using a one-step hydrothermal method for the treatment of Cr(VI) hexavalent chromium in wastewater. The material consisted of a combination of water hyacinth leaf protein concentrate (WHLPC) and GO, of which WHLPC accounted for about 90%. Sahraei et al. [80] prepared novel magnetic biosorbent hydrogel spheres using modified biopolymers, gum tragacanth (GT), poly(vinyl alcohol) (PVA), and GO, using the gel method to remove a variety of target pollutants, and they found that these spheres to be promising for a wide range of applications in the simultaneous removal of various pollutants from water. The preparation method is shown in Figure 7a. Yao et al. [95] synthesized tannic-acid-functionalized graphene (TA-rGO) in a simple and green method. A pH-responsive TA-functionalized graphene hydrogel (TRGAA) was synthesized though a random copolymerization reaction in the presence of TA-rGO using 2-acrylamido-2-methyl-1-propanesulfonic acid (AMPSA) and acrylic acid (AA) as raw materials, and the synthesis of TRGAA hydrogel is shown in Figure 7b.

Graphene aerogels are a class of monolithic materials with a well-developed pore structure. These 3D porous materials have the advantages of low density, low thermal conductivity, and well-developed pores. They also have an interconnected porous structure, high specific surface area, high pore volume, thermal and electrical conductivity, and environmentally friendly properties [96,97]. The methods to prepare them include the freeze-drying method, chemical reduction self-assembly method, and so on. Deng et al. [98] prepared a phosphoric-acid-modified aloe/CS aerogel (CS/AL-AP) though self-assembly with CS, phosphoric acid, and aloe vera powder, achieved by using vacuum freeze-drying. CS/AL-AP was prepared as shown in Figure 7c. Nundy et al. [99], meanwhile, fabricated 3D, porous, lamellar rGO aerogel to explore their potential for antimony removal from wastewater, where the process was the synthesis of rGO aerogels from graphite flakes using the modified Hummers method, as shown in Figure 7d.

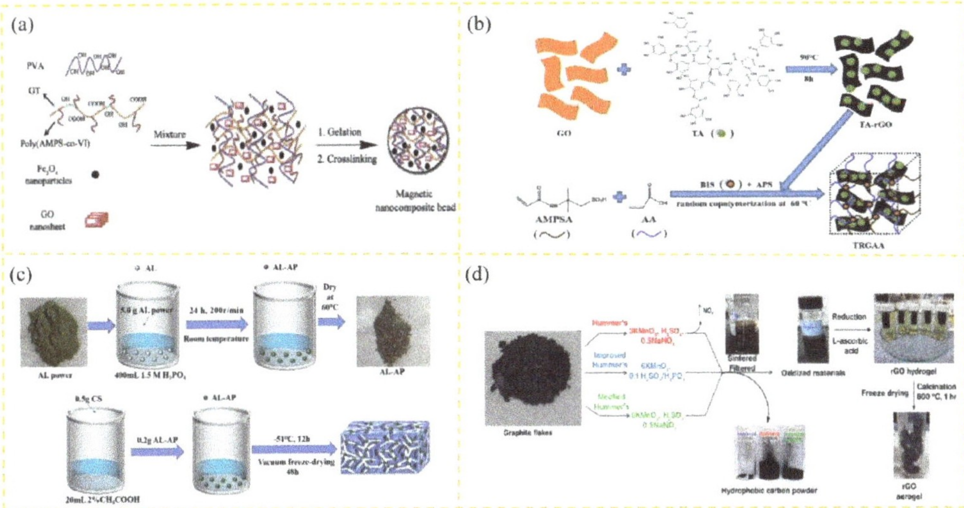

Figure 7. (**a**) Preparation of magnetic biosorbent hydrogel spheres. Reprinted with permission from reference [80]. Copyright 2017, Elsevier. (**b**) Synthesis of TRGAA hydrogel. Reprinted with permission from reference [95]. Copyright 2021, Elsevier. (**c**) Preparation of CS/AL-AP. Reprinted with permission from reference [98]. Copyright 2022, Springer Nature. (**d**) Synthesis of rGO aerogels from graphite flakes using a modified Hummers method. Reprinted with permission from reference [99]. Copyright 2021, Elsevier.

3. Adsorption of Inorganic Pollutants with Graphene-Based Adsorbents

3.1. Heavy Metals

The current environmental issue of heavy metal pollution, classified as an inorganic pollutant, is prevalent and has detrimental effects. The main heavy metal ions, including arsenic, chromium, lead, selenium, and mercury, are persistent and toxic substances that pose a significant threat to organisms and the environment. They enter the ecosystem through various industrial processes, mining activities, and inappropriate waste disposal [100]. Heavy metals are difficult to degrade, easy to accumulate, and have high toxicity after entering a water body. Heavy metals accumulate in animals and plants throughout the water body, and eventually, they pass to humans through the food chain. A large amount of heavy metal accumulation in organisms will affect their physiological activities, resulting in chronic poisoning and even death. Heavy metal pollution of water bodies is, thus, an environmental problem that needs to be solved urgently. If there is no effective, comprehensive treatment of polluted water bodies, the ecological environment will suffer serious damage and human health will be seriously harmed [101,102].

3.1.1. Arsenic

The high toxicity of arsenic renders it one of the prominent contaminants in various environmental pollutants. Arsenic can infiltrate the human body through the food chain or water, exerting toxic effects by influencing metabolic enzymes, inducing lipid peroxidation, causing genetic damage, and altering gene expression. The widespread concern over the detrimental effects caused by arsenic pollution necessitates urgent measures for its removal from various water bodies.

Various treatment techniques for arsenic removal have been investigated. These include physicochemical processes such as adsorption [103], precipitation [104], and membrane separation [105]. Among them, adsorption is considered an economical and simple operational technique for arsenic remediation. Various materials can be used as possible adsorbents for the removal of arsenic. Although commercial activated carbon has been used for As(V) adsorption, its adsorption capacity is low, reaching a maximum of only 4.96 $mg \cdot g^{-1}$ at pH 7 and 25 °C [106]. Elsewhere, GO has received attention as a potential adsorbent for arsenic removal. The presence of functional groups such as hydroxyl, carboxyl, and epoxy groups promotes the adsorption of arsenic on the GO surface. However, the use of GO faces certain challenges, such as self-agglomeration and difficulties in the post-processing separation of nanoparticles [107]. Therefore, GO has been modified with different materials to improve its performance. Meanwhile, iron-oxide-derived adsorbents have received great attention in arsenic removal due to their excellent arsenic adsorption properties compared to other types of adsorbents [108]. Das et al. [61] used the inorganic composite material GFeN for the adsorption of As(III) and As(V) with very high adsorption capacities of 306.10 ± 9.92 $mg \cdot g^{-1}$ and 431.41 ± 25.95 $mg \cdot g^{-1}$, respectively. These findings highlight the great potential of GFeN as a highly efficient arsenic removal material. Figure 8a illustrates the potential mechanism of arsenic removal by GFeN. Initially, arsenic ions in water make contact with the nanohybrid surface, forming an arsenic flux gradient between the solution and the adsorbent surface. Subsequently, electrostatic adsorption and surface complexation result in arsenic adsorption onto the GFeN surface. Furthermore, the arsenic absorbed on the GFeN surface is transformed and stabilized in the nanohybrid material. Figure 8b,c also illustrates that the arsenic removal from GFeN is not controlled by electrostatic forces alone; instead, surface complexation plays an important role, making it effective over a wide pH range. Qu et al. [109] investigated the adsorption removal performance of As(V) from simulated arsenic-containing wastewater using iron nitrate–graphene (FG) nanocomposites. The adsorption data demonstrated that, under optimal conditions, the removal efficiency of As(V) was as high as 99%, with the treated arsenic content reduced to a remarkably low level of 9.4 $\mu g \cdot L^{-1}$. As shown in Figure 8d,e, the main mechanism for the removal of As(V) from water by graphene involves the interaction of arsenic ions with carboxyl, hydroxyl, and other oxygen-containing functional groups on the graphene surface. In addition, electrostatic adsorption, ion exchange and complexation also contribute to this process. These surface hydroxyl groups enhance the electron density on the graphene surface, making it charged and forming hydrogen bonds with hydrated As(III) and As(V) ions. In an aqueous environment, these ions are held together by charge attraction and hydrogen bonding. As(V) removal maintained above 95% under both acidic and alkaline conditions, as shown in Figure 8f.

In addition to iron-derived adsorbents, there are other materials that have been used to improve the performance of graphene-based adsorbents and reduce the difficulty of post-processing separation for arsenic removal. Sherlala et al. [68] used CMGO for arsenic adsorption and showed that its adsorption capacity and removal efficiency were as high as 45 $mg \cdot g^{-1}$ and 61%, respectively. The adsorption mechanism is mainly through the complexation reaction between the surface functional groups of CMGO (e.g., $-NH_2$, $-OH$, $-OH_2$) and As(III). Such surface complexes can be formed with monodentate bonds or exosphere complexes. The superparamagnetic nature of the nanocomposites enables them to separate and recycle nanoparticles using external magnetic fields. Thus, the developed nanocomposites have potential for arsenic remediation.

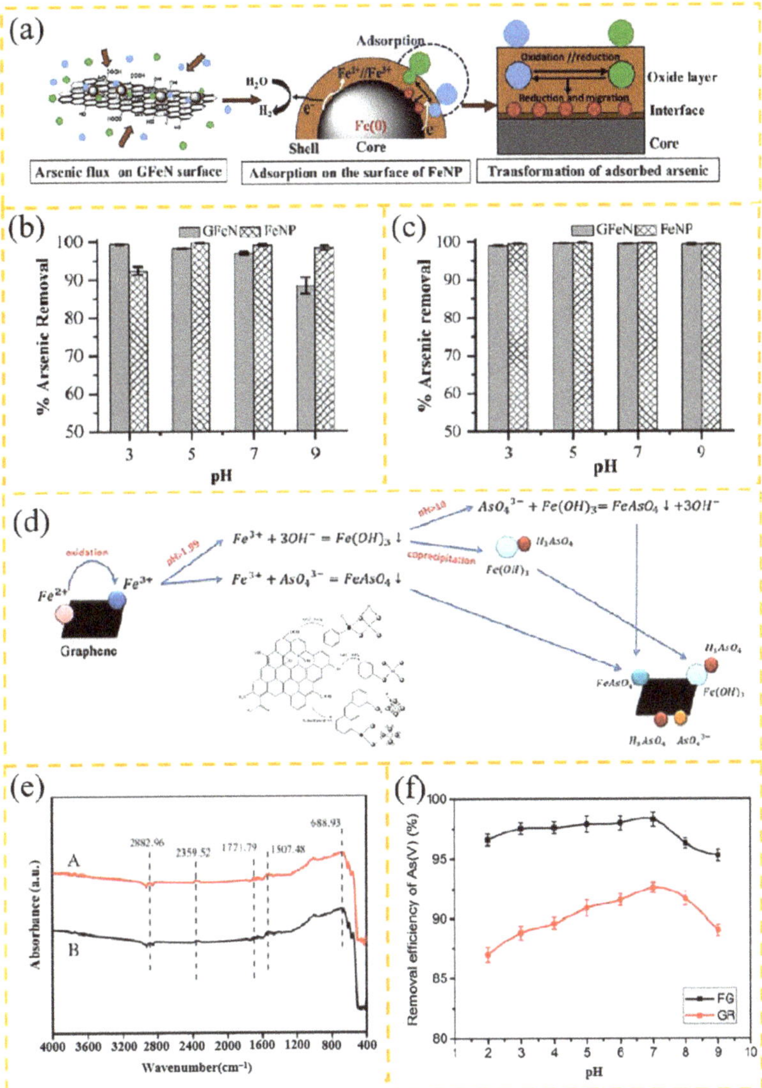

Figure 8. (**a**) The possible mechanism for arsenic removal by GFeN. Arsenic removal at different pH: (**b**) As(V) and (**c**) As(III). Reprinted with permission from reference [61]. Copyright 2020, Elsevier. (**d**) The removal mechanism of arsenic via ferric salt loaded on the surface of ferric nitrate-graphene (FG). (**e**) Fourier transform infrared spectra of the A fresh and B used ferric nitrate-graphene (FG) adsorbents. (**f**) Effect of pH on the adsorption of As(V). FG, ferric nitrate-graphene; GR, blank graphene. Reprinted with permission from reference [109]. Copyright 2019, John Wiley and Sons.

3.1.2. Chromium

Chromium exists in the environment mainly in the forms of chromium(III) and chromium(VI). The toxicity of chromium(VI) is higher than that of chromium(III) due to its elevated water solubility, fluidity, permeability, oxidation potential, and carcinogenic properties. According to the American Conference of Governmental Industrial Hygienists (ACGIH), the recommended concentrations for water-soluble and water-insoluble chromium(VI) compounds should not exceed 0.05 mg·m^{-3} and 0.01 mg·m^{-3}, respectively [110]. The presence of chromium(VI) in water poses significant risks to the kidneys, nervous system, lungs, and liver of mammals [111]. Given the severe toxicity associated with chromium(VI), it is imperative to prioritize the removal of this compound from water sources.

Various researchers have reported and investigated low-cost adsorbents for the removal of Cr(VI) from wastewater, such as orange peels [112], industrial wastes [113], biochar [114], etc. However, all these adsorbents have limited applicability due to their low adsorption capacity, poor selectivity, and lack of regeneration. Mondal et al. [115] concluded that graphene oxide can be used as an effective alternative to remove Cr(VI) from aqueous solutions. To make GO more appropriate for this application, it was chemically modified. Singh et al. [65] used the inorganic composite material ZnO–GO for the removal of Cr(VI) from aqueous solutions. The researchers delved into the adsorption characteristics of the material, including the kinetic, isothermal, and thermodynamic behaviors related to hexavalent chromium. The experimental data demonstrate that at a pH of 8.02, with an initial Cr(VI) concentration of 5 mg·L^{-1} and an adsorbent dose of 2 g·L^{-1}, the maximum adsorption efficiency is achieved. Additionally, the adsorption of Cr(VI) on the ZnO–GO surface follows a chemisorption process, as evidenced by the pseudo-second-order kinetic model. The utilization of ZnO–GO nanocomposites for Cr(VI) removal in water treatment demonstrates remarkable efficiency and feasibility, thereby having significant implications for the advancement of novel environmentally friendly materials.

There has been increasing attention paid to potential biosorbents for the removal of heavy metal ions in water treatment, such as CS [116], which possesses amine and hydroxyl groups that can remove heavy metal ions by forming stable metal chelates [117]. However, CS is only soluble in acidic solutions, which limits its application in the removal of heavy metal ions from aqueous media. Zhang et al. [70] addressed this limitation by crosslinking CS to adsorb Cr(VI) using the polymer composite GEC; the maximum adsorption of Cr(VI) by GEC under optimized conditions (pH 2, contact time 90 min) was 86.17 mg·g^{-1}. Based on the pH results, the adsorption mechanism can be mainly explained as follows: the NH_2 and OH reactive groups in GEC, which can be protonated in an acidic solution, give a positive charge to the GEC surface. Therefore, hexavalent chromium binds to GEC through the strong electrostatic attraction between the protonated groups of GEC and the negatively charged hexavalent chromium material.

The interest in the synthesis of graphene hydrogels for the removal of heavy metals has been growing in recent years. This is due to the high removal efficiency, reusability, and biocompatibility of hydrogels [118]. Hydrogels are 3D polymer networks capable of absorbing large amounts of water but insoluble in water. Interestingly, water hyacinth has been used as a precursor for the preparation of activated carbon, proteins, bioelectrodes, and hydrogels for different applications, including adsorbent materials for Cr(VI) removal [119]. Kumar et al. [120] used dried fenugreek root powder for Cr(VI) adsorption with a maximum adsorption capacity of only 1.28 mg/g. Worku et al. [121] used water-hyacinth-based activated carbon to adsorb Cr(VI), but its adsorption capacity was also low; a simple method to modify water hyacinth is still needed to obtain a higher adsorption capacity. Taye et al. [94] remediated Cr(VI) in wastewater with WHLPC/GO hydrogel. The experimental results showed that the adsorption of Cr(VI) was spontaneous and endothermic. The study of different factors revealed that the highest adsorption capacity of 322.00 mg·g^{-1} was achieved under optimum conditions, i.e., pH 1.0 and an equilibrium contact time of 420 min. The results of that study not only demonstrated the efficient adsorption capacity of the hydrogel materials but also highlighted their potential application in environmental treatment, especially in the reduction of toxic heavy metal emissions from industrial activities.

3.1.3. Lead

The presence of lead in water poses a significant threat to human health and various biological systems [122]. The US Environmental Protection Agency sets the acceptable concentration of Pb in drinking water at 15 µg·L^{-1} [123]. The treatment of Pb ions in wastewater is essential prior to their discharge into environmental water sources [124].

In recent years, graphene-based adsorbents have attracted researchers' attention as potential adsorbents for heavy metal removal from water and wastewater [125]. However, the current cost of pure graphite is still high. Therefore, recycling of graphite is very important. In an investigation aimed at finding a cost-effective method, it was proposed that waste-derived GO offers a better solution for the removal of toxic Pb from aqueous solutions [126]. The adsorption and removal of Pb(II) from water were investigated by Azam et al. [127], who prepared GO from graphite rods of waste dry cell batteries. The experimental results showed that the removal of Pb(II) (10 ppm) by GO was up to 98.87% at pH 4, a contact time of 20 min, and an adsorbent dosage of 0.25 g·L^{-1}, and the maximum adsorption capacity was calculated to be 55.80 mg·g^{-1} at 20 °C. The possible adsorption mechanism of Pb(II) on the GO surface in an aqueous solution is shown in Figure 9a. The adsorption of cationic heavy metals such as lead on the GO surface mainly involves electrostatic attraction between opposite charges, and the FTIR spectra of graphite and GO before and after Pb adsorption are shown in Figure 9b, confirming the mechanism.

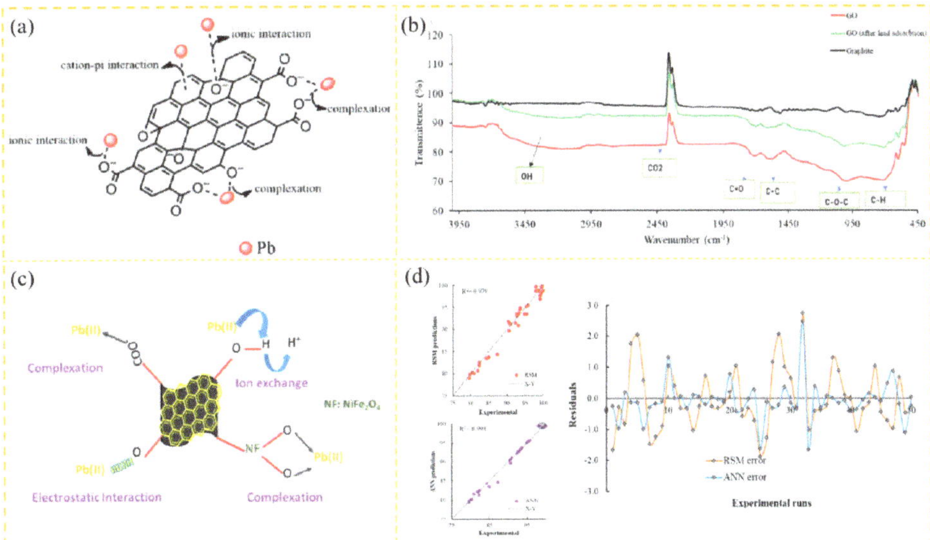

Figure 9. (**a**) The possible adsorption mechanism of Pb(II) on the GO surface from an aqueous solution. (**b**) FTIR spectrum of graphite powder, GO, and lead-adsorbed GO. Reprinted with permission from reference [127]. Copyright 2022, Elsevier. (**c**) Schematic representation of predominant mechanism for adsorption removal of Pb(II) with rGNF. (**d**) Comparison of performance of RSM and ANN predictions. Reprinted with permission from reference [124]. Copyright 2022, Elsevier.

Researchers have developed a number of novel GO composite adsorbents in order to increase the adsorption capacity. GO-functionalized binary metal oxides have been used as adsorbents due to their high stability, porosity, and high adsorption capacity for heavy metal ions. Mosleh et al. [62] used FeGO–TiLa to effectively remove harmful Pb(II) from water samples. The adsorption results showed that FeGO–TiLa removed up to 93% of Pb(II), which was significantly better than the results obtained for pure TiLa nanoparticles (81%) and magnetic GO (74%). Isothermal modeling analysis showed that the Langmuir model provided a better fit with the experimental data compared to the Freundlich model, thus indicating the existence of a monolayer physisorption mechanism. Kinetic studies also showed that the adsorption process follows a quasi-second-order rate equation. The adsorption of Pb(II) on FeGO flakes may be attributed to the coordination or complexation with oxygen-containing groups and non-covalent interactions between π–cations. Manganese oxides, moreover, have recently attracted a great deal of attention due to their low zero-charge potential, high surface area, negative charge-rich and strong surface activity, and various other properties. Lingamdinne et al. [128] investigated the adsorption mechanism of Pb(II) on these composites using composites of rGO and manganese oxide (Mn_3O_4). The adsorption of cationic heavy metals, such as Pb(II), from aqueous solutions onto the GO surface primarily involves electrostatic attraction between opposite charges (cation–π interaction and ion interaction) and surface complexation. Under optimized conditions (pH 4, adsorbent dose 0.5 $g \cdot L^{-1}$, contact time 60 min, temperature 25 ± 2 °C), the maximum adsorption capacity of the prepared rGO–Mn_3O_4 composite material for Pb(II) was determined to be 130.28 ± 0.52 $mg \cdot g^{-1}$, indicating a high efficiency in terms of adsorption performance. Furthermore, thermodynamic analysis revealed that the adsorption process is spontaneous and endothermic in nature.

The above studies did not use mathematical modeling techniques to optimize the intermittent adsorption process and the interactive effects of various variables on the removal efficiency. Mathematical and statistical methods are useful for planning experiments, detecting the effects of different independent process factors, and recommending ideal settings to obtain the best possible results. Narayana et al. [124] investigated the optimum values of independent process variables for maximum removal efficiency of Pb(II) from magnetic rGO reverse spinel nickel ferrite (rGNF) nanocomposites using response surface methodology (RSM) and artificial neural network (ANN). As shown in Figure 9c, surface complexation was identified as the main mechanism for the removal of Pb(II) via rGNF adsorption. The maximum removal efficiency of 99.8% was achieved when the initial concentration was 15 $mg \cdot L^{-1}$, the adsorbent dosage was 0.55 $g \cdot L^{-1}$, the treatment time was 100 min, the pH was maintained at 5, and the treatment temperature was 30 °C. The performance predicted by the RSM and ANN is shown in Figure 9d. It can be observed that both data-driven model predictions gave good predictions that were comparable to the experimental values; however, the ANN predictions were superior.

3.1.4. Copper and Cobalt

Electroplating wastewater mainly contains copper, cobalt, and other metals, which can cause a large amount of water pollution if not treated properly. When the human body ingests food contaminated with copper ions, acute copper ion poisoning may be triggered. Excessive intake of cobalt ions will also seriously jeopardize human health [129]. The adsorption of copper and cobalt thus plays an important role in environmental management. It may also contribute to the recycling of copper, cobalt, and nickel metals, which have irreplaceable roles as important non-ferrous metals in many fields in contemporary industrial development.

In order to increase the reusability and stability of graphene-based composite adsorbent materials for the adsorption of copper ions, GO was functionalized with Amberlite IRC 748 resin. Ma et al. [72] performed adsorption experiments of Cu(II) on an IRC748–GO composite. The adsorption process conformed to the quasi-secondary kinetic model, and the adsorption isotherm conformed to the Langmuir isotherm, with a maximum adsorption capacity of 127.22 $mg \cdot g^{-1}$ for Cu(II). The adsorption was investigated by using infrared spectroscopy, scanning electron microscopy, X-ray photoelectron spectroscopy, and an adsorption isotherm, and the results showed that the carboxyl groups in IRC748–GO composite produced effective complexation of Cu(II).

Functionalization of GO with inorganic complexes can also enhance the adsorption capacities of Cu(II) and Co(II). Ranjith et al. [130] adsorbed Cu(II) and Co(II) ions in water using zinc oxide nanorod–rGO (ZnO NR–rGO) nanocomposite. The results showed that the adsorbent was adsorbed at a solution adsorbent dosage of 10 mg, metal ion concentration of 10 mg/L, and pH of 6.0, and the maximum adsorption capacities for Cu(II) and Co(II) were 67.399 and 36.354 $mg \cdot g^{-1}$, respectively.

3.1.5. Selenium

Selenium pollution is a significant environmental concern, with selenium being present in water as a result of human activities such as coal combustion, mining operations, and crude oil processing. During the power generation process, solid waste and flue gas are released into the environment, leading to elevated concentrations of selenium in water in the forms of selenite (Se(IV)) and selenate (Se(VI)) [131]. It is estimated that approximately 80,000 tons of selenium are discharged into soil and water annually worldwide due to human activities. This subsequently affects animals, plants, and ecosystems, resulting in severe environmental and health consequences [132,133]. The recommended guidelines by the World Health Organization (WHO) and limits set by the Environmental Protection Agency (EPA) are 40 $\mu g \cdot L^{-1}$ and 50 $\mu g \cdot L^{-1}$, respectively [134,135]. Therefore, it is imperative to develop efficient technologies for selenium removal. By applying an adsorption technique, the pollution of water and soil with selenium can be reduced, and the ecosystem can be protected.

Although activated carbon as an adsorbent is applied for the adsorption of selenium from wastewater, it is not very efficient for selenium species investigated; it is reported that Se(IV) was adsorbed almost not at all (only 4%) at pH 6, and only 42% of Se(VI) was collected [136]. Activated carbon was modified, but the adsorption capacity remained low [137]. Recently, graphene oxide was found to selectively adsorb metal ions [138]. More studies were conducted to modify GO to adsorb metal ions. For instance, Xiao et al. [139] investigated the adsorption performance of selenium with polyamide-amide-functionalized GO (PAMAM–GO) material. The mapping of Se is illustrated in Figure 10a,b, indicating that Se was adsorbed on graphene oxide functionalized with a silane coupling agent. The maximum removal capacities for Se(VI) and Se(IV) were 77.9 $mg \cdot g^{-1}$ and 60.9 $mg \cdot g^{-1}$, respectively, at pH 6. It was shown that the chemically modified GO could effectively enhance the adsorption performance of selenium in water, especially at a lower pH, where its selective adsorption performance for Se(VI) is more prominent. As such, this chemically modified GO provides an effective material solution for the treatment of selenium-containing wastewater. According to analysis, one or two of the amine groups may interact with a selenium–oxygen anion, and the adsorption mechanism of PAMAM–GO on Se(IV) and Se(VI) is shown in Figure 10c. Lu et al. [140] used poly(allylamine)-modified magnetic GO (PAA–MGO) material for the adsorption removal of selenium from water, especially for Se(VI) and Se(IV). The adsorption results showed that the maximum adsorption capacities of PAA–MGO were 120.1 $mg \cdot g^{-1}$ for Se(IV) and 83.7 $mg \cdot g^{-1}$ for Se(VI) at pH 5.8. In addition, the effect of pH on selenium adsorption was investigated, and it was found that the selenium adsorption was favorable under acidic conditions, with an optimal pH of 3.1.

Currently, nano zero-valent iron (nZVI) seems a promising reductant and adsorbent for Se(IV) removal because of its large specific surface area and high reactivity [134]. Sun et al. [141] investigated the efficiency of the removal of selenate (Se(IV)) by this nZVI/rGO composite by loading GO prepared via the modified Hummers method on nZVI. The adsorption results showed that the removal ratio reached 99% when the loading ratio was 10%, the dose was 0.3 $g \cdot L^{-1}$, and the initial pH was 3. The influencing factors of nZVI/rGO on the removal ratio of Se(IV) are shown in Figure 10e,f. The isothermal adsorption was consistent with the Langmuir isothermal model, and the maximum adsorption capacity was as high as 173.53 $mg \cdot g^{-1}$. The process of Se(IV) removal by nZVI/rGO consisted of the following two main steps: (1) Se(IV) in aqueous solution was rapidly adsorbed on the surface of the material to form surface complexes. (2) Se(IV) first diffused on the surface of the material, underwent internal diffusion after a period of reaction, and finally reduced to Se(0). At the same time, Fe(0) was oxidized to Fe(II), and Fe(II) was oxidized to hydroxide precipitate covering the particle surface. The mechanism of Se(IV) removal by NZVI/rGO is shown in Figure 10d.

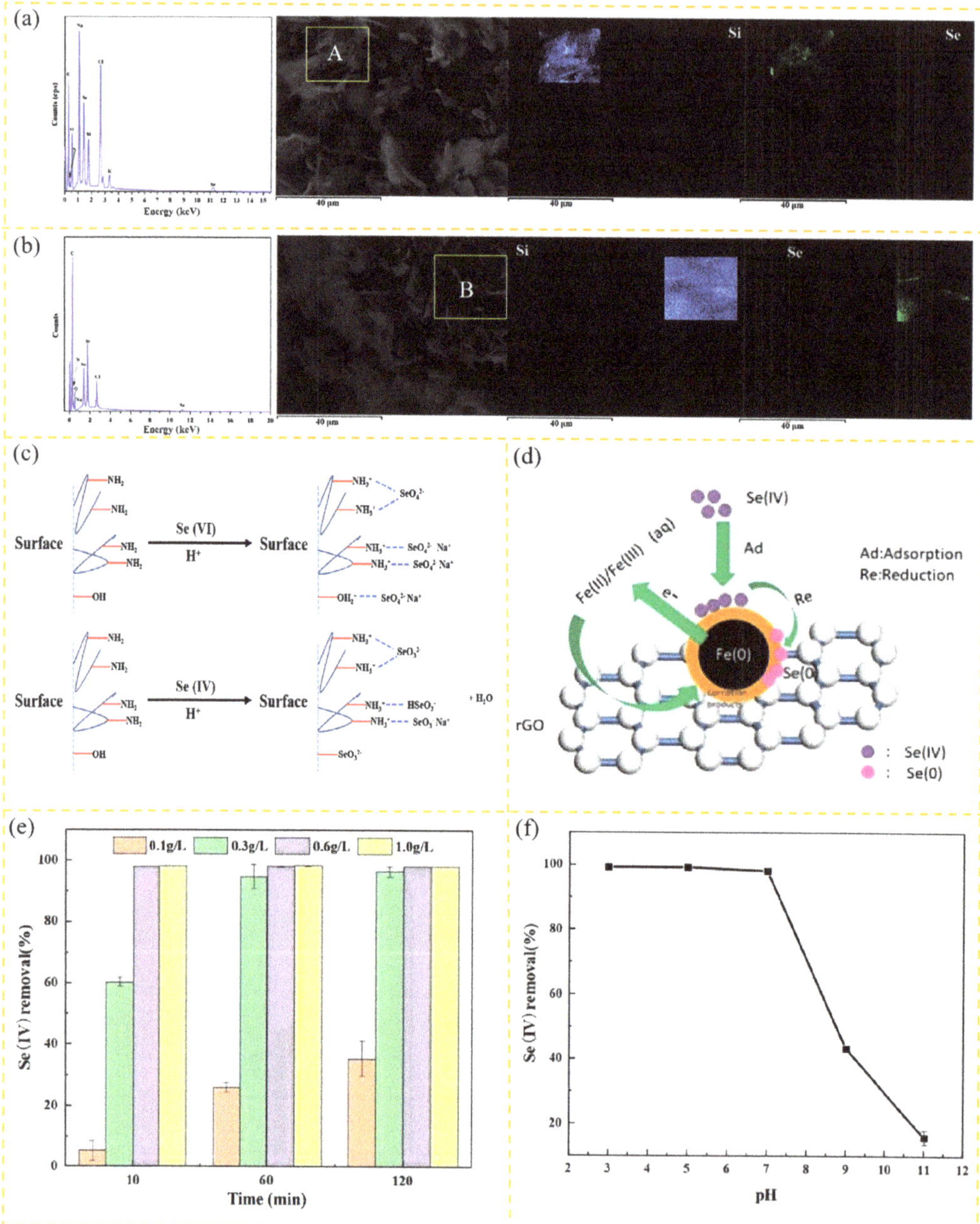

Figure 10. EDS spectra of PAMAM-GO after the adsorption of (**a**) selenate and (**b**) selenite ions. The scanned results of Si and Se in the yellow boxes (A and B) are shown on the right sides. (**c**) Schematic illustration of adsorption mechanisms for Se(IV) and Se(VI) on PAMAM-GO. Reprinted with permission from reference [139]. Copyright 2016, Elsevier. (**d**) Mechanism diagram of Se(IV) removal by nZVI/rGO. Influence of Se(IV) removal efficiency by nZVI/rGO: (**e**) contact time and dosage; (**f**) pH. Reprinted with permission from reference [141]. Copyright 2022, Springer Nature.

3.1.6. Summary of Adsorption of Heavy Metal Ions

A summary is given herein for a more intuitive comparison of the adsorption performances of various graphene-based adsorbents for heavy metal ions (see Table 1).

Table 1. Adsorption capacities of graphene-based adsorbents for selected heavy metal ions.

Adsorbent		Heavy Metal Ion	Adsorption Conditions	Adsorption Capacity/ Removal Efficiency	Reference
Material	Form				
CMGO	Powder	As(III)	pH = 7.3, 25 °C, 4 h	45 $mg \cdot g^{-1}$	[68]
Fe_3O_4–GO	Powder	As(III)	pH = 7, 25 °C, 12 h	85 $mg \cdot g^{-1}$	[142]
		As(V)	pH = 7, 25 °C, 12 h	38 $mg \cdot g^{-1}$	
Fe_3O_4–rGO	Powder	As(III)	pH = 7, 25 °C, 12 h	57 $mg \cdot g^{-1}$	[142]
		As(V)	pH = 7, 25 °C, 12 h	12 $mg \cdot g^{-1}$	
Fe_3O_4–GO	Aerogel	As(V)	Neutral pH, 25 °C, 12 h	40.048 $mg \cdot g^{-1}$	[143]
Fe_3O_4–rGO	Powder	As(III)	pH = 7, 30 °C, 2 h	0.81 $mg \cdot g^{-1}$	[144]
GO	Powder	As(III)	pH = 7, 30 °C, 2 h	0.70 $mg \cdot g^{-1}$	[144]
rGO	Powder	As(III)	pH = 7, 30 °C, 2 h	0.68 $mg \cdot g^{-1}$	[144]
Co–rGO	Powder	As(III)	pH = 7, 30 °C, 2 h	0.75 $mg \cdot g^{-1}$	[144]
GFeN	Powder	As(III)	pH = 7, 22 ± 2 °C	306.10 ± 9.92 $mg \cdot g^{-1}$	[61]
		As(V)	pH = 7, 22 ± 2 °C	431.41 ± 25.95 $mg \cdot g^{-1}$	
FG	Powder	As(V)	pH = 7, 35 °C	112.4 $mg \cdot g^{-1}$	[109]
TAIGO	Powder	Hg(II)	Natural pH, 15 min	23.81 $mg \cdot g^{-1}$	[145]
GO	Powder	Cr(VI)	pH = 4	1.222 $mg \cdot g^{-1}$	[115]
ZnO–GO	Powder	Cr(VI)	Neutral pH, 303 K	96%	[65]
GEC	Powder	Cr(VI)	pH = 2, 25 °C	86.17 $mg \cdot g^{-1}$	[70]
MCGO–IL	Powder	Cr(VI)	pH = 3, 30 °C	145.35 $mg \cdot g^{-1}$	[146]
GONF	Powder	Cr(III)	pH = 4, 298 ± 2 K	45.5 $mg \cdot g^{-1}$	[147]
		Pb(II)	pH = 5–6, 298 ± 2 K	25.0 $mg \cdot g^{-1}$	
WHLPC/GO	Hydrogel	Cr(VI)	pH = 1, 7 h	322.00 $mg \cdot g^{-1}$	[94]
AFGS	Sponge	Cr(VI)	pH > 4, 25 °C	166.46 $mg \cdot g^{-1}$	[87]
M–PAS–GO	Powder	Cr(VI)	pH = 3.2, 25 °C	46.48 $mg \cdot g^{-1}$	[148]
rGNF	Powder	Pb(II)	pH = 5, 30 °C, 100 min	99.8%	[124]
GO	Powder	Pb(II)	pH = 4, 25 °C, 20 min	98.87%	[127]
rGO–Mn_3O_4	Powder	Pb(II)	pH = 4, 25 ± 2 °C, 60 min	130.28 ± 0.52 $mg \cdot g^{-1}$	[128]
MCGO	Powder	Pb(II)	pH = 5, 1 h	50.23 $mg \cdot g^{-1}$	[149]
FeGO–TiLa	Powder	Pb(II)	pH = 5, 2 h	112 $mg \cdot g^{-1}$	[62]
$MSiO_2$–GO	Powder	Pb(II)	pH = 6, 25 °C	90.48%	[64]
Magnetic GO	Powder	Cd(II)	pH = 7, 45 °C	128.2 $mg \cdot g^{-1}$	[150]
		Pb(II)	pH = 7, 45 °C	385.1 $mg \cdot g^{-1}$	
ZnO NR–rGO	Powder	Co(II)	pH = 6	36.354 $mg \cdot g^{-1}$	[130]
		Cu(II)	pH = 6	67.399 $mg \cdot g^{-1}$	
AMGO	Powder	Cr(VI)	pH = 2	123.4 $mg \cdot g^{-1}$	[151]
TSCGO	Powder	Hg(II)	pH = 6, 298 K	933.2 $mg \cdot g^{-1}$	[152]
		Pb(II)	pH = 5, 298 K	890.4 $mg \cdot g^{-1}$	
GOFH	Powder	As(V)	pH = 4, 298 K	160 $mg \cdot g^{-1}$	[153]
		Cr(VI)	pH = 4, 298 K	66 $mg \cdot g^{-1}$	
IRC748–GO	Powder	Cu(II)	pH = 5.81, 30 °C	127.22 $mg \cdot g^{-1}$	[72]
CMC/SA/GO@Fe_3O_4	Powder	Cu(II)	pH = 5, 30 °C	55.96 $mg \cdot g^{-1}$	[154]
		Cd(II)	pH = 6, 30 °C	86.28 $mg \cdot g^{-1}$	
		Pb(II)	pH = 5, 30 °C	189.04 $mg \cdot g^{-1}$	
GO–EDTA–CS	Powder	Cu(II)	pH = 5.1, room temperature	130 ± 2.80 $mg \cdot g^{-1}$	[155]
		Hg(II)	pH = 5.1, room temperature	324 ± 3.30 $mg \cdot g^{-1}$	
GO/CS	Powder	Cu(II)	pH = 5, 313.15 K, 210 min	269 $mg \cdot g^{-1}$	[156]
GO/CS	Powder	Cu(II)	pH = 6, 20 °C	60.7 $mg \cdot g^{-1}$	[67]
		Cr(VI)	pH = 6, 20 °C	48.7 $mg \cdot g^{-1}$	
		Pb(II)	pH = 6, 20 °C	32.3 $mg \cdot g^{-1}$	

Table 1. *Cont.*

Adsorbent		Heavy Metal Ion	Adsorption Conditions	Adsorption Capacity/ Removal Efficiency	Reference
Material	Form				
mGO–PAMAM	Powder	Cu(II)	pH = 7, 25 °C	353.59 $mg \cdot g^{-1}$	[74]
		Cd(II)	pH = 6, 25 °C	435.85 $mg \cdot g^{-1}$	
		Pb(II)	pH = 7, 25 °C	326.729 $mg \cdot g^{-1}$	
GO–SA	Powder	Pb(II)	pH = 4, 25 °C	887.21 $mg \cdot g^{-1}$	[71]
		Zn(II)	pH = 5, 25 °C	161.25 $mg \cdot g^{-1}$	
		Cd(II)	pH = 7, 25 °C	139.62 $mg \cdot g^{-1}$	
GO–sand	Powder	Pb(II)	pH = 6, 298 K	289.68 $mg \cdot g^{-1}$	[157]
		Cr(III)	pH = 6, 298 K	258.87 $mg \cdot g^{-1}$	
GO–MBT	Powder	Pb(II)	pH = 6	116.959 $mg \cdot g^{-1}$	[158]
		Cd(II)	pH = 7	112.99 $mg \cdot g^{-1}$	
GO–CNC	Powder	Cr(VI)	pH = 5, 105 min	90.74%	[159]
		Ni(II)	pH = 8, 90 min	58.56%	
RGOS	Powder	Pb(II)	pH = 5.0, <10 min	689 $mg \cdot g^{-1}$	[160]
		Cu(II)	pH = 5.0, <10 min	59 $mg \cdot g^{-1}$	
		Ni(II)	pH = 5.5, <10 min	66 $mg \cdot g^{-1}$	
		Cd((II)	pH = 5.0, <10 min	267 $mg \cdot g^{-1}$	
		Cr(III)	pH = 5.5, <10 min	191 $mg \cdot g^{-1}$	
GRP–PPY–NP	Powder	Mn(II)	pH = 5.5, 90 min	89%	[161]
ZnO–Gr	Powder	Ni(II)	pH = 8.2, 25 °C	66.66 $mg \cdot g^{-1}$	[162]
Graphene	Powder	Ni(II)	pH = 8.2, 25 °C	3.78 $mg \cdot g^{-1}$	[162]
mp–CA/GO	Aerogel	Pb(II)	25 °C, 40 min	368.2 $mg \cdot g^{-1}$	[163]
		Cu(II)	25 °C, 40 min	98.1 $mg \cdot g^{-1}$	
		Cd(II)	25 °C, 40 min	183.6 $mg \cdot g^{-1}$	
GOF	Foam	Zn(II)	pH = 10	326.4 $mg \cdot g^{-1}$	[83]
		Cd(II)	pH = 10	252.5 $mg \cdot g^{-1}$	
		Pb(II)	pH = 10	381.3 $mg \cdot g^{-1}$	
		Fe(III)	pH = 10	587.6 $mg \cdot g^{-1}$	
rGO	Aerogel	Sb(III)	pH = 6 ± 0.5, 298 K	168.59 $mg \cdot g^{-1}$	[99]
		Sb(V)	pH = 6 ± 0.5, 298 K	206.72 $mg \cdot g^{-1}$	
TI/GO@ Fe_3O_4	Powder	Pb(II)	pH = 5, 293 K	461.00 $mg \cdot g^{-1}$	[164]
PAMAM–GO	Powder	Se(IV)	pH = 6	60.9 $mg \cdot g^{-1}$	[139]
		Se(VI)	pH = 6	77.9 $mg \cdot g^{-1}$	
PAA–MGO	Powder	Se(IV)	pH = 5.8	120.1 $mg \cdot g^{-1}$	[140]
		Se(VI)	pH = 5.8	83.7 $mg \cdot g^{-1}$	
NZVI/rGO	Powder	Se(IV)	pH = 3, 25 °C	173.53 $mg \cdot g^{-1}$	[141]
CTAB@GO	Powder	Se(IV)	pH = 2–3	139.95 $mg \cdot g^{-1}$	[165]
		Se(VI)	pH = 2–3	468.01 $mg \cdot g^{-1}$	

3.2. *Rare Metals*

The presence of rare metals is a particular concern in industrial wastewater treatment because they can cause serious environmental impacts and human health risks even at low concentrations. Accumulation of rare metals in the environment may lead to deterioration of water quality and affect the survival of aquatic organisms.

3.2.1. Radioactive Metals

A large amount of radioactive wastewater containing uranium, thorium, and other radioactive metals is generated during the mining and smelting of rare earth ores. These radioactive metals are difficult to reduce. Therefore, it is important to treat radioactive wastewater and reuse the treated water. Adsorption methods have attracted much attention because of the advantages of high selectivity, fast adsorption rate, and high capacity of adsorption materials such as graphene-based adsorption materials [166,167].

GO and its composites are promising materials for the adsorption of radioactive metals in water. Wu et al. [168] used persimmon-tannin-modified GO/CS (PGC) microspheres to investigate their effective adsorption behavior for U(VI) in aqueous solution. It was shown that the maximum adsorption capacity of PGC microspheres for U(VI) reached 199.13 mg·g^{-1} at 303 K, pH = 5, and initial uranium concentration of 10.00 mg/L, and the removal efficiency was 98.2%. Meanwhile, in order to make up for the lack of U(VI) adsorption on GO/CS, Zhou et al. [6] adsorbed U(VI) with PAO/GO/CS. Langmuir adsorption modeling showed that the maximum adsorption of U(VI) by PAO/GO/CS was 701.5, 821.7, and 1220.0 mg·g^{-1} at 293, 303, and 313 K, respectively. U(VI) was adsorbed spontaneously and predominantly chemisorbed onto oxygen-containing moieties. PAO/GO/CS could realize selective adsorption of common coexisting ions to U(VI) and excellent water separation performance.

It is well known that thorium and uranium often co-exist in some natural minerals such as monazite and thorium ores. Pan et al. [63] synthesized α–GOM_2 for the simultaneous removal of thorium/uranium ions from aqueous solutions. The maximum adsorption capacity of α–GOM_2 for Th(IV) was 497.5 mg·g^{-1} and 185.2 mg·g^{-1} for U(VI) in the single-component system (Th(IV) or U(VI)), which was higher than that of GO. In the binary-component system (Th(IV)/U(VI)), the saturated adsorption capacity on α–GOM_2 was 408.8 mg·g^{-1} for Th(IV) and 66.8 mg·g^{-1} for U(VI), which was also higher than that on GO. Based on the analysis of the XPS data, as shown in Figure 11a,b, it is believed that the adsorption process between Th(IV)/U(VI) and α–GOM_2 may involve four kinds of molecular interactions, as shown in Figure 11c, including ligand interactions, electrostatic interactions, cation–π interactions, and Lewis acid–base interactions. These results suggest that α-GOM_2 can be used as a potential adsorbent for the removal and separation of Th(IV)/U(VI) ions from aqueous solutions.

Produced by the fission of uranium 235, Cesium 137 is a radionuclide with a half-life of about 30 years. Due to its similarity in chemical properties to potassium ions, it can easily enter the ecosystem and is one of the hazardous substances in wastewater. The transition metal hexacyanoferrate (MHCF) has become a research hotspot because of its high selectivity for cesium ions, but nanoscale MHCF is difficult to recover. To overcome this problem, Kim et al. [169] used HCF–PEI–rGO composite membranes of GO grafted with polyethyleneimine (PEI) and potassium hexacyanoferrate (KCuHCF) nanoparticles for efficient and selective removal of cesium ions, and this composite was capable of efficiently treating radioactive materials. It was shown that the removal efficiency of the HCF–PEI–rGO membrane for Cs(I) exceeded 98%. In addition, the membrane showed stable cesium removal performance even under simulated general groundwater conditions. These results indicate that the HCF–PEI–rGO membrane can be effectively used for the rapid treatment of radiocaesium-contaminated water, which is promising for practical applications. Prussian blue has been investigated for the selective removal of radiocaesium from water due to its similar hydration radius with cesium ions. To address the difficulty of separating Prussian blue from treated solutions, Kadam et al. [170] utilized magnetic graphene oxide Prussian blue (PSMGPB) nanocomposites stabilized with its biopolymer pectin in order to enhance its properties and prevent aggregation and to improve the cesium adsorption efficiency. For PSMGPB nanocomposites, the adsorption capacity for Cs(I) was 1.609 mmol·g^{-1} at an optimal pH of 7 and an optimal adsorption temperature of 30 °C. This enhanced adsorption efficiency was attributed to the separation of pectin-stabilized GO sheets and the improved distribution of magnetite on the GO surface.

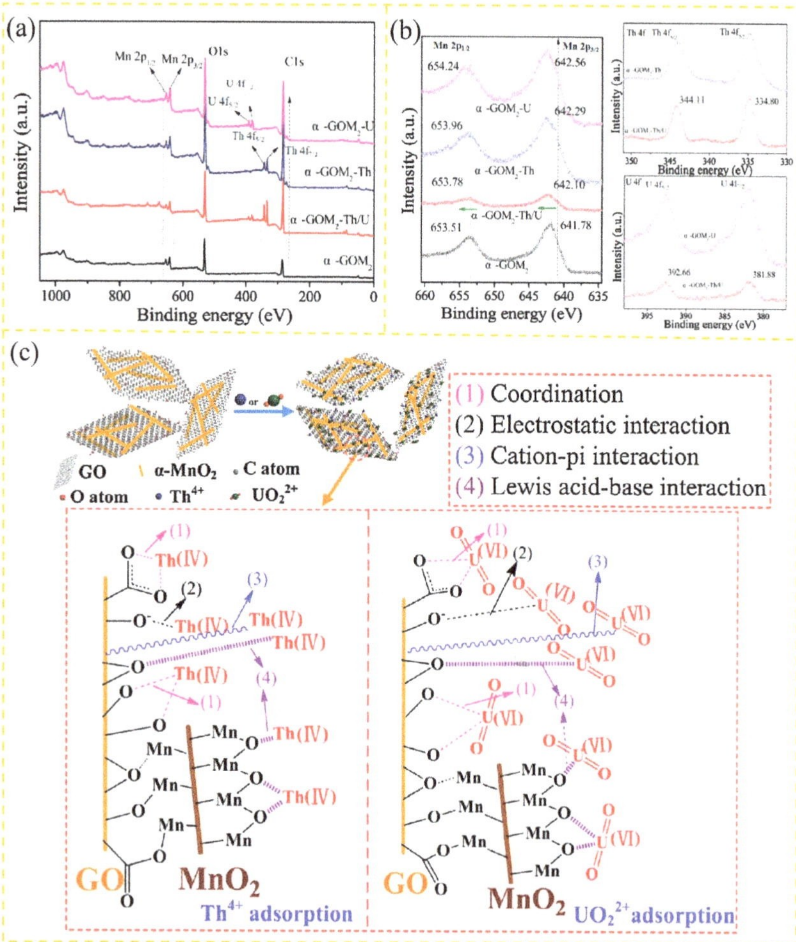

Figure 11. (**a**) XPS survey spectra and (**b**) Mn 2p, Th 4f, and U 4f peaks in the XPS spectra of α–GOM$_2$ before and after adsorbing single Th(IV), single U(VI), and Th(IV)/U(VI). The green arrows indicate the increase in Mn 2p binding energy of α–GOM$_2$ upon adsorption of Th(IV) or U(VI). (**c**) Schematic diagram of possible interaction between Th(IV) or U(VI) and α–GOM$_2$. The yellow arrow and the dotted line box indicate the possible interaction mechanisms in the adsorption of Th(IV) or U(VI) by α–GOM$_2$. Reprinted with permission from reference [63]. Copyright 2016, Elsevier.

3.2.2. Rare Earth Metals

Rare earth metals are mainly derived from ores, which are usually mined in specialized rare earth deposits and extracted through crushing, grinding, chemical treatment, and other steps. Rare earth metals generated during the mining process may enter surface water or groundwater through stormwater flushing or direct discharge. From an environmental perspective, the removal and recovery of rare earth metals can reduce dependence on natural resources and lead to sustainable resource utilization.

Cerium (Ce), as one of the major rare earth ionic pollutants, poses a threat to ecosystems and human health. GO and cellulose are two potential adsorption materials.GO has excellent adsorption properties due to its high specific surface area and abundant functional groups but at a higher cost. Cellulose is a renewable, low-cost natural polymer, but its adsorption capacity is limited by the limited number of functional groups. By combining

GO with modified cellulose, Huang et al. [171] could develop an efficient and low-cost adsorbent (CPAG) for the removal of Ce(III) from water. The results showed that CPAG adsorbed 271.8 mg·g^{-1} of Ce(III) with an adsorption temperature of 25 °C, the pH was 4.2, and the equilibrium contact adsorption time was 25 min. The adsorption mechanism was dominated by the complexation of ion exchange with ionized oxygen-containing functional groups. The composite material combines the advantages of both, improves the adsorption efficiency, reduces the cost, and provides a feasible solution for environmental protection.

Lanthanides, as important fission products, are prominent in environmental impact assessment and high-tech applications. Lanthanum and praseodymium, two of the most common elements, are widely used in various industrial fields, but they also raise concerns about environmental pollution. To address this issue, researchers have developed various adsorbents to remove these elements from water, including modified GO, functional ionic liquids, biomaterials, etc. Abu Elgoud et al. [172] obtained GO–imidazole–FeCN by combining 1-methylimidazole and epichlorohydrin onto graphene oxide, which was further treated with $K_3[Fe(CN)_6]$ composites showed high adsorption affinity for La(III) and Pr(III). In addition, the effects of different parameters on the adsorption process were investigated and the adsorbents were analyzed by various characterization means to optimize their performance and reduce the environmental risks. The experimental adsorption capacities of La(III) and Pr(III) were 785.55 and 867.67 mg·g^{-1}, respectively, and the GO–imidazole–FeCN composites exhibited excellent regeneration–reuse behavior.

3.3. *Other Inorganic Pollutants*

Inorganic pollutants in wastewater come from a wide variety of sources and have a significant impact on the environment. In addition to heavy metal ions, there are fluoride, phosphate, etc.

3.3.1. Fluorides

Improperly disposed of fluorinated wastes from industrial processes can enter water bodies through surface runoff or infiltration, and fluoride contained in agricultural drainage can also contaminate water sources through surface runoff. People who consume fluoride in excess can lead to the development of dental and skeletal fluorosis [173]. The presence of fluoride can disrupt the balance of aquatic ecosystems, so there is a need to remove fluoride from wastewater.

Current methods for removing fluoride from water, such as precipitation and Al^{3+} co-precipitation, are useful but produce toxic sludge. Therefore, Liu et al. [174] developed more environmentally friendly adsorbents, Zr–MCGO functionalized with Fe_3O_4/CS/GO (MCGO) and zirconium (IV) showed excellent adsorptive properties, was easy to separate, and could be operated over a wide pH range. When comparing the fluoride removal capacity of different adsorbents, it was found that the modified Zr–MCGO showed higher adsorption efficiency than the pristine MC and MCGO. The addition of zirconium significantly enhanced the adsorption capacity from 3.93 mg·g^{-1} to 5.78 mg·g^{-1}. Zr–MCGO outperformed Zr–MC, showing that the synergistic effect of GO and chitosan provided more adsorption sites. It was also shown that pH had a significant effect on the adsorption of fluoride, where the adsorption was better between pH 4.0 and 8.0, while high pH decreased the adsorption efficiency. Rashid et al. [175] studied the performance of a GO–CeO_2 nanocomposite for the adsorption removal of excess fluoride from drinking water. The adsorption data revealed that the maximum capacity of the nanocomplex for fluorine ion adsorption at pH 6.5 was 8.61 mg·g^{-1}, which increased to 16.07 mg·g^{-1} at pH 4.0. The experimental results showed that it played a major role in the rapid trapping of aqueous fluoride by ion exchange with ultrafast kinetic properties (<1 min), and the diagram of the mechanism of rapid capture of aqueous fluoride by ion exchange is shown in Figure 12a, and based on the adsorption behavior and XPS data, as shown in Figure 12b,c, it is proposed that the removal of fluoride by GO–CeO_2 is carried out in two steps.

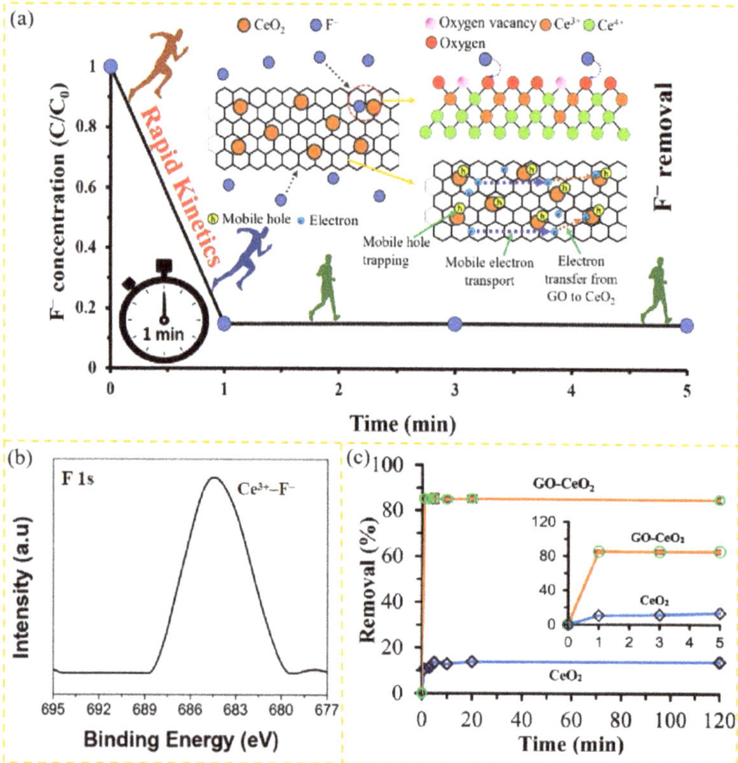

Figure 12. (**a**) Diagram of the mechanism of rapid capture of aqueous fluoride by ion exchange. (**b**) High-resolution XPS of F 1s spectrum after fluoride adsorption. (**c**) Fluoride removal by only CeO_2 NPs; the plot for fluoride removal by GO–CeO_2 is also shown for ease of comparison (Inset: Initial 5 min data zoomed in). Reprinted with permission from reference [175]. Copyright 2021, Elsevier.

3.3.2. Phosphates

Phosphates are widely available compounds in nature and human society, and they play an important role in many biological processes and industrial applications. However, phosphates from over-applied fertilizers enter rivers and lakes through farm drainage and surface runoff, and phosphate-containing detergents and other household cleaning products enter water bodies through municipal sewage systems. And phosphate is one of the major contributors to the eutrophication of water bodies, which leads to a decline in aquatic biodiversity and deterioration of water quality. In this context, control measures can be used, such as the adsorption and removal of phosphorus.

In order to effectively remove phosphate, Vasudevan et al. [176] tested the ability of graphene to adsorb phosphate from an aqueous solution by examining the experimental parameters, including pH, concentration, and temperature, and evaluated the associated adsorption isotherms, kinetics, and thermodynamic properties based on the experimental data. The experimental results showed that graphene is an excellent phosphate adsorbent with an adsorption capacity of up to 89.37 mg·g^{-1} under the experimental conditions of an initial phosphate concentration of 100 mg·L^{-1} and a temperature of 303 K. The results showed that graphene is an excellent phosphate adsorbent with a high adsorption capacity of up to 89.37 mg·g^{-1}.

To enhance the adsorption capacity as well as reusability of graphene adsorbent material, Eltaweil et al. [177] prepared a reusable lanthanum doped aminated graphene oxide/aminated chitosan microspheres (La-AmGO@AmCs) for the efficient removal of phosphate ions. The adsorption kinetic process conformed to the quasi-secondary kinetics and the proposed Langmuir model fitting, and the adsorption isotherm conformed to the Langmuir model fitting, with the maximum adsorption capacity of 125.00 $mg{\cdot}g^{-1}$. The removal rates of La-AmGO@AmCs microspheres were all above 86% after reuse five times. The adsorption mechanisms of LaAmGO@AmCs on phosphate ions include electrostatic interaction, intrasphere complexation, Lewis acid-base interaction, ion exchange, protonation, precipitation, and hydrogen bonding. The experimental results showed that the La-AmGO@AmCs microspheres could be used for the removal and recovery of phosphorus from wastewater. Wang et al. [178] researched the performance of a 3D graphene/La$(OH)_3$-nanorod aerogel for the removal and recovery of phosphate from wastewater. The adsorption data revealed that GLA-10 exhibited a maximum phosphorus adsorption capacity of 76.85 $mg{\cdot}g^{-1}$. The mechanism of phosphorus adsorption by GLA-10 can be divided into three stages, as shown in Figure 13a: (1) the –OH on the surface of La$(OH)_3$ is replaced by phosphate into the solution; (2) the exchanged phosphates form an inner-sphere ligand with La; and (3) the LaP coordination complex was transformed into $LaPO_4$-0.5H_2O crystals. The anomalous selectivity was further confirmed by comparing the XPS spectra before and after adsorption even in the presence of high concentrations of dissolved organic matter or various competing anions, as shown in Figure 13b,c.

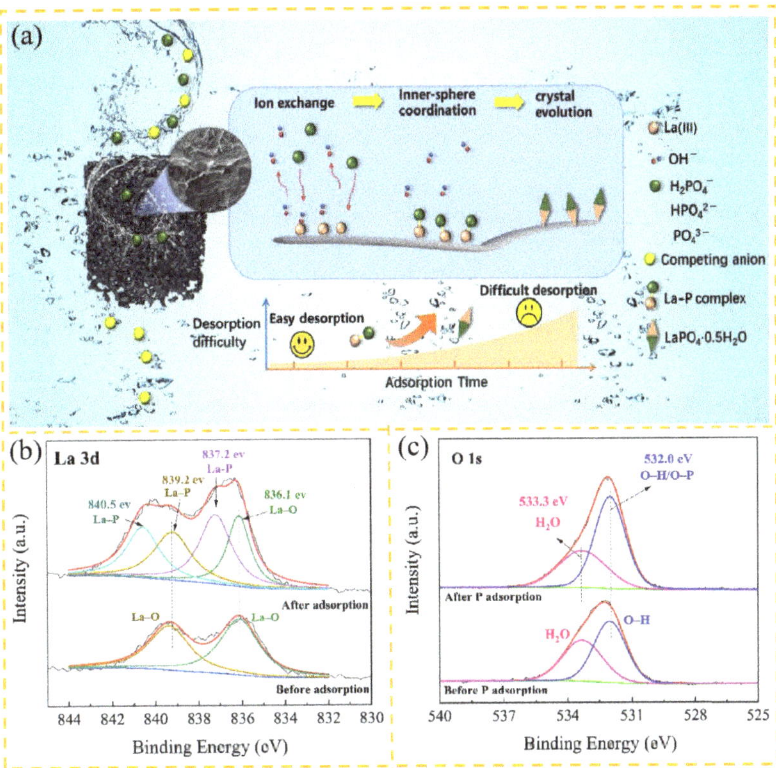

Figure 13. (**a**) Schematic illustration for the process of P adsorption by GLA-10. La 3d (**b**) and (**c**) O 1s XPS spectra of GLA-10 before and after P adsorption. Reprinted with permission from reference [178]. Copyright 2022, Elsevier.

4. Adsorption of Organic Pollutants with Graphene-Based Adsorbents

4.1. Organic Dyes

The presence of organic dyes in aquatic environments poses a significant threat to aquatic animals and plants [179]. Dye-containing wastewater obstructs light penetration, weakens photosynthesis in aquatic plants and algae, and reduces dissolved oxygen levels, which ultimately results in water eutrophication [180]. Dyes in wastewater are categorized into two main groups: cationic dyes and anionic dyes. When treating these dye wastewaters, different technologies and methods are required to target the removal of these dye ions. Many researchers have used different methods to remove organic dyes [181], and adsorption has been found to be easier than these methods and is thus promising for dye removal.

4.1.1. Organic Cationic Dyes

Cationic dyes such as rhodamine B (RhB), malachite green (MG), ethyl violet (EV), methylene blue (MB), nile blue (NB), and crystal violet (CV) are widely used in various industries such as food, textiles, cosmetics, paper, paint, printing, and others [182]. Due to their poor biodegradability, the discharge of dye-containing wastewater into water bodies leads to serious water pollution.

A carbon-based material with superhydrophobicity, porous graphene (PG), has been used as an effective adsorbent for dye removal [183]. Recently, a new technique was demonstrated for a greener approach to synthesizing PG, and the efficacy of this new green PG for the removal of seven emerging contaminants, such as RhB, was investigated for the first time [184]. The rGO–ZnS nanocomposites were used as adsorbent materials for two dyes, MG and EV [185]. The removal rate was rapid during the first 20 min. This fast initial adsorption process is due to the high number of available active sites on the nanocomposites. The adsorption data for both dyes fit well with the pseudo-second-order kinetic model, with adsorption capacities of 27.54 $mg{\cdot}g^{-1}$ and 20.04 $mg{\cdot}g^{-1}$ for MG and EV, respectively.

GO contains active surface oxygen-containing group pairs with a high capacity for the adsorption of organic dyes. Jayawardena et al. [186] investigated the adsorption performance of GO for the removal of MB from an aqueous solution by synthesizing GO with different degrees of oxidation. The results showed that the optimized GO materials had very high adsorption capacities, with the most effective samples having adsorption capacities of 2294 $mg{\cdot}g^{-1}$ at pH 7 and 2301 $mg{\cdot}g^{-1}$ at pH 12. By comparing the properties of the GO materials before and after adsorption, it was found that the removal of MB is affected by a number of factors, including the porosity of the materials, electrostatic attraction, hydrogen bonding, and π–π interaction. In addition, the best-performing GO materials showed excellent reproducibility, as they maintained a removal efficiency of more than 90% after five cycles. Li et al. [27] adsorbed MB using a magnetic humic acid/GO composite (MHAGO). The results of their adsorption experiments demonstrated that 30 mg of MHAGO adsorbent can treat 100 mL of aqueous solution containing 5.0 $mg{\cdot}L^{-1}$ MB at 45 °C, achieving a maximum MB adsorption capacity of approximately 59.00 $mg{\cdot}g^{-1}$. The effects of the initial MB concentration and pH on MB adsorption onto MHAGO are shown in Figure 14a,b. The mechanism for this adsorption by MHAGO is illustrated in Figure 14c, revealing a potential involvement of hydrogen bonds and π–π interactions during the process. As an innovative biological adsorbent, MHAGO exhibits exceptional performance in dye wastewater treatment, providing valuable insights for future research on recyclability and reuse.

In order to find effective adsorbents that can remove and adsorb dyes quickly, many research groups have investigated GO composites as adsorbents. Dang et al. [187] used carbon foam/CNTs/GO (CF/CNTs/GO) and investigated the adsorption properties of this material for MB. The experimental results showed that the prepared CF/CNTs/GO adsorbents exhibited remarkable removal efficiency (73%) and capacity (146 $mg{\cdot}g^{-1}$) for MB.

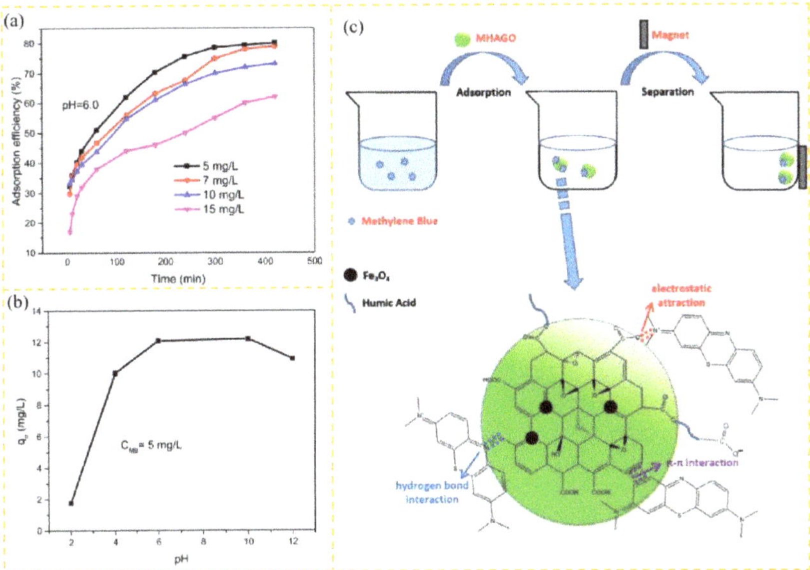

Figure 14. (**a**) Effect of contact time and initial MB concentration on adsorption efficiency of MHAGO for MB. (**b**) Effect of pH on equilibrium adsorption capacity of MHAGO for MB. (**c**) Schematic illustration of adsorption mechanisms for MHAGO. Reprinted with permission from reference [27]. Copyright 2021, Elsevier.

4.1.2. Organic Anionic Dyes

Anionic dyes such as Congo red (GR), orange G (OG), methyl orange (MO), and bromophenol blue (BB) have good water solubility, which allows them to penetrate the fibers uniformly during the dyeing process. Some anionic dyes are considered mutagenic and carcinogenic, and long-term exposure may increase the risk of cancer. Therefore, the development and adoption of effective dye wastewater treatment technologies is the key to reducing these hazards.

CR is a chemical substance with a complex aromatic composition that is not easily degraded. Rashad et al. [188] investigated the adsorption process of CR dye on the surface of mixed-phase copper oxide-graphene heterostructured nanocomposites. The results showed that the compound achieved 75% adsorption efficiency for CR. In order to reduce the adsorption cost, Zheng et al. [189] used CR as a dye pollutant model and obtained the raw material from used lithium-ion batteries, and investigated the effects of graphene aerogel input, initial concentration of pollutant, initial pH, temperature, and adsorption time on the adsorption of CR. The experimental results showed that the adsorption rate was as high as 99% under the optimal conditions. The adsorption process was multimolecular layer adsorption, and the quasi-second-order kinetic equation indicated that the adsorption might be chemisorption. The maximum adsorption capacity was 53.91 $mg{\cdot}g^{-1}$.

OG is a compound with acidic dye properties that is mainly used to color textiles, food, and pharmaceuticals. Since its molecule contains sulfonate groups that dissociate in water to form negatively charged anions, it can affect water quality and ecosystem balance when discharged directly into water bodies without treatment. GO has received much attention due to its excellent surface properties in the adsorption of dye wastewater. Yang et al. [190] used 3-aminopropyltriethoxysilane modified GO (AS–GO) to remove OG. The results showed that AS–GO had good adsorption capacity for OG. During the reaction process, the maximum adsorption capacity of OG was 576.6 $mg{\cdot}g^{-1}$ at an initial OG concentration of 100 $mg{\cdot}L^{-1}$, an initial adsorbent dose of 2.5 $g{\cdot}L^{-1}$, T = 313 K, and pH = 3. The reaction mechanism is shown in Figure 15a. The FTIR spectra proved the existence of π–π and hydrogen bonding interactions, as shown in Figure 15b. The XRD of AS–GO and AS–GO–OG showed no significant changes, indicating the stability of the adsorbent, as shown in Figure 15c. After four desorption–adsorption cycles, the removal rate remained above 86%, as shown in Figure 15d.

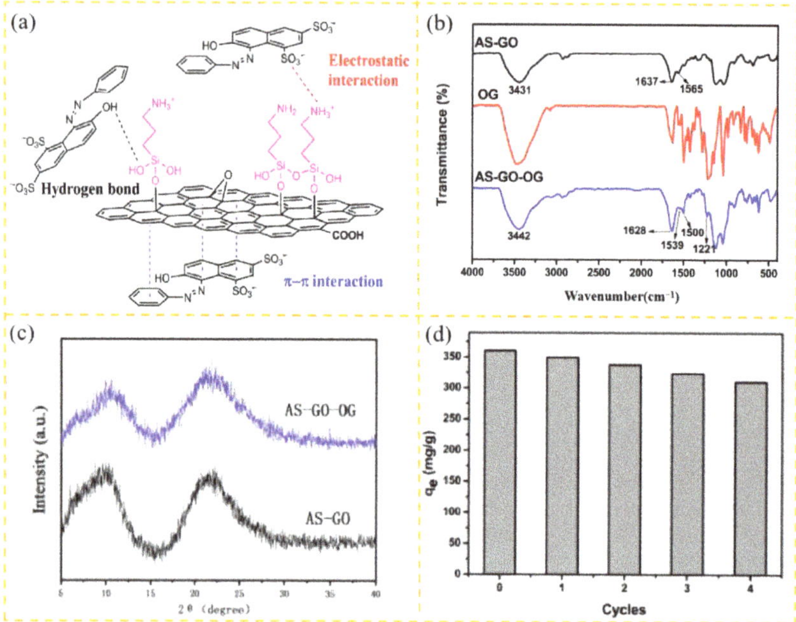

Figure 15. (**a**) Schematic illustration of adsorption mechanisms by AS−GO. (**b**) FTIR spectra of AS–GO, OG, and AS−GO−OG. (**c**) XRD spectra of AS−GO and AS−GO−OG. (**d**) Effect of regeneration cycle on equilibrium adsorption capacity of AS–GO for OG [190]. Copyright 2022, MDPI.

MO belongs to azo compounds, which may lead to color changes in water bodies after entering the water body, affecting not only water quality and aesthetics but also the survival of aquatic organisms. Labiadh et al. [191] investigated the performance of free-standing 3D graphene nano-edge as an effective adsorbent for the removal of MO from aqueous solution. This graphene material exhibits excellent MO absorption over a wide pH range from 2 to 11. It was found that the π–π interactions and hydrogen bonding existing between MO and graphene edges were responsible for the substantial adsorption of MO. Three kinetic models, Frendluich, Temkin, and Langmuir, were used to evaluate the experimental equilibrium data, among which the latter provided the best fit for the adsorption process, according to which the adsorption capacity of the monolayer was derived as 27.932 $mg{\cdot}g^{-1}$.

4.1.3. Summary of Adsorption of Organic Dyes

A summary is given herein for a more intuitive comparison of the adsorption performances of various graphene-based adsorbents for dye (see Table 2).

Table 2. Adsorption capacities of graphene-based adsorbents for selected dye.

Adsorbent		Organic Dye	Adsorption Conditions	Adsorption Capacity ($mg \cdot g^{-1}$)	Reference
Material	Form				
GO	Powder	MB	pH = 12	2301	[186]
MHAGO	Powder	MB	pH = 6.0, 45 °C	59.00	[27]
β-CD/GO	Powder	MB	30 °C, 1 h	76.4	[192]
RL–GO	Powder	MB	pH = 11, 318 K	581.40	[193]
GO–TSC–GO	Powder	MB	pH > 7, 25 °C, 120 min	596.642	[194]
GO	Powder	MB	pH > 7, 25 °C, 120 min	196.8	[194]
GO/DETA/$MnFe_2O_4$@SiO_2	Powder	MB	pH = 7, 25 °C	243.91	[195]
MFTGS	Sponge	MB	308 K, 10 min	224	[86]
CF/CNTs/GO	Foam	MB	C_0 = 100 $mg \cdot L^{-1}$	146	[187]
APAM/DTPA–CS/GO	Aerogel	MB	303 K, 95 min	652.99	[196]
GO/PEI	Aerogel	MB	pH = 10.5, 25 ± 1 °C	249.6	[197]
		MO	pH = 2.0, 25 ± 1 °C	331.0	
CS/GO	Aerogel	MB	pH = 8, 293 K	437.29	[198]
CGZ-3	Aerogel	MB	pH = 9, 298 K	427.35	[199]
M–rGO	Aerogel	MB	298 K	450.90	[200]
GO-MMT/SA	Aerogel	MB	60 min	150.66	[201]
TA–rGO	Hydrogel	MB	25 °C	1000	[95]
GO–EDTA–CS	Powder	MB	pH = 8.3, room temperature	141 ± 6.60 $mg \cdot g^{-1}$	[155]
		CV	pH = 8.3, room temperature	121 ± 3.50 $mg \cdot g^{-1}$	
CS/GO	Powder	RhB	pH = 4, 25 °C	858.0	[202]
rGO–ZnS	Powder	MG	25 °C	27.54	[185]
		EV	25 °C	20.04	
Ag–rGO	Powder	NB	pH = 6, 30 °C	36.8	[203]
rGO	Powder	NB	pH = 6, 30 °C	20	[203]
GO	Powder	NB	pH = 6, 30 °C	8.8	[203]
Ag_2O–Al_2O_3–ZrO_2–rGO	Powder	CR	30 °C, 90 min	333.32	[204]
CS/GO	Powder	CR	pH = 7, 300 K	182.5	[205]
		MB	pH = 7, 300 K	219.3	
Graphene	Aerogel	CR	pH = 2, 60 °C	53.91	[189]
CTAB–GO	Powder	CR	pH = 3, 298 K	2767	[206]
G/SnO_2	Powder	CR	pH = 2–4, 318 K	359.71	[207]
AS–GO	Powder	OG	pH = 3, 313 K	576.6	[190]
3D graphene	Powder	MO	pH = 2–11, 20 °C	27.932	[191]
GONs	Powder	MO	pH = 6.7, 298 K	179.21	[208]
		BB	pH = 6.7, 298 K	236.42	

4.2. *Pharmaceuticals*

4.2.1. Antibiotics

The main types of antibiotics found in the water environment include tetracycline, sulfonamides, and fluoroquinolones [209,210]. During the production, transportation, and usage of pharmaceutical antibiotics, a significant amount is released into the environment. Due to their stability and limited natural degradation, coupled with their continuous discharge from sewage treatment plants into the ecological environment, water pollution caused by antibiotic contamination has become increasingly severe. Such contamination can cause serious harm [211,212], while unnecessary ingestion of antibiotics by humans can result in toxicity [213,214]. Therefore, it is crucial to address antibiotic contamination. With the increasing social awareness of the need for environmental protection and the continuous progress of related technologies, the application of antibiotic adsorption technology has broad prospects and deserves continuous attention and in-depth research. Many researchers have used graphene-based adsorbents to adsorb antibiotics in water for environmental protection. The adsorption results for antibiotics by selected graphene-based adsorbents are shown in Table 3.

Table 3. Adsorption results of antibiotics by graphene-based adsorbents.

Adsorbent		Antibiotic	Adsorption Conditions	Adsorption Capacity/Removal Efficiency	Reference
Material	Form				
Graphene	Powder	Sulfamethoxazole	pH = 7, 298 K	181.3 $\mu g \cdot g^{-1}$	[215]
		Sulfamethazine	pH = 7, 298 K	196.3 $\mu g \cdot g^{-1}$	
		Sulfadiazine	pH = 7, 298 K	184.9 $\mu g \cdot g^{-1}$	
		Cefalexin	pH = 7, 298 K	207.2 $\mu g \cdot g^{-1}$	
		Olfloxain	pH = 7, 298 K	199.2 $\mu g \cdot g^{-1}$	
		Amoxicillin	pH = 7, 298 K	198.2 $\mu g \cdot g^{-1}$	
		Tetracycline	pH = 7, 298 K	200.2 $\mu g \cdot g^{-1}$	
GO	Powder	Sulfamethazine	pH = 5	240 $mg \cdot g^{-1}$	[216]
		Ciprofloxacin	pH = 9	379 $mg \cdot g^{-1}$	
GO	Powder	Doxycycline	pH = 6, 25 °C	116.5099 $mg \cdot g^{-1}$	[217]
		Ciprofloxacin	pH = 6, 25 °C	156.7634 $mg \cdot g^{-1}$	
		Tetracycline	pH = 6, 25 °C	80.980 $mg \cdot g^{-1}$	
GO	Powder	Chlortetracycline	pH = 5, 298 K	218.60 $mg \cdot g^{-1}$	[218]
		Tetracycline	pH = 5, 308 K	182.91 $mg \cdot g^{-1}$	
		Oxytetracycline	pH = 5, 308 K	176.62 $mg \cdot g^{-1}$	
SGO	Powder	Sparfloxacin	pH = 5.5, 25 °C	1428.57 $\mu mol \cdot g^{-1}$	[219]
GO/α–ATP	Powder	Tetracycline	pH = 5, 55 °C	38.73 $mg \cdot g^{-1}$	[220]
CS–GO	Powder	Metronidazole	Neutral pH, 298 K	29.76 $mg \cdot g^{-1}$	[221]
		Ciprofloxacin	Neutral pH, 298 K	102.04 $mg \cdot g^{-1}$	
CD–DGO	Powder	Sulfamethoxazole	pH = 2, 35 °C	144.2804 $mg \cdot g^{-1}$	[222]
		Sulfadiazine	pH = 2, 35 °C	152.2888 $mg \cdot g^{-1}$	
MGO	Powder	Erythromycin	pH = 3, 298 K	285.71 $mg \cdot g^{-1}$	[223]
TS–GO	Powder	Ciprofloxacin	pH = 5, 25 °C	769.23 $mg \cdot g^{-1}$	[224]
		Amoxicillin	pH = 3, 25 °C	625 $mg \cdot g^{-1}$	
GO–CNF	Aerogel	Doxycycline	25 °C	469.7 $mg \cdot g^{-1}$	[225]
		Chlortetracycline	25 °C	396.5 $mg \cdot g^{-1}$	
		Oxytetracycline	25 °C	386.5 $mg \cdot g^{-1}$	
		Tetracycline	25 °C	343.8 $mg \cdot g^{-1}$	
Alginate/rGO	Hydrogel	Tetracycline	pH = 8, 298 K	290.70 $mg \cdot g^{-1}$	[226]
		Ciprofloxacin	pH = 8, 298 K	344.83 $mg \cdot g^{-1}$	

Adsorption by activated carbon, graphene, and GO is one of the most attractive methods to remove antibiotics [84,227], but the pH has a strong influence on the adsorption of antibiotics. Li et al. [218] utilized GO as the main material to explore the adsorption performance of GO on tetracycline antibiotics under different pH conditions, and the experimental results showed that the pH is an important factor affecting the adsorption performance of GO. The functional groups on the surface of GO may change at a high pH, which affects its interaction with antibiotics. The adsorption results showed that the maximum adsorption capacity of GO for the three antibiotics was achieved at pH 5, with the highest adsorption capacity of 167.0 $mg \cdot g^{-1}$ for chloro-tetracycline. This finding is of great significance for understanding and optimizing the ability of GO to remove antibiotics from water in practical water treatment processes. To achieve better adsorption, GO polymers have been used as antibiotic adsorbents. Yu et al. [222] prepared a novel ternary composite, β-cyclodextrin/dopamine hydrochloride–GO (CD–DGO), and investigated its adsorption properties on sulfonamide antibiotics (sulfonamides). The effects of different doses of the composite, the ratio of the components, and the pH of the solution on the adsorption capacity were experimentally evaluated. The results showed that the maximum adsorption capacities of CD–DGO for sulfamethoxazole (SMX) and sulfadiazine (SDZ) were 144 $mg \cdot g^{-1}$ and 152 $mg \cdot g^{-1}$ at pH = 2, respectively.

For easier separation of nanomaterials from water after adsorption, the preparation of nanomaterial hydrogels and polymer composite hydrogels proved to be useful. Zhuang et al. [226] investigated the performance of adsorption for the removal of antibiotics from water using modified alginate/graphene bi-network porous hydrogels. It was shown that hydrogen bonding had a greater effect on adsorption than carboxyl groups. The adsorption results revealed maximum capacities of 290.70 mg·g^{-1} for tetracycline and 344.83 mg·g^{-1} for ciprofloxacin in the modified hydrogel. Antibiotic adsorption is not only of great significance for protecting the environment and promoting sustainable development but also plays a key role in enhancing economic benefits, safeguarding public health, promoting scientific research, and strengthening environmental monitoring and management.

4.2.2. Other Organic Pharmaceuticals

In addition to antibiotics, there are other medications whose dangers should not be overlooked, such as sulfamethoxazole, carbamazepine, ketoprofen, valsartan, diclofenac, and amphetamines (ATDs). Most of these drugs originate from synthesis or extraction processes in the pharmaceutical industry. Entry into water sources through improper handling or discharge may lead to contamination. Removal of these drugs is necessary to ensure that water quality is safe and meets regulatory requirements.

Granular activated carbon (GAC) is widely used for the removal of active pharmaceutical ingredients (APIs), but these adsorbents suffer from degraded performance and expensive regeneration due to the presence of natural organic macromolecules (NOMs) [228,229]. A stable, self-regenerating adsorbent, metal oxide nanocomposite (S–MGC) containing TiO_2 and SiO_2 combined with 3D GO, which adsorbs APIs and is regenerated by light, was used by Fu et al. Removal of APIs in a variety of environments, including five model APIs: sulfamethoxazole, carbamazepine, ketoprofen, valsartan, and diclofenac. In the absence of TiO_2/SiO_2, the adsorption performance of 3D GO (CGB) was superior to that of GAC, while S–MGCs further enhanced the adsorption capacity of CGB.

ATDs are harmful in a variety of ways, primarily including damage to the blood system, nervous system, skin, and liver. Cao et al. [230] used deep eutectic solvent-functionalized composites (ZMG–DES) as adsorbents for the adsorption of three typical ATDs (MDMA, MAM, and AM). The adsorption schematic is shown in Figure 16a. The results of batch experiments showed that the maximum adsorption of MDMA (933.652 μg·g^{-1}) was 2.3 and 2.8 times higher than that of MAM (412.849 μg·g^{-1} and AM (328.652 μg·g^{-1}), respectively. The affected in different environments as shown in Figure 16b–d. ZMG–DES exhibited a fast adsorption rate (within 1 h), good affinity (96.76%, 74.33%, and 61.35% removal, respectively), and excellent reusability (five cycles) for MDMA, MAM, and AM. Quantum chemical theory and Multiwfn were used to study the weak interactions, especially the differences in H-bonding. In addition, mechanisms such as π–π, van der Waals forces, hydrophobic interactions, and electrostatic attraction were involved in the adsorption.

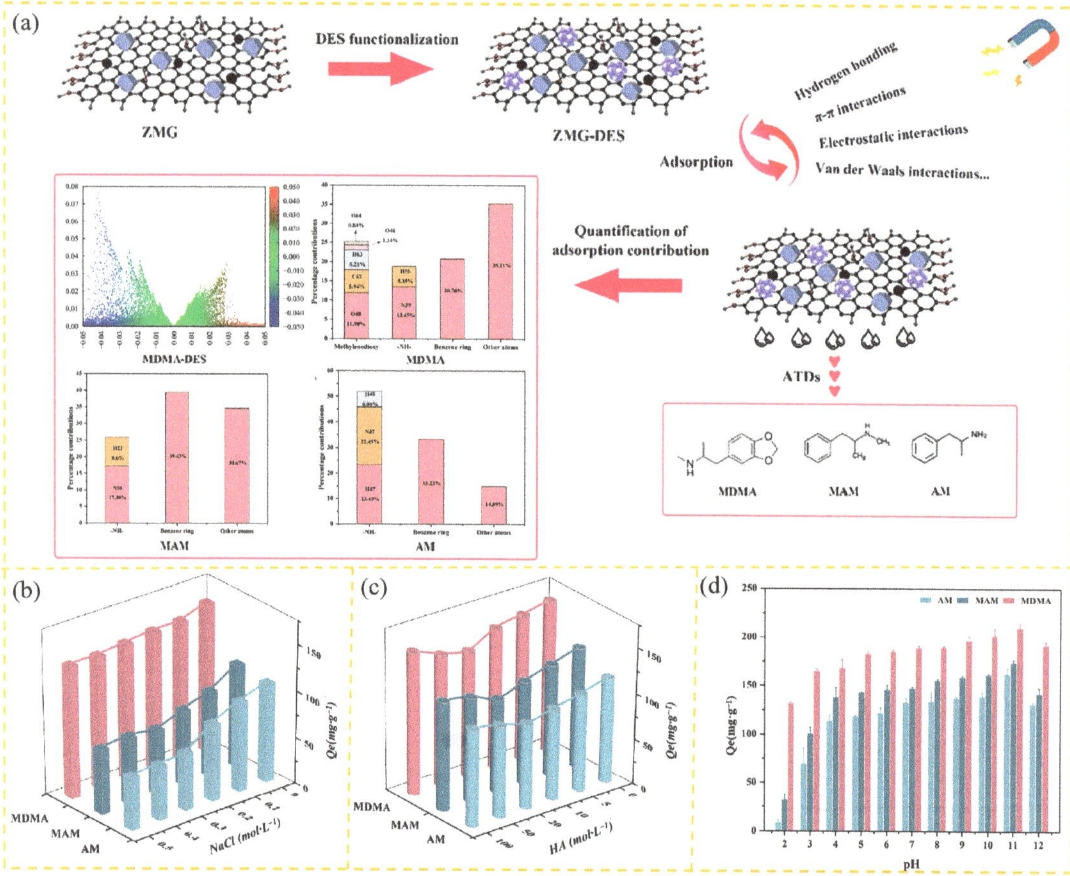

Figure 16. (**a**) Schematic diagram of ZMG–DES adsorbed ATDs. Effects of (**b**) NaCl concentration, (**c**) HA concentration, (**d**) solution pH of three ATD adsorption on ZMG–DES. Reprinted with permission from reference [230]. Copyright 2024, Elsevier.

4.3. Aromatic Pollutants

Most aromatic compounds are recognized as toxic pollutants, among which phenolic compounds such as phenol, nitrobenzene, bisphenol A, and triclosan, as well as polycyclic aromatic hydrocarbons such as anthracene (ANT), 2-methyl anthraquinone (2-MAQ), quinoline, are of great concern.

Phenolic compounds are highly toxic and can cause skin burns, respiratory irritation, and poisoning in humans. They seriously pollute water bodies, disrupt the ecological balance, and affect food safety. Phenol is toxic and tends to accumulate in the food chain, inducing heart disease and leukemia, thus posing an irreversible risk to public health [231]. In order to develop a recyclable and efficient adsorbent for the adsorption of phenol, Cui et al. [232] used 3D nitrogen-doped graphene oxide aerogel (NGOA) for the adsorption of phenol and the static saturated adsorption capacity for phenol reached 108.19 $mg \cdot g^{-1}$, which was significantly higher than that of most adsorbents. Importantly, monolithic NGOA can be easily separated and recovered after adsorption saturation. Finally, the adsorption mechanism of phenol on NGOA was proposed, as shown in Figure 17a. The adsorption of phenol on NGOA mainly consists of porous adsorption, π–π interactions, hydrogen bonding, and electrostatic gravitational forces, which are physical adsorption. Nitrophenols are highly toxic and may cause central nervous system damage and organ dysfunction with prolonged exposure, requiring stringent safety measures to minimize their impact on human health and the environment. In order to remove nitrobenzene, Severo et al. [233] used a 3D graphene sponge material (WS-Graphene) and used it to adsorb *o*- and *p*-nitrophenol from wastewater, as shown in Figure 17b. When the concentration of both adsorbents was fixed at 700 $mg \cdot L^{-1}$, the adsorption process of both nitrophenols was very rapid, reaching equilibrium in less than 5 min. In addition, the maximum adsorption capacities were 1842 $mg \cdot g^{-1}$ for *o*-nitrophenol and 879.6 $mg \cdot g^{-1}$ for *p*-nitrophenol at 30 °C. For the removal of bisphenol A (BPA) and triclosan (TCS), Wang et al. [234] investigated the adsorption behavior of BPA and TCS on graphene and compared it with activated carbon, and the adsorption schematic is shown in Figure 17c. The maximum adsorption capacities of BPA on both graphene and activated carbon reached about 2 $mg \cdot g^{-1}$. In contrast, the strong partitioning capacity of TCS on graphene indicated that graphene material can be used for the removal of TCS from wastewater. This phenomenon should be attributed to the establishment of electrostatic repulsion between anionic BPA (or TCS) molecules and graphene (or activated carbon) surfaces at higher pH conditions.

Polycyclic aromatic hydrocarbons (PAHs) are a class of compounds consisting of two or more benzene rings attached to each other. They are widely present in the environment, mainly from incomplete combustion of fossil fuels and wood, spills of petroleum products, etc. PAHs are acutely and chronically toxic to aquatic organisms and are capable of affecting growth, reproduction, and the immune system of living organisms. Song et al. [235] used nitrogen-doped rGO (NRGO) for the removal of PAHs and their oxygen derivatives in aqueous solutions (OPAHs). NRGO showed significant adsorption efficiency for ANT and 2-MAQ. At 298 K and pH = 7, the adsorption capacity of NRGO was 5.77 $mg \cdot g^{-1}$ for ANT and 9.29 $mg \cdot g^{-1}$ for 2-MAQ, which increased more than 2.76 and 2.29 times, respectively, compared with GO and 1.4 times compared with rGO. Based on the characterization results and adsorption experiments, intra-particle diffusion, hydrophobic effects, and π–π interactions may be simultaneously involved in the adsorption process. This work provides ideas for the application of NRGO in the environment and the possibility of removing low concentrations of PAHs and OPAHs from wastewater.

Quinoline is a nitrogen-containing heterocyclic compound, and its presence in wastewater is potentially harmful in various ways. After entering the water body, quinoline may persist in the aqueous environment for a long period and cause persistent effects on the water quality due to its chemical stability. An effective method for the removal of quinoline is adsorption. Kang et al. [236] worked on the adsorption properties of a nitrogen-doped hollow carbon nanospheres/graphene composite aerogel (HCNS/NGA) for quinoline. The results of the adsorption study revealed that HCNS/NGA exhibited a high adsorption capacity for quinoline, reaching 138.37 ± 2.58 $mg \cdot g^{-1}$ at 298 K, demonstrating significant potential for practical applications. The adsorption of quinoline molecules on HCNS/NGA may be the result of the synergistic effect of multiple mechanisms, among which physical adsorption may be dominant, and the mechanism is shown in Figure 17d.

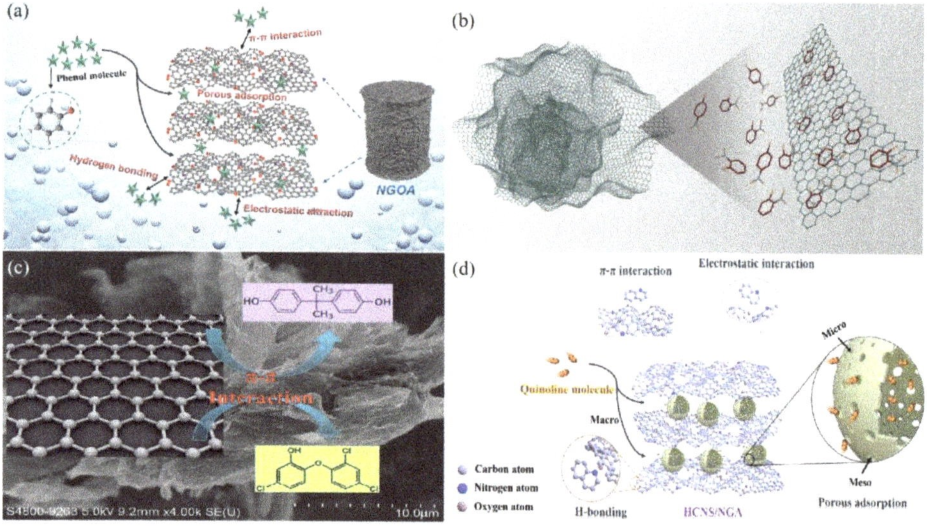

Figure 17. (**a**) The adsorption mechanism of phenol on NGOA. Reprinted with permission from reference [232]. Copyright 2023, Elsevier. (**b**) Schematic diagram of nitrophenol adsorption by WS-Graphene. Reprinted with permission from reference [233]. Copyright 2023, Elsevier. (**c**) Schematic diagram of BPA and TCS adsorption by graphene [234]. Copyright 2017, ACS. (**d**) Schematic illustration of the possible interaction between HCNS/NGA and quinoline molecules. Reprinted with permission from reference [236]. Copyright 2021, Elsevier.

5. Conclusions and Prospects

Graphene, as a two-dimensional material, exhibits considerable research and application potential in the field of adsorbent materials due to its unique physical and chemical properties, such as the high hydrophobicity of graphene materials, good adsorption affinity for organic pollutants, and huge specific surface area. In particular, GO contains abundant oxygen-containing functional groups and hydroxyl and epoxy groups, which provide good adsorption of pollutants in adsorbent water, and its adsorption performance and recycling rate can be improved though chemical or physical modification. There are two main methods for graphene synthesis: bottom-up methods and top-down methods. This paper describes the main preparation methods of graphene and its derivatives, as well as the main preparation methods of 3D graphene. The rapid development of the preparation technology of graphene and its derivatives has enabled a variety of high-performance graphene-based adsorbents to effectively remove toxic pollutants from wastewater, such as heavy metals, rare metals, organic dyes, and pharmaceuticals, with fast kinetics and excellent adsorption performance. Further functionalization of these graphene-based adsorbents through the introduction of specific functional groups or the incorporation of other nanomaterials is expected to significantly improve their performance in removing heavy metals and organic pollutants. This paper further summarized the adsorption results obtained in recent years regarding graphene-based adsorbents on inorganic and organic pollutants, as well as the adsorption mechanisms in recent years.

In the future, continuous production of graphene-based adsorbents using low-cost raw materials has the potential to reduce their cost in the future. Large-scale preparation of cost-effective graphene-based adsorbents for wastewater treatment still requires further research. The application of graphene-based adsorbents in wastewater treatment plants is expected to be a breakthrough for future research. The research directions on graphene-based adsorbents will be directed toward developing more environmentally friendly methods for graphene preparation, reducing the utilization of hazardous chemicals, minimizing production costs, and enhancing the sustainability of graphene. By precisely manipulating the layer count, pore size distribution, and surface properties of graphene, it is possible to design graphene-based materials with enhanced selectivity and adsorption capacities. Innovative functionalization strategies, such as biomolecular modification and magnetic nanoparticle incorporation, are being explored to augment the performance of graphene-based materials in specific applications. A comprehensive understanding of the interaction mechanism between graphene-based materials and pollutants offers theoretical guidance for the development of more efficient adsorbents. Graphene-based adsorbents hold significant potential for application in the field of environmental purification. Through continuous technological innovation and extensive research, they are expected to address current environmental issues and make substantial contributions to safeguarding human health and preserving the ecological environment.

Author Contributions: Conceptualization, S.W.; writing—original draft preparation, G.L. and R.D.; writing—review and editing, Z.C., C.L., J.X. and X.M.; visualization, G.L.; project administration, S.W.; funding acquisition, S.W. All authors have read and agreed to the published version of the manuscript.

Funding: This research was funded by the National Key Research and Development Program of China (No. 2023YFC2908203), the Fundamental Research Funds for the Central Universities of Central South University, China (No. 2022ZZTS0492), and the Natural Science Foundation of Hunan Province, China (No. 2018JJ2484).

Data Availability Statement: The data presented in this study are available on request from the corresponding author.

Conflicts of Interest: The authors declare that they have no known competing financial interests or personal relationships that could have appeared to influence the work reported in this paper.

References

1. Sarath, N.G.; Puthur, J.T. Heavy metal pollution assessment in a mangrove ecosystem scheduled as a community reserve. *Wetl. Ecol. Manag.* **2021**, *29*, 719–730. [CrossRef]
2. Liu, M.; Zhang, L.; Wang, M.; Wang, X.; Cui, H.; Wei, J.; Li, X. The role of metal-organic frameworks in removing emerging contaminants in wastewater. *J. Clean. Prod.* **2023**, *429*, 139526. [CrossRef]
3. Wójcik, G. Sorption behaviors of light lanthanides(III) (La(III), Ce(III), Pr(III), Nd(III)) and Cr(III) using nitrolite. *Materials* **2020**, *13*, 2256. [CrossRef] [PubMed]
4. Vergara-Araya, M.; Oeltze, H.; Radeva, J.; Roth, A.G.; Göbbert, C.; Niestroj-Pahl, R.; Dähne, L.; Wiese, J. Operation of hybrid membranes for the removal of pharmaceuticals and pollutants from water and wastewater. *Membranes* **2022**, *12*, 502. [CrossRef]
5. Duan, G.; Cao, Z.; Zhong, H.; Ma, X.; Wang, S. Highly efficient poly(6-acryloylamino-N-hydroxyhexanamide) resin for adsorption of heavy metal ions. *J. Environ. Manag.* **2022**, *308*, 114631. [CrossRef]
6. Zhou, S.; Xie, Y.; Zhu, F.; Gao, Y.; Liu, Y.; Tang, Z.; Duan, Y. Amidoxime modified chitosan/graphene oxide composite for efficient adsorption of U(VI) from aqueous solutions. *J. Environ. Chem. Eng.* **2021**, *9*, 106363. [CrossRef]
7. Tang, X.; Xue, H.; Li, J.; Wang, S.; Yu, J.; Zeng, T. Degradation of bisphenol A by nitrogen-rich ZIF-8-derived carbon materials-activated peroxymonosulfate. *Toxics* **2024**, *12*, 359. [CrossRef]
8. Cui, Y.; Li, S.; Yu, N.; Yu, X.; Ji, X.; Wang, L. Highly biocompatible hemoglobin-stabilized gold nanoparticles for an enhanced catalytic reduction of 4-nitrophenol. *Inorganics* **2024**, *12*, 136. [CrossRef]
9. Benalia, M.C.; Youcef, L.; Bouaziz, M.G.; Achour, S.; Menasra, H. Removal of heavy metals from industrial wastewater by chemical precipitation: Mechanisms and sludge characterization. *Arab. J. Sci. Eng.* **2022**, *47*, 5587–5599. [CrossRef]
10. Wang, L.; Luo, Y.; Li, H.; Yu, D.; Wang, Y.; Wang, W.; Wu, M. Preparation and selective adsorption of surface-imprinted microspheres based on hyperbranched polyamide-functionalized sodium alginate for the removal of Sb(III). *Colloids Surf. A Physicochem. Eng. Asp.* **2020**, *585*, 124106. [CrossRef]
11. Wang, S.; Li, J.; Narita, H.; Tanaka, M. Solvent extraction equilibrium modeling for the separation of ammonia, nickel(II), and copper(II) from the loaded LIX84-I. *Miner. Eng.* **2021**, *172*, 107132. [CrossRef]
12. Pei, J.; Huang, L.; Jiang, H.; Liu, H.; Liu, X.; Hu, X. Inhibitory effect of hydrogen ion on the copper ions separation from acid solution across graphene oxide membranes. *Sep. Purif. Technol.* **2019**, *210*, 651–658. [CrossRef]
13. Wang, J.; Yang, F.; Wang, S.; Zhong, H.; Wu, Z.-k.; Cao, Z.-f. Reactivation of nano-Fe_3O_4/diethanolamine/rGO catalyst by using electric field in Fenton reaction. *J. Taiwan Inst. Chem. Eng.* **2019**, *99*, 113–122. [CrossRef]
14. Jiang, C.-L.; Wang, R.; Chen, X.; Zheng, L.-G.; Cheng, H. Preparation of chitosan modified fly ash under acid condition and its adsorption mechanism for Cr(VI) in water. *J. Cent. South Univ.* **2021**, *28*, 1652–1664. [CrossRef]

15. Shi, T.-S.; Jiang, F.; Wang, P.; Yue, T.; Sun, W. Deep purification of As(V) in drinking water by silica gel loaded with FeOOH and MnO_2. *J. Cent. South Univ.* **2021**, *28*, 1692–1706. [CrossRef]
16. Qiu, S.; Yan, L.; Jing, C. Simultaneous removal of arsenic and antimony from mining wastewater using granular TiO_2: Batch and field column studies. *J. Environ. Sci.* **2019**, *75*, 269–276. [CrossRef] [PubMed]
17. Guo, X.; Wu, Z.; He, M.; Meng, X.; Jin, X.; Qiu, N.; Zhang, J. Adsorption of antimony onto iron oxyhydroxides: Adsorption behavior and surface structure. *J. Hazard. Mater.* **2014**, *276*, 339–345. [CrossRef]
18. Ighalo, J.O.; Iwuozor, K.O.; Igwegbe, C.A.; Adeniyi, A.G. Verification of pore size effect on aqueous-phase adsorption kinetics: A case study of methylene blue. *Colloids Surf. A Physicochem. Eng. Asp.* **2021**, *626*, 127119. [CrossRef]
19. Alshabib, M.; Oluwadamilare, M.A.; Tanimu, A.; Abdulazeez, I.; Alhooshani, K.; Ganiyu, S.A. Experimental and DFT investigation of ceria-nanocomposite decorated AC derived from groundnut shell for efficient removal of methylene-blue from wastewater effluent. *Appl. Surf. Sci.* **2021**, *536*, 147749. [CrossRef]
20. Yu, W.; Xu, J.; Li, J.; Zhu, S.; Xie, J.; Zhou, Z.; Wang, B.; Li, J.; Chen, K. Hollow structured kapok fiber-based hierarchical porous biocarbons for ultrahigh adsorption of organic dyes. *Ind. Eng. Chem. Res.* **2022**, *61*, 4114–4124. [CrossRef]
21. Chen, Q.; Zhang, Q.; Yang, Y.; Wang, Q.; He, Y.; Dong, N. Synergetic effect on methylene blue adsorption to biochar with gentian violet in dyeing and printing wastewater under competitive adsorption mechanism. *Case Stud. Therm. Eng.* **2021**, *26*, 101099. [CrossRef]
22. Park, C.; Engel, E.S.; Crowe, A.; Gilbert, T.R.; Rodriguez, N.M. Use of carbon nanofibers in the removal of organic solvents from water. *Langmuir* **2000**, *16*, 8050–8056. [CrossRef]
23. Mishakov, I.V.; Bauman, Y.I.; Brzhezinskaya, M.; Netskina, O.V.; Shubin, Y.V.; Kibis, L.S.; Stoyanovskii, V.O.; Larionov, K.B.; Serkova, A.N.; Vedyagin, A.A. Water purification from chlorobenzenes using heteroatom-functionalized carbon nanofibers produced on self-organizing Ni-Pd catalyst. *J. Environ. Chem. Eng.* **2022**, *10*, 107873. [CrossRef]
24. Dehghani, Z.; Ostovari, F.; Sharifi, S. A comparison of the crystal structure and optical properties of reduced graphene oxide and aminated graphene nanosheets for optoelectronic device applications. *Optik* **2023**, *274*, 170551. [CrossRef]
25. Kim, J.B.; Koo, S.H.; Kim, I.H.; Kim, J.T.; Kim, J.G.; Jayaraman, B.; Lim, J.; Kim, S.O. Characteristic dual-domain composite structure of reduced graphene oxide and its application to higher specific capacitance. *Chem. Eng. J.* **2022**, *446*, 137390. [CrossRef]
26. Bryan, M.Y.K.; Chai, P.V.; Law, J.Y.; Mahmoudi, E. Graphene oxide-chitosan composite material as adsorbent in removing methylene blue dye from synthetic wastewater. *Mater. Today Proc.* **2022**, *64*, 1587–1596. [CrossRef]
27. Li, D.; Hua, T.; Yuan, J.; Xu, F. Methylene blue adsorption from an aqueous solution by a magnetic graphene oxide/humic acid composite. *Colloids Surf. A Physicochem. Eng. Asp.* **2021**, *627*, 127171. [CrossRef]
28. Hua, Y.; Li, F.; Hu, N.; Fu, S.-Y. Frictional characteristics of graphene oxide-modified continuous glass fiber reinforced epoxy composite. *Compos. Sci. Technol.* **2022**, *223*, 109446. [CrossRef]
29. Jun, S.-Y.; Park, S.H.; Sohn, M.K.; Kim, S.; Lee, J.M.; Kong, D.S.; Lee, T.-Y.; Jung, J.H.; Kim, M.-S.; Yoo, S.; et al. Reduction time effect on the dielectric characteristics of reduced-graphene-oxide--encapsulated barium titanate powder fillers. *Carbon* **2022**, *199*, 23–32. [CrossRef]
30. Anandaraj, C.; Venkatapathy, R.; Bharath Sabarish, V.C.; Kalaivani, P.; Durairajan, A.; Graça, M.P.; Valente, M.A.; Gajendiran, J.; Gokul Raj, S.; Ramesh Kumar, G. Structural characteristics, optical, electrical and electrochemical sensing properties of graphene and multi walled carbon nanotube admixtured bismuth iron oxide composite ceramics. *Ceram. Int.* **2021**, *47*, 28042–28049. [CrossRef]
31. Hu, H.; Wen, W.; Ou, J.Z. Construction of adsorbents with graphene and its derivatives for wastewater treatment: A review. *Environ. Sci. Nano* **2022**, *9*, 3226–3276. [CrossRef]
32. Wang, Y.; Gai, Z.; Guo, F.; Zhang, M.; Zhang, L.; Xia, G.; Chai, X.; Ren, Y.; Zhang, X.; Jiang, X. A diamond/graphene/diamond eectrode for waste water treatment. *Nanomaterials* **2023**, *13*, 3043. [CrossRef] [PubMed]
33. Faysal Hossain, M.D.; Akther, N.; Zhou, Y. Recent advancements in graphene adsorbents for wastewater treatment: Current status and challenges. *Chin. Chem. Lett.* **2020**, *31*, 2525–2538. [CrossRef]
34. Asif, F.C.; Saha, G.C. Graphene-like carbon structure synthesis from biomass pyrolysis: A critical review on feedstock–process–properties relationship. *C* **2023**, *9*, 31. [CrossRef]
35. Geim, A.K.; Novoselov, K.S. The rise of graphene. *Nat. Mater.* **2007**, *6*, 183–191. [CrossRef] [PubMed]
36. Laraba, S.R.; Luo, W.; Rezzoug, A.; Zahra, Q.u.a.; Zhang, S.; Wu, B.; Chen, W.; Xiao, L.; Yang, Y.; Wei, J.; et al. Graphene-based composites for biomedical applications. *Green Chem. Lett. Rev.* **2022**, *15*, 724–748. [CrossRef]
37. Brzhezinskaya, M.; Irzhak, A.; Irzhak, D.; Kang, T.W.; Kononenko, O.; Matveev, V.; Panin, G.; Roshchupkin, D. Direct growth of graphene film on piezoelectric $La_3Ga_{5.5}Ta_{0.5}O_{14}$ crystal. *Phys. Status Solidi (RRL)—Rapid Res. Lett.* **2016**, *10*, 639–644. [CrossRef]
38. Kononenko, O.; Brzhezinskaya, M.; Zotov, A.; Korepanov, V.; Levashov, V.; Matveev, V.; Roshchupkin, D. Influence of numerous Moiré superlattices on transport properties of twisted multilayer graphene. *Carbon* **2022**, *194*, 52–61. [CrossRef]
39. Santhiran, A.; Iyngaran, P.; Abiman, P.; Kuganathan, N. Graphene synthesis and its recent advances in applications—A review. *C* **2021**, *7*, 76. [CrossRef]
40. Jana, A.; Scheer, E.; Polarz, S. Synthesis of graphene–transition metal oxide hybrid nanoparticles and their application in various fields. *Beilstein J. Nanotechnol.* **2017**, *8*, 688–714. [CrossRef]
41. Pedrazzetti, L.; Gibertini, E.; Bizzoni, F.; Russo, V.; Lucotti, A.; Nobili, L.; Magagnin, L. Graphene growth on electroformed copper substrates by atmospheric pressure CVD. *Materials* **2022**, *15*, 1572. [CrossRef] [PubMed]

42. Sun, L.; Chen, B.; Wang, W.; Li, Y.; Zeng, X.; Liu, H.; Liang, Y.; Zhao, Z.; Cai, A.; Zhang, R.; et al. Toward epitaxial growth of misorientation-free graphene on Cu(III) foils. *ACS Nano* **2022**, *16*, 285–294. [CrossRef]
43. Olorunkosebi, A.A.; Eleruja, M.A.; Adedeji, A.V.; Olofinjana, B.; Fasakin, O.; Omotoso, E.; Oyedotun, K.O.; Ajayi, E.O.B.; Manyala, N. Optimization of graphene oxide through various Hummers' methods and comparative reduction using green approach. *Diam. Relat. Mater.* **2021**, *117*, 108456. [CrossRef]
44. Vacacela Gomez, C.; Guevara, M.; Tene, T.; Villamagua, L.; Usca, G.T.; Maldonado, F.; Tapia, C.; Cataldo, A.; Bellucci, S.; Caputi, L.S. The liquid exfoliation of graphene in polar solvents. *Appl. Surf. Sci.* **2021**, *546*, 149046. [CrossRef]
45. Tan, H.; Navik, R.; Liu, Z.; Xiang, Q.; Zhao, Y. Scalable massive production of defect-free few-layer graphene by ball-milling in series with shearing exfoliation in supercritical CO_2. *J. Supercrit. Fluids* **2022**, *181*, 105496. [CrossRef]
46. Edison, T.N.J.I.; Atchudan, R.; Karthik, N.; Chandrasekaran, P.; Perumal, S.; Arunachalam, P.; Raja, P.B.; Sethuraman, M.G.; Lee, Y.R. Electrochemically exfoliated graphene sheets as electrode material for aqueous symmetric supercapacitors. *Surf. Coat. Technol.* **2021**, *416*, 127150. [CrossRef]
47. Rabchinskii, M.K.; Sysoev, V.V.; Glukhova, O.E.; Brzhezinskaya, M.; Stolyarova, D.Y.; Varezhnikov, A.S.; Solomatin, M.A.; Barkov, P.V.; Kirilenko, D.A.; Pavlov, S.I.; et al. Guiding graphene derivatization for the on-chip multisensor arrays: From the synthesis to the theoretical background. *Adv. Mater. Technol.* **2022**, *7*, 2101250. [CrossRef]
48. Marcano, D.C.; Kosynkin, D.V.; Berlin, J.M.; Sinitskii, A.; Sun, Z.; Slesarev, A.; Alemany, L.B.; Lu, W.; Tour, J.M. Improved synthesis of graphene oxide. *ACS Nano* **2010**, *4*, 4806–4814. [CrossRef]
49. Peng, L.; Xu, Z.; Liu, Z.; Wei, Y.; Sun, H.; Li, Z.; Zhao, X.; Gao, C. An iron-based green approach to 1-h production of single-layer graphene oxide. *Nat. Commun.* **2015**, *6*, 5716. [CrossRef]
50. Ye, C.; Wang, G.; Yuan, H.; Li, J.; Ni, K.; Pan, F.; Guo, M.; Wu, Y.; Ji, H.; Zhang, F.; et al. Microfluidic oxidation of graphite in two minutes with capability of real-time monitoring. *Adv. Mater.* **2022**, *34*, 2107083. [CrossRef]
51. Zhu, Y.; Kong, G.; Pan, Y.; Liu, L.; Yang, B.; Zhang, S.; Lai, D.; Che, C. An improved Hummers method to synthesize graphene oxide using much less concentrated sulfuric acid. *Chin. Chem. Lett.* **2022**, *33*, 4541–4544. [CrossRef]
52. Zhou, A.; Yu, T.; Liang, X.; Yin, S. H_2O_2-free strategy derived from Hummers method for preparing graphene oxide with high oxidation degree. *FlatChem* **2023**, *38*, 100487. [CrossRef]
53. Sahila Grace, A.; Littis Malar, G.S.P. Synthesis and characterization of graphene oxide from coconut husk ash. *Orient. J. Chem.* **2020**, *36*, 348–352. [CrossRef]
54. Amir Faiz, M.S.; Che Azurahanim, C.A.; Yazid, Y.; Suriani, A.B.; Siti Nurul Ain, M.J. Preparation and characterization of graphene oxide from tea waste and it's photocatalytic application of TiO_2/graphene nanocomposite. *Mater. Res. Express* **2020**, *7*, 015613. [CrossRef]
55. Sujiono, E.H.; Zurnansyah; Zabrian, D.; Dahlan, M.Y.; Amin, B.D.; Samnur; Agus, J. Graphene oxide based coconut shell waste: Synthesis by modified Hummers method and characterization. *Heliyon* **2020**, *6*, e04568. [CrossRef]
56. Tohamy, H.-A.S.; El-Sakhawy, M.; Kamel, S. Development of magnetite/graphene oxide hydrogels from agricultural wastes for water treatment. *J. Alloys Compd.* **2022**, *10*, 1889–1909. [CrossRef]
57. Harres, A.; Garcia, W.J.S.; Salles, T.R.; Bruckmann, F.S.; Sulzenco, J.B.; Schneider, A.D.; Rhoden, C.R.B. Magnetic properties of graphene oxide decorated with magnetite nanoparticles. *Diam. Relat. Mater.* **2023**, *138*, 110238. [CrossRef]
58. Shen, J.; Shi, M.; Ma, H.; Yan, B.; Li, N.; Ye, M. Hydrothermal synthesis of magnetic reduced graphene oxide sheets. *Mater. Res. Bull.* **2011**, *46*, 2077–2083. [CrossRef]
59. Wang, X.; Yu, J. Application of Fe_3O_4/graphene oxide composite for the separation of Cs(I) and Sr(II) from aqueous solution. *J. Radioanal. Nucl. Chem.* **2015**, *303*, 807–813. [CrossRef]
60. Chong, S.; Zhang, G.; Tian, H.; Zhao, H. Rapid degradation of dyes in water by magnetic Fe^0/Fe_3O_4/graphene composites. *J. Environ. Sci.* **2016**, *44*, 148–157. [CrossRef]
61. Das, T.K.; Sakthivel, T.S.; Jeyaranjan, A.; Seal, S.; Bezbaruah, A.N. Ultra-high arsenic adsorption by graphene oxide iron nanohybrid: Removal mechanisms and potential applications. *Chemosphere* **2020**, *253*, 126702. [CrossRef]
62. Mosleh, N.; Joolaei Ahranjani, P.; Parandi, E.; Rashidi Nodeh, H.; Nawrot, N.; Rezania, S.; Sathishkumar, P. Titanium lanthanum three oxides decorated magnetic graphene oxide for adsorption of lead ions from aqueous media. *Environ. Res.* **2022**, *214*, 113831. [CrossRef]
63. Pan, N.; Li, L.; Ding, J.; Li, S.; Wang, R.; Jin, Y.; Wang, X.; Xia, C. Preparation of graphene oxide-manganese dioxide for highly efficient adsorption and separation of Th(IV)/U(VI). *J. Hazard. Mater.* **2016**, *309*, 107–115. [CrossRef]
64. Bagbi, Y.; Solanki, P.R. Fabrication of mesoporous silica nanoparticle-decorated graphene oxide sheets for the effective removal of lead (Pb^{2+}) from water. *ACS Omega* **2023**, *9*, 304–316. [CrossRef]
65. Singh, S.; Anil, A.G.; Khasnabis, S.; Kumar, V.; Nath, B.; Adiga, V.; Kumar Naik, T.S.S.; Subramanian, S.; Kumar, V.; Singh, J.; et al. Sustainable removal of Cr(VI) using graphene oxide-zinc oxide nanohybrid: Adsorption kinetics, isotherms and thermodynamics. *Environ. Res.* **2022**, *203*, 111891. [CrossRef]
66. Li, X.; Wang, Z.; Li, Q.; Ma, J.; Zhu, M. Preparation, characterization, and application of mesoporous silica-grafted graphene oxide for highly selective lead adsorption. *Chem. Eng. J.* **2015**, *273*, 630–637. [CrossRef]
67. Li, L.; Zhao, L.; Ma, J.; Tian, Y. Preparation of graphene oxide/chitosan complex and its adsorption properties for heavy metal ions. *Green Process. Synth.* **2020**, *9*, 294–303. [CrossRef]

68. Sherlala, A.I.A.; Raman, A.A.A.; Bello, M.M.; Buthiyappan, A. Adsorption of arsenic using chitosan magnetic graphene oxide nanocomposite. *J. Environ. Manag.* **2019**, *246*, 547–556. [CrossRef] [PubMed]
69. Xia, H.; Ren, Q.; Lv, J.; Wang, Y.; Feng, Z.; Li, Y.; Wang, C.; Liu, Y.; Wang, Y. Hydrothermal fabrication of phytic acid decorated chitosan-graphene oxide composites for efficient and selective adsorption of uranium (VI). *J. Environ. Chem. Eng.* **2023**, *11*, 110760. [CrossRef]
70. Zhang, L.; Luo, H.; Liu, P.; Fang, W.; Geng, J. A novel modified graphene oxide/chitosan composite used as an adsorbent for Cr(VI) in aqueous solutions. *Int. J. Biol. Macromol.* **2016**, *87*, 586–596. [CrossRef] [PubMed]
71. Abd-Elhamid, A.I.; Elgoud, E.M.A.; Aly, H.F. Alginate modified graphene oxide for rapid and effective sorption of some heavy metal ions from an aqueous solution. *Cellulose* **2022**, *29*, 6231–6245. [CrossRef]
72. Ma, X.; Duan, G.-y.; Huang, J.-q.; Yang, J.; Cao, Z.-f.; Wang, S. Preparation of graphene oxide/polyiminodiacetic acid resin as a high-performance adsorbent for Cu(II). *J. Cent. South Univ.* **2023**, *30*, 3881–3896. [CrossRef]
73. Arshad, F.; Selvaraj, M.; Zain, J.; Banat, F.; Abu Haija, M. Polyethylenimine modified graphene oxide hydrogel composite as an efficient adsorbent for heavy metal ions. *Sep. Purif. Technol.* **2019**, *209*, 870–880. [CrossRef]
74. Peer, F.E.; Bahramifar, N.; Younesi, H. Removal of Cd (II), Pb (II) and Cu (II) ions from aqueous solution by polyamidoamine dendrimer grafted magnetic graphene oxide nanosheets. *J. Taiwan Inst. Chem. Eng.* **2018**, *87*, 225–240. [CrossRef]
75. Ma, Y.-X.; Xing, D.; Shao, W.-J.; Du, X.-Y.; La, P.-Q. Preparation of polyamidoamine dendrimers functionalized magnetic graphene oxide for the adsorption of Hg(II) in aqueous solution. *J. Colloid Interface Sci.* **2017**, *505*, 352–363. [CrossRef]
76. Nardecchia, S.; Carriazo, D.; Ferrer, M.L.; Gutiérrez, M.C.; del Monte, F. Three dimensional macroporous architectures and aerogels built of carbon nanotubes and/or graphene: Synthesis and applications. *Chem. Soc. Rev.* **2013**, *42*, 794–830. [CrossRef]
77. Ren, R.-P.; Li, W.; Lv, Y.-K. A robust, superhydrophobic graphene aerogel as a recyclable sorbent for oils and organic solvents at various temperatures. *J. Colloid Interface Sci.* **2017**, *500*, 63–68. [CrossRef]
78. Wang, Y.; Xie, W.; Liu, H.; Gu, H. Hyperelastic magnetic reduced graphene oxide three-dimensional framework with superb oil and organic solvent adsorption capability. *Adv. Compos. Hybrid Mater.* **2020**, *3*, 473–484. [CrossRef]
79. Tiwari, J.N.; Mahesh, K.; Le, N.H.; Kemp, K.C.; Timilsina, R.; Tiwari, R.N.; Kim, K.S. Reduced graphene oxide-based hydrogels for the efficient capture of dye pollutants from aqueous solutions. *Carbon* **2013**, *56*, 173–182. [CrossRef]
80. Sahraei, R.; Sekhavat Pour, Z.; Ghaemy, M. Novel magnetic bio-sorbent hydrogel beads based on modified gum tragacanth/graphene oxide: Removal of heavy metals and dyes from water. *J. Clean. Prod.* **2017**, *142*, 2973–2984. [CrossRef]
81. Yu, B.; Bai, Y.; Ming, Z.; Yang, H.; Chen, L.; Hu, X.; Feng, S.; Yang, S.-T. Adsorption behaviors of tetracycline on magnetic graphene oxide sponge. *Mater. Chem. Phys.* **2017**, *198*, 283–290. [CrossRef]
82. Jayanthi, S.; KrishnaRao Eswar, N.; Singh, S.A.; Chatterjee, K.; Madras, G.; Sood, A.K. Macroporous three-dimensional graphene oxide foams for dye adsorption and antibacterial applications. *RSC Adv.* **2016**, *6*, 1231–1242. [CrossRef]
83. Lei, Y.; Chen, F.; Luo, Y.; Zhang, L. Synthesis of three-dimensional graphene oxide foam for the removal of heavy metal ions. *Chem. Phys. Lett.* **2014**, *593*, 122–127. [CrossRef]
84. Wu, R.; Yu, B.; Liu, X.; Li, H.; Wang, W.; Chen, L.; Bai, Y.; Ming, Z.; Yang, S.-T. One-pot hydrothermal preparation of graphene sponge for the removal of oils and organic solvents. *Appl. Surf. Sci.* **2016**, *362*, 56–62. [CrossRef]
85. Bagoole, O.; Rahman, M.M.; Shah, S.; Hong, H.; Chen, H.; Al Ghaferi, A.; Younes, H. Functionalized three-dimensional graphene sponges for highly efficient crude and diesel oil adsorption. *Environ. Sci. Pollut. Res.* **2018**, *25*, 23091–23105. [CrossRef]
86. Maimaiti, T.; Hu, R.; Yuan, H.; Liang, C.; Liu, F.; Li, Q.; Lan, S.; Yu, B.; Yang, S.-T. Magnetic Fe_3O_4/TiO_2/graphene sponge for the adsorption of methylene blue in aqueous solution. *Diam. Relat. Mater.* **2022**, *123*, 108811. [CrossRef]
87. Sun, Y.; Yu, F.; Li, L.; Ma, J. Adsorption-reduction synergistic effect for rapid removal of Cr (VI) ions on superelastic NH_2-graphene sponge. *Chem. Eng. J.* **2021**, *421*, 129933. [CrossRef]
88. Myung, Y.; Jung, S.; Tung, T.T.; Tripathi, K.M.; Kim, T. Graphene-based aerogels derived from biomass for energy storage and environmental remediation. *ACS Sustain. Chem. Eng.* **2019**, *7*, 3772–3782. [CrossRef]
89. Kaushik, J.; Gunture; Tripathi, K.M.; Singh, R.; Sonkar, S.K. Thiourea-functionalized graphene aerogel for the aqueous phase sensing of toxic Pb(II) metal ions and H_2O_2. *Chemosphere* **2022**, *287*, 132105. [CrossRef] [PubMed]
90. Lee, B.; Lee, S.; Lee, M.; Jeong, D.H.; Baek, Y.; Yoon, J.; Kim, Y.H. Carbon nanotube-bonded graphene hybrid aerogels and their application to water purification. *Nanoscale* **2015**, *7*, 6782–6789. [CrossRef] [PubMed]
91. Li, S.; Xu, J.; Yao, G.; Liu, H. Self-adhesive, self-healable, and triple-responsive hydrogel doped with polydopamine as an adsorbent toward methylene blue. *Ind. Eng. Chem. Res.* **2019**, *58*, 17075–17087. [CrossRef]
92. Pandey, S.; Do, J.Y.; Kim, J.; Kang, M. Fast and highly efficient removal of dye from aqueous solution using natural locust bean gum based hydrogels as adsorbent. *Int. J. Biol. Macromol.* **2020**, *143*, 60–75. [CrossRef]
93. Gutekunst, S.B.; Siemsen, K.; Huth, S.; Moehring, A.; Hesseler, B.; Timmermann, M.; Paulowicz, I.; Mishra, Y.K.; Siebert, L.; Adelung, R.; et al. 3D hydrogels containing interconnected microchannels of subcellular size for capturing human pathogenic acanthamoeba castellanii. *ACS Biomater. Sci. Eng.* **2019**, *5*, 1784–1792. [CrossRef]
94. Taye, A.; Yifru, A.; Getachew, N.; Mehretie, S.; Admassie, S. Adsorption of hexavalent chromium using water hyacinth leaf protein concentrate/graphene oxide hydrogel. *Environ. Monit. Assess.* **2023**, *195*, 1342. [CrossRef]
95. Yao, G.; Liu, X.; Zhang, G.; Han, Z.; Liu, H. Green synthesis of tannic acid functionalized graphene hydrogel to efficiently adsorb methylene blue. *Colloids Surf. A Physicochem. Eng. Asp.* **2021**, *625*, 126972. [CrossRef]

96. Phiri, J.; Johansson, L.-S.; Gane, P.; Maloney, T. A comparative study of mechanical, thermal and electrical properties of graphene-, graphene oxide- and reduced graphene oxide-doped microfibrillated cellulose nanocomposites. *Compos. Part B Eng.* **2018**, *147*, 104–113. [CrossRef]
97. Cao, X.; Zhang, J.; Chen, S.; Varley, R.J.; Pan, K. 1D/2D nanomaterials synergistic, compressible, and response rapidly 3D graphene aerogel for piezoresistive sensor. *Adv. Funct. Mater.* **2020**, *30*, 2003618. [CrossRef]
98. Deng, M.; Zhao, L.; Wang, Z.; Yang, P.; Sun, Y. Preparation of phosphoric-modified aloe vera/chitosan aerogels and their efficient adsorption of U(VI). *Environ. Sci. Pollut. Res.* **2023**, *30*, 33229–33242. [CrossRef] [PubMed]
99. Nundy, S.; Ghosh, A.; Nath, R.; Paul, A.; Tahir, A.A.; Mallick, T.K. Reduced graphene oxide (rGO) aerogel: Efficient adsorbent for the elimination of antimony (III) and (V) from wastewater. *J. Hazard. Mater.* **2021**, *420*, 126554. [CrossRef] [PubMed]
100. Sreedevi, P.R.; Suresh, K.; Jiang, G. Bacterial bioremediation of heavy metals in wastewater: A review of processes and applications. *J. Water Process Eng.* **2022**, *48*, 102884. [CrossRef]
101. Bodrud-Doza, M.; Islam, S.M.D.-U.; Hasan, M.T.; Alam, F.; Haque, M.M.; Rakib, M.A.; Asad, M.A.; Rahman, M.A. Groundwater pollution by trace metals and human health risk assessment in central west part of Bangladesh. *Groundw. Sustain. Dev.* **2019**, *9*, 100219. [CrossRef]
102. Manzoor, Q.; Shahab, M.R.; Sajid, A.; Yaseen, H.M.; Alqahtani, F.O.; Malik, Q.M.; Nazir, A.; Arif, K.; Iqbal, M. Eco-benign preparation of biosorbent using momordica charantia for the efficient removal of Cr(VI) ions from wastewater. *Z. Für Phys. Chem.* **2022**, *236*, 1461–1491. [CrossRef]
103. Ghosal, P.S.; Kattil, K.V.; Yadav, M.K.; Gupta, A.K. Adsorptive removal of arsenic by novel iron/olivine composite: Insights into preparation and adsorption process by response surface methodology and artificial neural network. *J. Environ. Manag.* **2018**, *209*, 176–187. [CrossRef]
104. da Silva, E.B.; Mussoline, W.A.; Wilkie, A.C.; Ma, L.Q. Arsenic removal and biomass reduction of As-hyperaccumulator Pteris vittata: Coupling ethanol extraction with anaerobic digestion. *Sci. Total Environ.* **2019**, *666*, 205–211. [CrossRef]
105. Criscuoli, A.; Figoli, A. Pressure-driven and thermally-driven membrane operations for the treatment of arsenic-contaminated waters: A comparison. *J. Hazard. Mater.* **2019**, *370*, 147–155. [CrossRef]
106. Nieto-Delgado, C.; Gutiérrez-Martínez, J.; Rangel-Méndez, J.R. Modified activated carbon with interconnected fibrils of iron-oxyhydroxides using Mn^{2+} as morphology regulator, for a superior arsenic removal from water. *J. Environ. Sci.* **2019**, *76*, 403–414. [CrossRef]
107. Su, H.; Ye, Z.; Hmidi, N. High-performance iron oxide-graphene oxide nanocomposite adsorbents for arsenic removal. *Colloids Surf. A Physicochem. Eng. Asp.* **2017**, *522*, 161–172. [CrossRef]
108. Tang, S.C.N.; Lo, I.M.C. Magnetic nanoparticles: Essential factors for sustainable environmental applications. *Water Res.* **2013**, *47*, 2613–2632. [CrossRef]
109. Qu, G.; Li, R.; Zhou, Y.; Wu, B.; Cai, Y.; Ning, P. Preparation of ferric nitrate-graphene nanocomposite and its adsorption of arsenic(V) from simulated arsenic-containing wastewater. *Appl. Organomet. Chem.* **2019**, *33*, e5221. [CrossRef]
110. Ren, L.; Xu, J.; Zhang, Y.; Zhou, J.; Chen, D.; Chang, Z. Preparation and characterization of porous chitosan microspheres and adsorption performance for hexavalent chromium. *Int. J. Biol. Macromol.* **2019**, *135*, 898–906. [CrossRef]
111. Pavesi, T.; Moreira, J.C. Mechanisms and individuality in chromium toxicity in humans. *J. Appl. Toxicol.* **2020**, *40*, 1183–1197. [CrossRef]
112. Ben Khalifa, E.; Rzig, B.; Chakroun, R.; Nouagui, H.; Hamrouni, B. Application of response surface methodology for chromium removal by adsorption on low-cost biosorbent. *Chemom. Intell. Lab. Syst.* **2019**, *189*, 18–26. [CrossRef]
113. Panda, H.; Tiadi, N.; Mohanty, M.; Mohanty, C.R. Studies on adsorption behavior of an industrial waste for removal of chromium from aqueous solution. *S. Afr. J. Chem. Eng.* **2017**, *23*, 132–138. [CrossRef]
114. Govindaraju, K.; Vinu, R.; Gautam, R.; Vasantharaja, R.; Niranjan, M.; Sundar, I. Microwave-assisted torrefaction of biomass kappaphycus alvarezii based biochar and magnetic biochar for removal of hexavalent chromium Cr(VI) from aqueous solution. *Biomass Convers. Biorefinery* **2024**, *14*, 3643–3653. [CrossRef]
115. Mondal, N.K.; Chakraborty, S. Adsorption of Cr(VI) from aqueous solution on graphene oxide prepared from graphite: Equilibrium, kinetic and thermodynamic studies. *Appl. Water Sci.* **2020**, *10*, 61. [CrossRef]
116. Heidari, A.; Younesi, H.; Mehraban, Z.; Heikkinen, H. Selective adsorption of Pb(II), Cd(II), and Ni(II) ions from aqueous solution using chitosan–MAA nanoparticles. *Int. J. Biol. Macromol.* **2013**, *61*, 251–263. [CrossRef]
117. Kołodyńska, D. Chitosan as an effective low-cost sorbent of heavy metal complexes with the polyaspartic acid. *Chem. Eng. J.* **2011**, *173*, 520–529. [CrossRef]
118. Jia, B.; Li, G.; Cao, E.; Luo, J.; Zhao, X.; Huang, H. Recent progress of antibacterial hydrogels in wound dressings. *Mater. Today Bio* **2023**, *19*, 100582. [CrossRef]
119. Shang, M.-R.; Liu, Y.-G.; Liu, S.-B.; Zeng, G.-M.; Tan, X.-F.; Jiang, L.-H.; Huang, X.-X.; Ding, Y.; Guo, Y.-M.; Wang, S.-F. A novel graphene oxide coated biochar composite: Synthesis, characterization and application for Cr(VI) removal. *RSC Adv.* **2016**, *6*, 85202–85212. [CrossRef]
120. Kumar, P.; Chauhan, M.S. Adsorption of chromium (VI) from the synthetic aqueous solution using chemically modified dried water hyacinth roots. *J. Environ. Chem. Eng.* **2019**, *7*, 103218. [CrossRef]
121. Worku, Z.; Tibebu, S.; Nure, J.F.; Tibebu, S.; Moyo, W.; Ambaye, A.D.; Nkambule, T.T.I. Adsorption of chromium from electroplating wastewater using activated carbon developed from water hyacinth. *BMC Chem.* **2023**, *17*, 85. [CrossRef] [PubMed]

122. Wang, J.; Chen, Y.; Sun, T.; Saleem, A.; Wang, C. Enhanced removal of Cr(III)-EDTA chelates from high-salinity water by ternary complex formation on DETA functionalized magnetic carbon-based adsorbents. *Ecotoxicol. Environ. Saf.* **2021**, *209*, 111858. [CrossRef] [PubMed]
123. Kinuthia, G.K.; Ngure, V.; Beti, D.; Lugalia, R.; Wangila, A.; Kamau, L. Levels of heavy metals in wastewater and soil samples from open drainage channels in Nairobi, Kenya: Community health implication. *Sci. Rep.* **2020**, *10*, 8434. [CrossRef]
124. Narayana, P.L.; Lingamdinne, L.P.; Karri, R.R.; Devanesan, S.; AlSalhi, M.S.; Reddy, N.S.; Chang, Y.-Y.; Koduru, J.R. Predictive capability evaluation and optimization of Pb(II) removal by reduced graphene oxide-based inverse spinel nickel ferrite nanocomposite. *Environ. Res.* **2022**, *204*, 112029. [CrossRef]
125. Yu, J.-G.; Yu, L.-Y.; Yang, H.; Liu, Q.; Chen, X.-H.; Jiang, X.-Y.; Chen, X.-Q.; Jiao, F.-P. Graphene nanosheets as novel adsorbents in adsorption, preconcentration and removal of gases, organic compounds and metal ions. *Sci. Total Environ.* **2015**, *502*, 70–79. [CrossRef] [PubMed]
126. Akhavan, O.; Bijanzad, K.; Mirsepah, A. Synthesis of graphene from natural and industrial carbonaceous wastes. *RSC Adv.* **2014**, *4*, 20441–20448. [CrossRef]
127. Azam, M.G.; Kabir, M.H.; Shaikh, M.A.A.; Ahmed, S.; Mahmud, M.; Yasmin, S. A rapid and efficient adsorptive removal of lead from water using graphene oxide prepared from waste dry cell battery. *J. Water Process Eng.* **2022**, *46*, 102597. [CrossRef]
128. Lingamdinne, L.P.; Godlaveeti, S.K.; Angaru, G.K.R.; Chang, Y.-Y.; Nagireddy, R.R.; Somala, A.R.; Koduru, J.R. Highly efficient surface sequestration of Pb^{2+} and Cr^{3+} from water using a Mn_3O_4 anchored reduced graphene oxide: Selective removal of Pb^{2+} from real water. *Chemosphere* **2022**, *299*, 134457. [CrossRef]
129. Mu, R.; Liu, B.; Chen, X.; Wang, N.; Yang, J. Adsorption of Cu (II) and Co (II) from aqueous solution using lignosulfonate/chitosan adsorbent. *Int. J. Bio. Macromol.* **2020**, *163*, 120–127. [CrossRef]
130. Ranjith, K.S.; Manivel, P.; Rajendrakumar, R.T.; Uyar, T. Multifunctional ZnO nanorod-reduced graphene oxide hybrids nanocomposites for effective water remediation: Effective sunlight driven degradation of organic dyes and rapid heavy metal adsorption. *Chem. Eng. J.* **2017**, *325*, 588–600. [CrossRef]
131. Santos, S.; Ungureanu, G.; Boaventura, R.; Botelho, C. Selenium contaminated waters: An overview of analytical methods, treatment options and recent advances in sorption methods. *Sci. Total Environ.* **2015**, *521*, 246–260. [CrossRef] [PubMed]
132. Awual, M.R.; Yaita, T.; Suzuki, S.; Shiwaku, H. Ultimate selenium(IV) monitoring and removal from water using a new class of organic ligand based composite adsorbent. *J. Hazard. Mater.* **2015**, *291*, 111–119. [CrossRef] [PubMed]
133. Qureshi, S.S.; Memon, S.A.; Rafi ul, Z.; Ram, N.; Saeed, S.; Mubarak, N.M.; Karri, R.R. Rapid adsorption of selenium removal using iron manganese-based micro adsorbent. *Sci. Rep.* **2022**, *12*, 17207. [CrossRef]
134. Qi, Z.; Liu, R.; Joshi, T.P.; Peng, J.; Qu, J. Highly efficient removal of selenite by electrolysis-assisted nano-zerovalent iron (nZVI): Implication for corrosion and reduction. *Chem. Eng. J.* **2021**, *405*, 126564. [CrossRef]
135. Okonji, S.O.; Dominic, J.A.; Pernitsky, D.; Achari, G. Removal and recovery of selenium species from wastewater: Adsorption kinetics and co-precipitation mechanisms. *J. Water Process Eng.* **2020**, *38*, 101666. [CrossRef]
136. Latva, S.; Peräniemi, S.; Ahlgrén, M. Study of metal-loaded activated charcoals for the separation and determination of selenium species by energy dispersive X-ray fluorescence analysis. *Anal. Chim. Acta* **2003**, *478*, 229–235. [CrossRef]
137. Dobrowolski, R.; Otto, M. Preparation and evaluation of Fe-loaded activated carbon for enrichment of selenium for analytical and environmental purposes. *Chemosphere* **2013**, *90*, 683–690. [CrossRef]
138. Jankovský, O.; Šimek, P.; Klímová, K.; Sedmidubský, D.; Pumera, M.; Sofer, Z. Highly selective removal of Ga^{3+} ions from Al^{3+}/Ga^{3+} mixtures using graphite oxide. *Carbon* **2015**, *89*, 121–129. [CrossRef]
139. Xiao, W.; Yan, B.; Zeng, H.; Liu, Q. Dendrimer functionalized graphene oxide for selenium removal. *Carbon* **2016**, *105*, 655–664. [CrossRef]
140. Lu, Z.; Yu, J.; Zeng, H.; Liu, Q. Polyamine-modified magnetic graphene oxide nanocomposite for enhanced selenium removal. *Sep. Purif. Technol.* **2017**, *183*, 249–257. [CrossRef]
141. Sun, F.; Zhu, Y.; Liu, X.; Chi, Z. Highly efficient removal of Se(IV) using reduced graphene oxide-supported nanoscale zero-valent iron (nZVI/rGO): Selenium removal mechanism. *Environ. Sci. Pollut. Res.* **2023**, *30*, 27560–27569. [CrossRef] [PubMed]
142. Yoon, Y.; Park, W.K.; Hwang, T.-M.; Yoon, D.H.; Yang, W.S.; Kang, J.-W. Comparative evaluation of magnetite–graphene oxide and magnetite-reduced graphene oxide composite for As(III) and As(V) removal. *J. Hazard. Mater.* **2016**, *304*, 196–204. [CrossRef]
143. Ye, Y.; Yin, D.; Wang, B.; Zhang, Q. Synthesis of three-dimensional Fe_3O_4 graphene aerogels for the removal of Arsenic Ions from Water. *J. Nanomater.* **2015**, *2015*, 864864. [CrossRef]
144. Singh, N.; Naseh, M.F.; Ansari, J.R.; Sarkar, T.; Datta, A. Removal of aqueous arsenic (III) by graphene-based systems at micro-trace level. *Carbon Lett.* **2023**, *33*, 233–243. [CrossRef]
145. Chen, H.; Liu, F.; Cai, C.; Wu, H.; Yang, L. Removal of Hg^{2+} from desulfurization wastewater by tannin-immobilized graphene oxide. *Environ. Sci. Pollut. Res.* **2022**, *29*, 17964–17976. [CrossRef] [PubMed]
146. Li, L.; Luo, C.; Li, X.; Duan, H.; Wang, X. Preparation of magnetic ionic liquid/chitosan/graphene oxide composite and application for water treatment. *Int. J. Biol. Macromol.* **2014**, *66*, 172–178. [CrossRef] [PubMed]
147. Lingamdinne, L.P.; Koduru, J.R.; Choi, Y.-L.; Chang, Y.-Y.; Yang, J.-K. Studies on removal of Pb(II) and Cr(III) using graphene oxide based inverse spinel nickel ferrite nano-composite as sorbent. *Hydrometallurgy* **2016**, *165*, 64–72. [CrossRef]

148. Huang, Y.; Huang, W.; Chen, Y.; Sun, J.; Liang, M.; Guo, Y.; Liu, X.; Liu, M.; Wei, Y.; Wei, J.; et al. Effective removal of hexavalent chromium from aqueous solutions using quaternary ammonium-functionalized magnetic graphene oxide composites. *Separations* **2023**, *10*, 463. [CrossRef]
149. Fan, L.; Luo, C.; Sun, M.; Li, X.; Qiu, H. Highly selective adsorption of lead ions by water-dispersible magnetic chitosan/graphene oxide composites. *Colloids Surf. B Biointerfaces* **2013**, *103*, 523–529. [CrossRef]
150. Bao, S.; Yang, W.; Wang, Y.; Yu, Y.; Sun, Y. One-pot synthesis of magnetic graphene oxide composites as an efficient and recoverable adsorbent for Cd(II) and Pb(II) removal from aqueous solution. *J. Alloys Compd.* **2020**, *381*, 120914. [CrossRef]
151. Zhao, D.; Gao, X.; Wu, C.; Xie, R.; Feng, S.; Chen, C. Facile preparation of amino functionalized graphene oxide decorated with Fe3O4 nanoparticles for the adsorption of Cr(VI). *Appl. Surf. Sci.* **2016**, *384*, 1–9. [CrossRef]
152. Wu, W.P.; Ge, H.C. Preparation of amino-silane and thiosemicarbazide modified graphene oxide composite for high uptake adsorption of Hg(II) and Pb(II). *J. Alloys Compd.* **2024**, 2312839. [CrossRef]
153. Joya-Cárdenas, D.R.; Rodríguez-Caicedo, J.P.; Corona-Rivera, M.A.; Saldaña-Robles, N.; Damián-Ascencio, C.E.; Saldaña-Robles, A. Removal of As(V) in the presence of Cr(VI) in contaminated water from the Bajio region of Mexico using ferrihydrite-functionalized graphene oxide (GOFH): A case study. *Emerg. Contam.* **2024**, *10*, 100312. [CrossRef]
154. Wu, Z.; Deng, W.; Zhou, W.; Luo, J. Novel magnetic polysaccharide/graphene oxide @Fe_3O_4 gel beads for adsorbing heavy metal ions. *Carbohydr. Polym.* **2019**, *216*, 119–128. [CrossRef]
155. Verma, M.; Lee, I.; Oh, J.; Kumar, V.; Kim, H. Synthesis of EDTA-functionalized graphene oxide-chitosan nanocomposite for simultaneous removal of inorganic and organic pollutants from complex wastewater. *Chemosphere* **2022**, *287*, 132385. [CrossRef]
156. Zhang, H.; Liu, X.; Han, J.; Niu, W.; Wang, B.; Wu, Z.; Wei, Z.; Zhu, Y.; Guo, Q.; Wang, X. Acid-resistant chitosan/graphene oxide adsorbent for Cu^{2+} removal: The role of mixed cross-linking and amino-functionalized. *Int. J. Biol. Macromol.* **2024**, *273*, 133096. [CrossRef] [PubMed]
157. Azfar Shaida, M.; Daniyal; Ali, S.; Saalim Badar, M.; Salim Mahtab, M.; Umar Khan, M.; Ullah Khan, S.; Ahmad, I.; Atika; Haq Farooqi, I.; et al. Synthesis, characterization, and breakthrough modeling of low-cost graphene oxide-sand supported adsorbent filter for the separation of lead and chromium ions. *J. Mol. Liq.* **2024**, *403*, 124611. [CrossRef]
158. Bhardwaj, M.; Tewari, S.; Kumari, N.; Bhardwaj, A.; Misra, N.; Shukla, S.; Dwivedi, J.; Sharma, S. Adsorption of Pb(II) and Cd(II) by functionalized graphene oxide (GO-MBT): Mechanisms, antibacterial activity, and DFT studies. *Ind. Eng. Chem. Res.* **2024**, *164*, 112464. [CrossRef]
159. Elumalai, N.S.; Jaisankar, S.M.; Kumaran, C. Utilization of graphene oxide-chitosan nanocomposite for the removal of heavy metals: Kinetics, isotherm, and error analysis. *Water Conserv. Sci. Eng.* **2024**, *9*, 9. [CrossRef]
160. Zhang, C.-Z.; Chen, B.; Bai, Y.; Xie, J. A new functionalized reduced graphene oxide adsorbent for removing heavy metal ions in water via coordination and ion exchange. *Sep. Sci. Technol.* **2018**, *53*, 2896–2905. [CrossRef]
161. Attia, N.F.; Diab, M.A.; Attia, A.S.; El-Shahat, M.F. Greener approach for fabrication of antibacterial graphene-polypyrrole nanoparticle adsorbent for removal of Mn^{2+} from aqueous solution. *Synth. Met.* **2021**, *282*, 107132. [CrossRef]
162. Hadadian, M.; Goharshadi, E.K.; Fard, M.M.; Ahmadzadeh, H. Synergistic effect of graphene nanosheets and zinc oxide nanoparticles for effective adsorption of Ni(II) ions from aqueous solutions. *ACS Biomater. Sci. Eng.* **2018**, *124*, 239. [CrossRef]
163. Pan, L.; Wang, Z.; Yang, Q.; Huang, R. Efficient removal of lead, copper and cadmium ions from water by a porous calcium alginate/graphene oxide composite aerogel. *Nanomaterials* **2018**, *8*, 957. [CrossRef] [PubMed]
164. Cao, Z.-f.; Wen, X.; Wang, J.; Yang, F.; Zhong, H.; Wang, S.; Wu, Z.-k. In situ nano-Fe_3O_4/triisopropanolamine functionalized graphene oxide composites to enhance Pb^{2+} ions removal. *Colloids Surf. A Physicochem. Eng. Asp.* **2019**, *561*, 209–217. [CrossRef]
165. Ibrahim, A.I.; Onaizi, S.A.; Vohra, M.S. Novel CTAB functionalized graphene oxide for selenium removal: Adsorption results and ANN & RSM modeling. *Emergent Mater.* **2024**, *7*, 547–564. [CrossRef]
166. Liu, H.; Zhou, Y.; Yang, Y.; Zou, K.; Wu, R.; Xia, K.; Xie, S. Synthesis of polyethylenimine/graphene oxide for the adsorption of U(VI) from aqueous solution. *Appl. Surf. Sci.* **2019**, *471*, 88–95. [CrossRef]
167. Nan, Y.; Wang, J.; Chang, X.; Shao, K.; Lin, Y.; Qian, L.; Li, Z.; Hu, P. Functionalized graphene oxide/sodium alginate beads with ion responsiveness for uranium trapping. *Carbohydr. Polym.* **2023**, *300*, 120259. [CrossRef]
168. Wu, F.; Huang, H.; Sun, X.; Xie, S.; Yuan, H.; Liu, Y.; Guo, Y. Persimmon tannin-modified graphene oxide/chitosan microsphere for removing U(VI) in rare earth wastewater. *J. Radioanal. Nucl. Chem.* **2023**, *332*, 3617–3633. [CrossRef]
169. Kim, Y.; Eom, H.H.; Kim, Y.K.; Harbottle, D.; Lee, J.W. Effective removal of cesium from wastewater via adsorptive filtration with potassium copper hexacyanoferrate-immobilized and polyethyleneimine-grafted graphene oxide. *Chemosphere* **2020**, *250*, 126262. [CrossRef]
170. Kadam, A.A.; Jang, J.; Lee, D.S. Facile synthesis of pectin-stabilized magnetic graphene oxide Prussian blue nanocomposites for selective cesium removal from aqueous solution. *Bioresour. Technol.* **2016**, *216*, 391–398. [CrossRef]
171. Huang, T.; Qu, J.; Tan, L.; Yao, R.; Jiao, W.; Wang, Y.; Lin, T.; Hao, Y.; Yang, H.; Yang, H.; et al. Efficient adsorption of Ce (III) using cellulose graft poly (2-acrylamido-2-methyl-1-propanesulfonic acid)/graphene oxide composite. *Colloids Surf. A Physicochem. Eng. Asp.* **2024**, *682*, 132981. [CrossRef]
172. Abu Elgoud, E.M.; Abd-Elhamid, A.I.; Aly, H.F. Modification of graphene oxide with imidazolium-based ionic liquid for significant sorption of La(III) and Pr(III) from aqueous solutions. *Appl. Water Sci.* **2023**, *13*, 152. [CrossRef]
173. Gbadebo, A.M. Groundwater fluoride and dental fluorosis in southwestern Nigeria. *Environ. Geochem. Health* **2012**, *34*, 597–604. [CrossRef] [PubMed]

174. Liu, M.; Zang, Z.; Zhang, S.; Ouyang, G.; Han, R. Enhanced fluoride adsorption from aqueous solution by zirconium (IV)-impregnated magnetic chitosan graphene oxide. *Int. J. Biol. Macromol.* **2021**, *182*, 1759–1768. [CrossRef]
175. Rashid, U.S.; Das, T.K.; Sakthivel, T.S.; Seal, S.; Bezbaruah, A.N. GO-CeO_2 nanohybrid for ultra-rapid fluoride removal from drinking water. *Sci. Total Environ.* **2021**, *793*, 148547. [CrossRef]
176. Vasudevan, S.; Lakshmi, J. The adsorption of phosphate by graphene from aqueous solution. *RSC Adv.* **2012**, *2*, 5234–5242. [CrossRef]
177. Eltaweil, A.S.; Ibrahim, K.; El-Monaem, E.M.A.; El-Subruiti, G.M.; Omer, A.M. Phosphate removal by Lanthanum-doped aminated graphene oxide@aminated chitosan microspheres: Insights into the adsorption mechanism. *J. Clean. Prod.* **2023**, *385*, 135640. [CrossRef]
178. Wang, P.; Li, L.; Tian, Y.; Sun, L.; Zhan, W.; Chen, S.; Zhang, J.; Zuo, W. Three-dimensional graphene/La$(OH)_3$-nanorod aerogel adsorbent by self-assembly process for enhanced removal and recovery of phosphate in wastewater. *Sci. Total Environ.* **2022**, *809*, 152124. [CrossRef]
179. He, J.; Du, Y.-e.; Bai, Y.; An, J.; Cai, X.; Chen, Y.; Wang, P.; Yang, X.; Feng, Q. Facile formation of anatase/rutile TiO_2 nanocomposites with enhanced photocatalytic activity. *Molecules* **2019**, *24*, 2996. [CrossRef]
180. Shooto, N.D.; Thabede, P.M.; Bhila, B.; Moloto, H.; Naidoo, E.B. Lead ions and methylene blue dye removal from aqueous solution by mucuna beans (velvet beans) adsorbents. *J. Environ. Chem. Eng.* **2020**, *8*, 103557. [CrossRef]
181. Cao, Z.-f.; Wen, X.; Chen, P.; Yang, F.; Ou, X.-l.; Wang, S.; Zhong, H. Synthesis of a novel heterogeneous fenton catalyst and promote the degradation of methylene blue by fast regeneration of Fe^{2+}. *Colloids Surf. A Physicochem. Eng. Asp.* **2018**, *549*, 94–104. [CrossRef]
182. Raees, A.; Jamal, M.; Ahmed, I.; Silanpaa, M.; Saad Algarni, T. Synthesis and characterization of CeO_2/CuO nanocomposites for photocatalytic degradation of methylene blue in visible light. *Coatings* **2021**, *11*, 305. [CrossRef]
183. Tabish, T.A.; Memon, F.A.; Gomez, D.E.; Horsell, D.W.; Zhang, S. A facile synthesis of porous graphene for efficient water and wastewater treatment. *Sci. Rep.* **2018**, *8*, 1817. [CrossRef]
184. Joshi, B.; Khalil, A.M.E.; Zhang, S.; Memon, F.A. Investigating the potential of greener-porous graphene for the treatment of organic pollutants in wastewater. *C* **2023**, *9*, 97. [CrossRef]
185. Naeem, H.; Tofil, H.M.; Soliman, M.; Hai, A.; Zaidi, S.H.H.; Kizilbash, N.; Alruwaili, D.; Ajmal, M.; Siddiq, M. Reduced graphene oxide-zinc sulfide nanocomposite decorated with silver nanoparticles for wastewater treatment by adsorption, photocatalysis and antimicrobial action. *Molecules* **2023**, *28*, 926. [CrossRef] [PubMed]
186. Jayawardena, R.; Eldridge, D.S.; Malherbe, F. Sonochemical synthesis of improved graphene oxide for enhanced adsorption of methylene blue. *Colloids Surf. A Physicochem. Eng. Asp.* **2022**, *650*, 129587. [CrossRef]
187. Dang, A.; Yuan, Z.; Liu, X.; Ma, S.; Yang, Y.; Zada, A.; Gao, Y.; Liu, Y.; Li, T.; Han, Y. Graphene oxide mediated carbon foam/CNTs composites for highly efficient adsorption of methylene blue and mechanism insight. *Ceram. Int.* **2023**, *49*, 36970–36978. [CrossRef]
188. Rashad, M.; Helali, S.; Shaalan, N.M.; Albalawi, A.E.; Alatawi, N.S.; Al-Faqiri, B.; Al-Belwi, M.M.; Alsharari, A.M. Dual studies of photo degradation and adsorptions of Congo red in wastewater on graphene–copper oxide heterostructures. *Materials* **2023**, *16*, 3721. [CrossRef]
189. Zheng, Y.; Zhou, X.; Luo, H.; Ling, H.; Mo, W.; Fang, H.; Shen, C.; Lei, J.; Sun, M.; Li, J. Efficient removal of Congo red with graphene aerogel derived from recycled anode of lithium-ion battery. *Int. J. Environ. Sci. Technol.* **2021**, *18*, 3995–4006. [CrossRef]
190. Yang, Z.; He, C.; Liao, W.; Zhang, X.; Liu, W.; Zou, B. Adsorption of orange G in liquid solution by the amino functionalized GO. *Separations* **2022**, *9*, 391. [CrossRef]
191. Labiadh, L.; Kamali, A.R. 3D graphene nanoedges as efficient dye adsorbents with ultra-high thermal regeneration performance. *Appl. Surf. Sci.* **2019**, *490*, 383–394. [CrossRef]
192. Yang, Z.; Liu, X.; Liu, X.; Wu, J.; Zhu, X.; Bai, Z.; Yu, Z. Preparation of β-cyclodextrin/graphene oxide and its adsorption properties for methylene blue. *Colloids Surf. B Biointerfaces* **2021**, *200*, 111605. [CrossRef]
193. Wu, Z.; Zhong, H.; Yuan, X.; Wang, H.; Wang, L.; Chen, X.; Zeng, G.; Wu, Y. Adsorptive removal of methylene blue by rhamnolipid-functionalized graphene oxide from wastewater. *Water Res.* **2014**, *67*, 330–344. [CrossRef]
194. Bu, J.; Yuan, L.; Zhang, N.; Liu, D.; Meng, Y.; Peng, X. High-efficiency adsorption of methylene blue dye from wastewater by a thiosemicarbazide functionalized graphene oxide composite. *Diam. Relat. Mater.* **2020**, *101*, 107604. [CrossRef]
195. Banaei, A.; Saadat, A.; Javadi, R.; Pargolghasemi, P. Preparation magnetic graphene oxide/diethylenetriamine composite for removal of methylene blue from aqueous solutions. *Sci. Rep.* **2024**, *14*, 15457. [CrossRef] [PubMed]
196. Liu, X.; Jing, K.; Peng, S.; Shi, Q.; Liu, H. Facile preparation of graphene oxide-based composite aerogel to efficiently adsorb methylene blue. *Colloids Surf. A Physicochem. Eng. Asp.* **2024**, *681*, 132754. [CrossRef]
197. Zhao, Q.; Zhu, X.; Chen, B. Stable graphene oxide/poly(ethyleneimine) 3D aerogel with tunable surface charge for high performance selective removal of ionic dyes from water. *Chem. Eng. J.* **2018**, *334*, 1119–1127. [CrossRef]
198. Xu, W.; Li, Y.; Wang, H.; Du, Q.; Li, M.; Sun, Y.; Cui, M.; Li, L. Study on the adsorption performance of casein/graphene oxide aerogel for methylene blue. *ACS Omega* **2021**, *6*, 29243–29253. [CrossRef] [PubMed]
199. Fu, Y.; Wang, Z.; Tian, T.; Li, G.; Gu, J.; Zhou, J.; Dong, S. Preparation of carboxymethyl cellulose/graphene oxide/ZIF-8 aerogels for efficient methylene blue adsorption. *Colloids Surf. A Physicochem. Eng. Asp.* **2024**, *696*, 134338. [CrossRef]

200. Zhou, S.; Yin, J.; Ma, Q.; Baihetiyaer, B.; Sun, J.; Zhang, Y.; Jiang, Y.; Wang, J.; Yin, X. Montmorillonite-reduced graphene oxide composite aerogel (M−rGO): A green adsorbent for the dynamic removal of cadmium and methylene blue from wastewater. *Sep. Purif. Technol.* **2022**, *296*, 121416. [CrossRef]
201. E, T.; Ma, D.; Yang, S.; Hao, X. Graphene oxide-montmorillonite/sodium alginate aerogel beads for selective adsorption of methylene blue in wastewater. *J. Alloys Compd.* **2020**, *832*, 154833. [CrossRef]
202. Li, Y.; Ni, M.; He, Q.; Li, X.; Zhang, W.; Wang, H. Adsorption of rhodamine B and Cr^{3+} ion onto graphene/chitosan composite. *J. Comput. Methods Sci. Eng.* **2021**, *21*, 927–938. [CrossRef]
203. Natasha; Khan, A.; Rahman, U.U.; Sadaf; Yaseen, M.; Abumousa, R.A.; Khattak, R.; Rehman, N.; Bououdina, M.; Humayun, M. Effective removal of nile blue dye from wastewater using silver-decorated reduced graphene oxide. *ACS Omega* **2024**, *9*, 19461–19480. [CrossRef] [PubMed]
204. Sharma, G.; AlGarni, T.S.; Kumar, P.S.; Bhogal, S.; Kumar, A.; Sharma, S.; Naushad, M.; Alothman, Z.A.; Stadler, F.J. Utilization of Ag_2O-Al_2O_3-ZrO_2 decorated onto rGO as adsorbent for the removal of Congo red from aqueous solution. *Environ. Res.* **2021**, *197*, 111179. [CrossRef]
205. Dissanayake, N.S.L.; Pathirana, M.A.; Wanasekara, N.D.; Mahltig, B.; Nandasiri, G.K. Removal of methylene blue and congo red using a chitosan–graphene oxide-electrosprayed functionalized polymeric nanofiber membrane. *Nanomaterials* **2023**, *13*, 1350. [CrossRef]
206. Su, J.; He, S.; Zhao, Z.; Liu, X.; Li, H. Efficient preparation of cetyltrimethylammonium bromide-graphene oxide composite and its adsorption of Congo red from aqueous solutions. *Colloids Surf. A Physicochem. Eng. Asp.* **2018**, *554*, 227–236. [CrossRef]
207. Mahmood, T.; Noreen, U.; Ali, R.; Ullah, A.; Naeem, A.; Aslam, M. Adsorptive removal of Congo red from aqueous phase using graphene–tin oxide composite as a novel adsorbent. *Int. J. Environ. Sci. Technol.* **2022**, *19*, 10275–10290. [CrossRef]
208. Li, Y.; Liu, H.; Nie, R.; Li, Y.; Li, Q.; Lei, Y.; Guo, M.; Zhang, Y. Highly efficient adsorption of anionic dyes on a porous graphene oxide nanosheets/chitosan composite aerogel. *Ind. Crops Prod.* **2024**, *220*, 119146. [CrossRef]
209. Dong, J.; Chen, Q.; Zhang, J.; Wang, Z.; Cai, J.; Yan, H.; Chen, C. Effects of rainfall events on behavior of tetracycline antibiotics in a receiving river: Seasonal differences in dominant processes and mechanisms. *Sci. Total Environ.* **2019**, *692*, 511–518. [CrossRef]
210. Hu, X.; Zhou, Q.; Luo, Y. Occurrence and source analysis of typical veterinary antibiotics in manure, soil, vegetables and groundwater from organic vegetable bases, northern China. *Environ. Pollut.* **2010**, *158*, 2992–2998. [CrossRef]
211. Zhao, H.; Liu, X.; Cao, Z.; Zhan, Y.; Shi, X.; Yang, Y.; Zhou, J.; Xu, J. Adsorption behavior and mechanism of chloramphenicols, sulfonamides, and non-antibiotic pharmaceuticals on multi-walled carbon nanotubes. *J. Hazard. Mater.* **2016**, *310*, 235–245. [CrossRef] [PubMed]
212. Liu, P.; Liu, W.-J.; Jiang, H.; Chen, J.-J.; Li, W.-W.; Yu, H.-Q. Modification of bio-char derived from fast pyrolysis of biomass and its application in removal of tetracycline from aqueous solution. *Bioresour. Technol.* **2012**, *121*, 235–240. [CrossRef]
213. Gao, Y.; Li, Y.; Zhang, L.; Huang, H.; Hu, J.; Shah, S.M.; Su, X. Adsorption and removal of tetracycline antibiotics from aqueous solution by graphene oxide. *J. Colloid Interface Sci.* **2012**, *368*, 540–546. [CrossRef]
214. Moussavi, G.; Hossaini, Z.; Pourakbar, M. High-rate adsorption of acetaminophen from the contaminated water onto double-oxidized graphene oxide. *Chem. Eng. J.* **2016**, *287*, 665–673. [CrossRef]
215. Peng, B.; Chen, L.; Que, C.; Yang, K.; Deng, F.; Deng, X.; Shi, G.; Xu, G.; Wu, M. Adsorption of antibiotics on graphene and biochar in aqueous solutions induced by π-π interactions. *Sci. Rep.* **2016**, *6*, 31920. [CrossRef]
216. Chen, H.; Gao, B.; Li, H. Removal of sulfamethoxazole and ciprofloxacin from aqueous solutions by graphene oxide. *J. Hazard. Mater.* **2015**, *282*, 201–207. [CrossRef] [PubMed]
217. Rostamian, R.; Behnejad, H. A comprehensive adsorption study and modeling of antibiotics as a pharmaceutical waste by graphene oxide nanosheets. *Ecotoxicol. Environ. Saf.* **2018**, *147*, 117–123. [CrossRef]
218. Li, Z.; Jiang, H.; Wang, X.; Wang, C.; Wei, X. Effect of pH on adsorption of tetracycline antibiotics on graphene oxide. *Int. J. Environ. Res. Public Health* **2023**, *20*, 2448. [CrossRef] [PubMed]
219. Shaha, C.K.; Mahmud, M.A.A.; Saha, S.; Karmaker, S.; Saha, T.K. Efficient removal of sparfloxacin antibiotic from water using sulfonated graphene oxide: Kinetics, thermodynamics, and environmental implications. *Heliyon* **2024**, *10*, e33644. [CrossRef]
220. Song, X.; Shui, B.; Wang, Y.; Zhou, J.; Wang, S.; Wu, N. Adsorption performance of GO-doped activated ATP composites towards tetracycline. *RSC Adv.* **2022**, *12*, 19917–19928. [CrossRef]
221. Parashar, D.; Harafan, A.; Achari, G.; Kumar, M. Ciprofloxacin and metronidazole adsorption on chitosan-modified graphene oxide as single-compound and binary mixtures: Kinetics, isotherm, and sorption mechanism. *J. Hazard. Toxic Radioact. Waste* **2023**, *27*, 04022042. [CrossRef]
222. Yu, H.; Zheng, K.; Xu, X.; Liu, X.; Zhao, B.; Ding, H.; Yu, Z.; Deng, C. Preparation of β-cyclodextrin/dopamine hydrochloride-graphene oxide and its adsorption properties for sulfonamide antibiotics. *Environ. Sci. Pollut. Res.* **2022**, *29*, 70192–70201. [CrossRef] [PubMed]
223. Fatholahi, P.; Salehzadeh, H.; Hosseini, K.; Wantala, K.; Shivaraju, H.P.; Jenkins, D.; Shahmoradi, B. Removal of erythromycin antibiotic from the aqueous media using magnetic graphene oxide nanoparticles. *Desalination Water Treat.* **2023**, *303*, 142–150. [CrossRef]
224. Bahar, P.; Hassani, A.H.; Panahi, H.A.; Moniri, E. Application of modified graphene oxide with thermosensitive polymers for adsorption of antibiotics from synthetic contaminated water. *Dalton Trans.* **2021**, *210*, 281–295. [CrossRef]

225. Wang, J.; Yao, Q.; Sheng, C.; Jin, C.; Sun, Q. One-step preparation of graphene oxide/cellulose nanofibril hybrid aerogel for adsorptive removal of four kinds of antibiotics. *J. Nanomater.* **2017**, *2017*, 5150613. [CrossRef]
226. Zhuang, Y.; Yu, F.; Ma, J.; Chen, J. Enhanced adsorption removal of antibiotics from aqueous solutions by modified alginate/graphene double network porous hydrogel. *J. Colloid Interface Sci.* **2017**, *507*, 250–259. [CrossRef]
227. Karaolia, P.; Michael-Kordatou, I.; Hapeshi, E.; Drosou, C.; Bertakis, Y.; Christofilos, D.; Armatas, G.S.; Sygellou, L.; Schwartz, T.; Xekoukoulotakis, N.P.; et al. Removal of antibiotics, antibiotic-resistant bacteria and their associated genes by graphene-based TiO_2 composite photocatalysts under solar radiation in urban wastewaters. *Appl. Catal. B-Environ.* **2018**, *224*, 810–824. [CrossRef]
228. Tang, L.; Ma, X.Y.; Wang, Y.; Zhang, S.; Zheng, K.; Wang, X.C.; Lin, Y. Removal of trace organic pollutants (pharmaceuticals and pesticides) and reduction of biological effects from secondary effluent by typical granular activated carbon. *Sci. Total Environ.* **2020**, *749*, 141611. [CrossRef]
229. Gouveia, T.I.A.; Mota, I.H.; Silva, A.M.T.; Alves, A.; Santos, M.S.F. Are cytostatic drugs in surface waters a potential threat? *Sci. Total Environ.* **2022**, *853*, 158559. [CrossRef]
230. Cao, S.; Liu, Y.; Ming, D.; Tian, J.; You, J.; Chen, Z. Evaluation of the difference in adsorption of amphetamine-type drugs on deep eutectic solvent-functionalized graphene oxide/ZIF-67 composite: Experiment and theoretical calculations. *Environ. Res.* **2024**, *249*, 118356. [CrossRef]
231. Al-Ghouti, M.A.; Sayma, J.; Munira, N.; Mohamed, D.; Da'na, D.A.; Qiblawey, H.; Alkhouzaam, A. Effective removal of phenol from wastewater using a hybrid process of graphene oxide adsorption and UV-irradiation. *Environ. Technol. Innov.* **2022**, *27*, 102525. [CrossRef]
232. Cui, Y.; Kang, W.; Hu, J. Effectiveness and mechanisms of the adsorption of phenol from wastewater onto N-doped graphene oxide aerogel. *J. Water Process Eng.* **2023**, *53*, 103665. [CrossRef]
233. Severo, L.S.; Thue, P.S.; Lima, D.R.; Didó, C.A.; Vasconcellos, M.A.Z.; Armas, L.E.G.; Lima, E.C.; Benvenutti, E.V.; de Menezes, E.W. 3D graphene sponge biomass-derived with high surface area applied as adsorbent for nitrophenols. *J. Environ. Chem. Eng.* **2023**, *11*, 109924. [CrossRef]
234. Wang, F.; Lu, X.; Peng, W.; Deng, Y.; Zhang, T.; Hu, Y.; Li, X.-y. Sorption behavior of bisphenol A and triclosan by graphene: Comparison with activated carbon. *ACS Omega* **2017**, *2*, 5378–5384. [CrossRef] [PubMed]
235. Song, T.; Tian, W.; Qiao, K.; Zhao, J.; Chu, M.; Du, Z.; Wang, L.; Xie, W. Adsorption behaviors of polycyclic aromatic hydrocarbons and oxygen derivatives in wastewater on n-doped reduced graphene oxide. *Sep. Purif. Technol.* **2021**, *254*, 117565. [CrossRef]
236. Kang, W.; Cui, Y.; Yang, Y.; Guo, M.; Zhao, Z.; Wang, X.; Liu, X. Preparation of nitrogen-doped hollow carbon nanosphere/graphene composite aerogel for efficient removal of quinoline from wastewater. *J. Hazard. Mater.* **2021**, *417*, 126160. [CrossRef]

MDPI AG
Grosspeteranlage 5
4052 Basel
Switzerland
Tel.: +41 61 683 77 34

C Editorial Office
E-mail: carbon@mdpi.com
www.mdpi.com/journal/carbon

www.ingramcontent.com/pod-product-compliance
Lightning Source LLC
Chambersburg PA
CBHW061846210926
51850CB00006B/980
9783725852673